HISTORY AND SCIENCE OF KNOTS

SERIES ON KNOTS AND EVERYTHING

Editor-in-charge: Louis H. Kauffman *(Univ. of Illinois, Chicago)*

The Series on Knots and Everything: is a book series polarized around the theory of knots. Volume 1 in the series is Louis H Kauffman's Knots and Physics.

One purpose of this series is to continue the exploration of many of the themes indicated in Volume 1. These themes reach out beyond knot theory into physics, mathematics, logic, linguistics, philosophy, biology and practical experience. All of these outreaches have relations with knot theory when knot theory is regarded as a pivot or meeting place for apparently separate ideas. Knots act as such a pivotal place. We do not fully understand why this is so. The series represents stages in the exploration of this nexus.

Details of the titles in this series to date give a picture of the enterprise.

Published:*

Vol. 1: Knots and Physics (3rd Edition)
by L. H. Kauffman

Vol. 2: How Surfaces Intersect in Space — An Introduction to Topology (2nd Edition)
by J. S. Carter

Vol. 3: Quantum Topology
edited by L. H. Kauffman & R. A. Baadhio

Vol. 4: Gauge Fields, Knots and Gravity
by J. Baez & J. P. Muniain

Vol. 5: Gems, Computers and Attractors for 3-Manifolds
by S. Lins

Vol. 6: Knots and Applications
edited by L. H. Kauffman

Vol. 7: Random Knotting and Linking
edited by K. C. Millett & D. W. Sumners

Vol. 8: Symmetric Bends: How to Join Two Lengths of Cord
by R. E. Miles

Vol. 9: Combinatorial Physics
by T. Bastin & C. W. Kilmister

Vol. 10: Nonstandard Logics and Nonstandard Metrics in Physics
by W. M. Honig

Vol. 11: History and Science of Knots
edited by J. C. Turner & P. van de Griend

*The complete list of the published volumes in the series can also be found at
http://www.worldscientific.com/series/skae

K&E Series on Knots and Everything — Vol. 11

HISTORY AND SCIENCE OF KNOTS

editors

J C Turner
University of Waikato, New Zealand

P van de Griend
Aarhus University, Denmark

World Scientific
Singapore • New Jersey • London • Hong Kong

Published by

World Scientific Publishing Co. Pte. Ltd.

5 Toh Tuck Link, Singapore 596224

USA office: 27 Warren Street, Suite 401-402, Hackensack, NJ 07601

UK office: 57 Shelton Street, Covent Garden, London WC2H 9HE

Library of Congress Cataloging-in-Publication Data

History and science of knots / editors, J. C. Turner, P. van de Griend.
p. cm. -- (K & E series on knots and everything : vol. 11)
Includes index.
ISBN-13 978-981-02-2469-1 -- ISBN-10 981-02-2469-9
1. Knots and splices -- History. 2. Knot theory. I. Turner, J. C. (John Christopher), 1928– . II. Griend, P. C. van de. III. Series.
VM533.H57 1995
623.88'22--dc20 95-42872
CIP

British Library Cataloguing-in-Publication Data

A catalogue record for this book is available from the British Library.

CONTENTS

Part V. DECORATIVE KNOTS AND OTHER ASPECTS

PREFACE

Knots are more numerous than the stars; and equally mysterious and beautiful ... John Turner (1988)

The importance of twines, cords, strings and ropes in the lives of humankind is immense. All civilizations have depended upon them for their safety, and for many other activities and uses which have contributed towards their well-being. These activities range from house, bridge and boat building to weaving and cloth production; from construction of fishing knots and nets to maintenance of record-keeping systems; from making of articles of apparel to decorative braiding of bags, belts and wall-hangings.

For much of mankind's history, knots and knot lore were closely linked with magic, medicine and religious beliefs. Knots have even served as bases for mathematical systems (by the Mayans, for example), before writing skills were available; and of course, string games and other recreational uses were and are legion. All of these activities have been practised since time immemorial, in one form or another, and by all races. They are still practised today, throughout the world. Furthermore, it is safe to say that they will go on being practised until the day arrives that humankind no longer exists.

The original intention for this book was that a collection of essays should be gathered, to form a useful, entertaining and authoritative account of the history of many of these activities. We drew up a list of topics that might be included—anything that involved a 'crossing of strings' was considered—and we came to realise that the whole field of knots and knotting applications simply could not be covered properly in a single book. To do the subject full justice an encyclopaedia would be needed. We had to be selective.

The Contents page shows the topics that we finally were able to include. Notable and regrettable omissions from the list are ropemaking and rope tools, basketry, weaving, knitting, tools for knotting, and uses of knots in magic, medicine and religion. Much more could have been included, too, on the special uses of knots made by many of the great civilizations that have flourished in world history. We were able to include a chapter on Chinese knots; but we are silent on the knotting practices of the Japanese, the Indians, the Arabs, the Babylonians, and so on. We knew no authorities on knots of these races.

Nevertheless, the eighteen essays in the book cover a very wide range of knot topics, and each deals with historical aspects of its own topic. The remote history of our subject is generally shrouded in mystery and mythology; for example, the art of making fish nets is said to have begun with Aphrodite (a Greek goddess of love and beauty), who rose from the sea each morning and gathered women around her to teach them how to make nets and to weave and spin*. Our book begins with a search for less imaginative explanations of how mankind began to make and use knots. Our desire has been to establish the subject as much as possible on a scientific base. In truth, real evidence on which to base the story of knots only arises when archaeological finds are made, of artifacts which either still have rope and knots attached to them, or which can reasonably be said to have been associated with use of rope, twine or knots. Our first four Chapters deal with matters of this kind; many archaeological discoveries pertaining to knots are described, and speculations made about them.

Other sources of knot-history are the occasional references to the subject matter which have been found in early writings of the ancients; and knowledge gleaned from wall paintings in caves, heiroglyphs (for example, see Fig. 1), clay tablets, papyri, quipus, tomb finds, pottery, carvings, and so on. However, it has been noted that clear depictions of knots are extremely rare on ancient paintings or earthenware; a reason mooted for this is that knots bore magical or religious significance, and artists were afraid to portray them too faithfully.

Fig. 1. The ancient Egyptian hieroglyph for the word 'tjeset' = knot

Indeed, the oldest *written* source on the subject (containing the only known descriptions of knots before the 18th century) is the *Iatrikon Synagogus*, a medical treatise compiled by a Greek physician, Oribasius of Pergamum, in the 4th century A.D. He states that he derives his information from Heraklas, who is believed to have been a physician who flourished about A.D. 100. The knots recorded in this book were those used by physicians for slings, or as parts of slings, during operations or in the treatment of bone fractures. They are carefully described by C. L. Day in *The Art of Knotting and Splicing* (U.S. Naval Institute, Annapolis, Maryland, 1970).

*In Norse mythology it is said to have been Loki, the god of mischief and destruction, who taught mankind the art of netting.

It seems that it was not until the late 17th century and just beyond, when mathematics flourished and scientific reasoning began to lead the minds of great thinkers, that knots became worthy elements of study, in searches for rational nautical knowledge. In the mid–18th century the grand encyclopedia of Diderot and D'Alembert appeared, and knots and their uses became recorded as part of man's heritage. It was about this time that shipbuilding was becoming more than an art: methodical thinking was being given to hull shapes, strengths of spars, and ropes and riggings. Studies of principles of seamanship began to be made too. By the end of the 18th century, books such as Hutchinson's *A Treatise on Practical Seamanship, with Hints and Remarks Relating Hereto: designed to contribute something towards fixing rules upon philosophical and rational principle* (1777), and Steel's *Elements and Practice of Rigging and Seamanship* (1794), had appeared. Textbooks bearing instruction on seamen's knots found ready sale, as the need for trained ship's officers grew rapidly with the great expansion of the British Navy during the period of the Napoleonic Wars. As is so often observed in history, advances of a science are greatest in times of war.

In view of the venerable and extensive use of knots over the long millennia, it is somewhat surprising that they did not become a topic deemed worthy of *mathematical* study until the late eighteenth century. No doubt there had been many attempts in the past to discover the most efficient kinds of knot, and the best kinds of rope to use, in specific applications—for example in seamanship. But it seems that before about A.D. 1771 no-one saw any possibility* of modelling knots mathematically; one certainly couldn't apply Euclidean geometry to do so—and that generally was firmly believed to be the mathematics of God's plan for the Universe and all that it contained.

It took the genius of a man like Gauss, one of the greatest mathematicians and astronomers the world has known, to break this taboo. In 1794 he prepared sketches of thirteen knots (Fig. 2), with English names, perhaps copied from an English book of seafarers' knots; this note, together with other papers bearing sketches of knots, was found amongst his papers after his death in 1855. It is clear from his notes that he considered the study of knots to be an important task for mathematicians to attempt; and he encouraged some of his students to engage in that study. He himself published only one paper in the field.

The story of these beginnings of topological knot theory, and of its gradual development into what it has now become, a major branch of Topology, is told in Part IV, Chapter 11.

---

*In 1771, A. T. Vandermonde (1735–1796) published a paper in *Mémoires de l'Académie Royale des Sciences, Paris*, in which he suggests that 'the craftsman who fashions a braid, a net, or some knots will be concerned, not with questions of measurement, but with those of position: what he sees there is the manner in which the threads are interlaced.' The title of his paper was 'Remarques sur les problèmes de situation.'

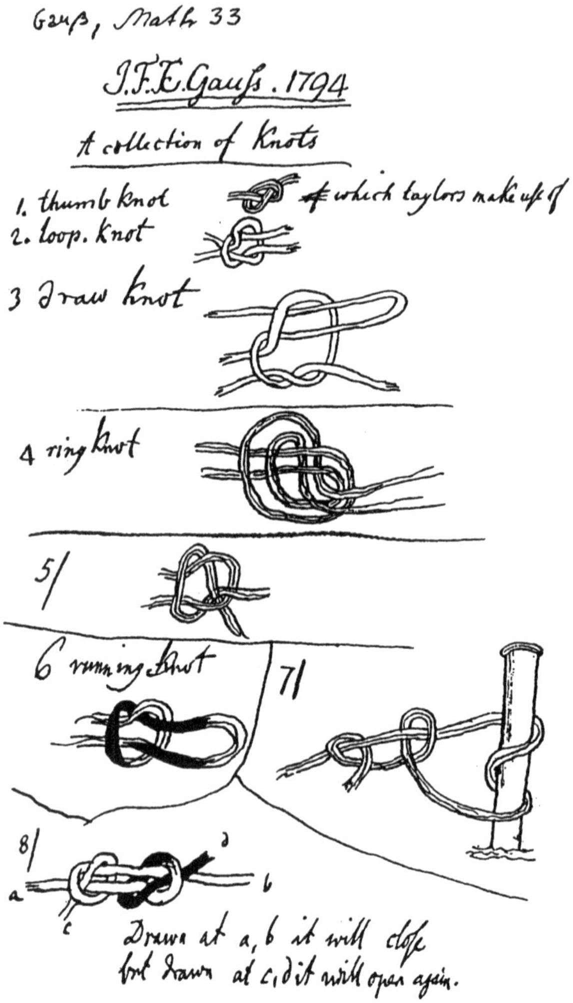

Fig. 2. Copy of a page from the notebooks of C. F. Gauss (1777–1855)

There has been considerable discussion amongst authors and editors of this book as to whether the term 'Science' can legitimately be applied to describe the study of knots. As Professor C. E. M. Joad would have said, when speaking on the British radio programme *Brains Trust*, during World War II: "It all depends on what you mean by 'Science'." We have reached no agreement! Chapter 12 discusses this issue, and also outlines some non-topological knot theories that have emerged recently.

We have organised the eighteen chapters into five sets, or Parts, to impose what we believe to be a useful ordering upon them.

Part I contains four essays which deal with evidences of knots in **Prehistory and Antiquity**. First, evidence of knotting from the Pleistocene period is presented and discussed: speculations are made on what knots would be the first to be used, and how they might have been thought of in the first place. Then evidence from various archaeological sites in Europe is treated. Finally, discoveries of cordwork and knots on articles found at sites of ancient Egyptian settlements are described.

Part II is entitled **Non-European Traditions**. It contains essays on the Peruvian Quipu, Chinese knotting and Inuit (Eskimo) Knots.

Working Knots is the title of Part III; its two essays deal with seamen's and fishermen's knots, and *life support knots*. The last name was coined by the author of the essay, who groups under its banner a variety of specialist knots and ropes used for activities such as tree and rock climbing, caving, and mountaineering; he has studied the histories of development of these knotting applications.

As noted above, Part IV is devoted to the relatively recent history of attempts to study properties of knots, to model them mathematically, and to place them into some kind of a theoretical framework; we place these five essays under the tentative heading **Towards a Science of Knots?**

Finally, in Part V we have grouped, under the title **Decorative Knots and Other Aspects**, five essays which deal with historical aspects of different types of decorative knotting—the uses of knots as adornments, as signs in heraldry, and for the making of attractive clothing and other articles of aesthetic value.

As far as we are aware, this is the first time that a comprehensive study of the history of knots and knotting has been attempted. Normally the topic is dealt with selectively, in a very few pages, at the beginnings of manuals of knots, instruction books for lace, macramé or other forms of knotting, or else at the ends of texts on knot theory. We hope that our book will provide something of interest to all who have occasion to use or study knots. Perhaps one of our readers will be encouraged to build upon this collection of essays, and produce the encyclopaedia that is really necessary if the whole field of the history of knots is to be covered.

We believe that knots in all of their forms constitute a subject most worthy of serious treatment; indeed, in Part IV the argument is put that they already form the basis of a science. This science will grow steadily as theoretical modelling is aimed at a wider range of knot problems, and as more applications arise in other sciences.

The story is told of how in ancient Greece, at Gordium, the chief city of Phrygia, Alexander the Great attempted to unravel the Gordian knot which held fast a famous chariot bound in cords made of the rind of a cornel tree. According to tradition, anyone succeeding in this task would become ruler of the world. Alexander, frustrated by the difficulty of the knot, took his sword and cut it asunder. Perhaps this story can be adapted for those who are engaged in developing the newly emerging knot science. The problems posed by knots are legion; and the rewards for solving them are potentially great. Solvers will surely fall under the spell of the magic of knots. And they will meet with many setbacks and frustrations. However, they are unlikely to find sharp solutions, like Alexander's sword: they must forge ever more powerful mathematical tools to overcome their difficulties.

Acknowledgements

We wish to thank all those who have helped to bring this book into being. Our special thanks go to Charles Warner, for the great help he gave in the design of the book, and for his careful editing of many of the chapters. Heartfelt thanks are due to Bill Rogers, of the University of Waikato, New Zealand, who has given much help with the computer type-setting; without his advice, and his ever ready willingness to solve our LaTeX computing problems, it would have taken us very much longer to put the book into camera-ready form.

John Turner gives warm thanks to his colleague Georg Schaake, for the very many pleasant hours spent together discussing Schaake's theory of braiding, and working on publishing and research projects. He acknowledges the use in Chapter 12 of many diagrams and quotations from the books and pamphlets of Schaake, Turner *et al.*

With regard to copyright questions, every author was asked to take care to acknowledge any quotations used, making adequate references to their sources. If they used diagrams from research journals or books, they were asked to write to the copyright holders for permission to do so, if deemed necessary. It is evident that in a book such as this, all of our authors will have drawn on very many sources for much of their information. We wish to acknowledge, collectively, our debt to all these sources.

Finally, we thank the twelve authors who between them produced the eighteen essays which comprise this book. We thank them for their scholarship, their authorship and their patient help in bringing the essays to their final chapter forms.

The authors submitted biographical notes on their lives and work, and these are gathered together in a section entitled **About the Authors**. It may be found at the end of the book.

Pieter van de Griend
John Turner
1996

PART I. PREHISTORY AND ANTIQUITY

The four chapters in Part I deal with evidence of knots being used by mankind, and other animals, from prehistoric times up to ancient periods of European and Egyptian civilizations.

The first chapter attempts a chronology of the advent of cordage and knot technology over the vast range of time from the Lower Palaeolithic Age (two and a half million years ago) to the Neolithic Age and beyond, when modern *Homo sapiens* appeared some five or six thousand years ago. Most of the evidence is circumstantial, drawn from sparse knowledge of art, dwellings, beads and clothing, tools and weapons, and sea voyages made by various peoples who left their marks in this long and far-distant period.

The second chapter discusses the question *Why knot?* The author speculates on what types of knot were the first to be used, and how they might have been thought of in the first place.

The third and fourth chapters are written by archaeologists who have examined evidence, direct or circumstantial, of knots and rope found at many excavation sites. Chapter 3 deals with knotting evidence found at sites in Denmark and Switzerland (Swiss Neolithic lake dwellings); and with the knots and equipment found with the Ice Man, who was discovered in 1991, in a glacier just south of the Italian-Austrian border. Chapter 4 contains diagrams of findings of knots, plaits, basketry and rope, found by the author during ten expeditions to four different archaelogical sites in Egypt. The earliest site dates from 1350 B.C. at the time of Pharao Amenhotep IV.

CHAPTER 1

PLEISTOCENE KNOTTING

Charles Warner and Robert G. Bednarik

Introduction

Cordage of some sort, and the knots needed for its use, can be safely assumed to have played a crucial role in the earliest technological development of humans. The lack of actual remains of the materials to which knot technology was applied is not really relevant to its history, as we shall see. We have not a single solid piece of evidence from the Pleistocene, the period of the Ice Ages between about two million and ten thousand years ago, that knots were made, and yet many things would have needed binding, joining or attaching. Cordage of some kind (sinews, thongs, hair, vegetable fibres and so on, as single filaments or twisted or plaited into rope) and knots, in the broadest sense, may be safely assumed to have been necessary as fastenings for many of these functions. Some could have been attained by the interlocking of forked sticks and the like, or by the use of such adhesives as might have been available, such as gums and resins. Even these, however, would have been made more effective and reliable if supplemented with cordage and knots.

To understand the reason for the lack of finds in the early record, it is not sufficient simply to remind ourselves that such materials were inevitably organic, and thus subject to decay. Many types of perishable materials, including wooden objects and bark, have survived for hundreds of thousands of years [15, e.g. pp. 47, 60]. What we know of the composition, mode of occurrence, spatial distribution and statistics of archaeological materials is not necessarily representative of the material culture of a given society. It relates much more to such factors as the manner of use and deposition of materials, chemical and physical conditions at the site of deposition, archaeological techniques of data recovery, and all factors involved in selective survival. If a satisfactory picture of a material culture is to be gained, account must also be taken of scholarly biases and competence of investigations, even the language and manner in which the evidence is published; and a multitude of other factors. The study of the interplay of such factors is known as *taphonomy*, and it is this interplay which determines what is used for the formation of our models of what happened in the past. The need to compensate for these complexities has only recently become apparent and many traditional archaeological studies were made without consideration of the taphonomy of the objects of study. So while our claims of the very early use of cordage and knots must be based on the archaeological evidence that is available, we will expound this record in a creative, taphonomically guided, fashion.

There is much indirect evidence for the use of cordage in the Pleistocene, of two basic types. Firstly, there are instances where the use of strings, ropes or thongs is actually demonstrated by such things as representations in art or wear marks on artefacts, even though no actual remains of cordage are available, and where it thus appears inevitable that knots of some sort were used. Secondly, there are many instances where the use of binding or tying material is implied but not proven. The further we go back in time, the more the second, indirect and less reliable, kind of evidence persists; and eventually the record peters out at some point in time. Beyond that, there is no relevant information at all, but this absence does not necessarily indicate the advent of cordage and knot technology: it may still be a taphonomic phenomenon. Consequently, the simplest and most correct answer to the question, when did the use of knots begin, is that this beginning is somewhere in the unfathomable depths of the Lower Palaeolithic period (about 2 500 000 to 250 000 years ago); no doubt the question will never be more accurately answered.

It is to be expected that knotting, as with other techniques of early mankind, began in a very unsophisticated way. Only a very small number of different knots would have been used, though no doubt haphazard conglomerations of several such simple basic knots would have occurred. Progressive improvements in technique would tend to accelerate with time. Technical skills enlarge the habitable range, providing more opportunity for greater versatility so that the momentum of cultural change increases. The available archaeological evidence suggests that hundreds of thousands of years could pass, in the first million or so years of our history, before there were any detectable changes in a material culture. That period had reduced to only a few thousand years by the end of the last Ice Age; and now, as we all know, we are living in a world obviously different from that even of our parents. The technology of cordage and the knots used with it is still developing, but, perhaps because of the increasing availability of substitute fastenings, the rate of change may no longer be accelerating.

There have been some exciting recent findings in archaeology and anthropology concerning the dispersal of peoples, ancient *(Homo habilis* or *Homo erectus)*, archaic *(Homo sapiens)*, or modern *(Homo sapiens sapiens)*, around various parts of the world, and about the number of species or subspecies involved and their relations to modern humans. These findings have not yet been integrated with all previous ones; some are unexpected, such as that anatomically modern humans appear to have arrived in Australia twenty to thirty thousand years before they arrived in Europe. The findings will have to be confirmed and extended before any clear picture can emerge.

However, this turmoil does not seem to apply to the dates of the various artefacts discussed here, though our ideas on just who was responsible for them may have to be modified. And it is far from certain how much technique was transmitted between the various populations of *Homo*, and how much developed independently when new people arrived.

Modern Apes

Modern apes cannot be considered as necessarily similar to our remote ancestors, but it is of interest to know that they have some very elementary skills at knotting and ropework. Sanderson [27] noted that some wild gorillas made complete knots, mostly Granny Knots [3, #1405] but also (rarely) Reef Knots [3, #1402] when constructing their nests. On the other hand, Schaller [28] claimed he saw no interlacing, weaving or knot-tying whatever, though examination of his photographs and diagrams suggests that some Half Knots [3, #47], have been formed. The two gorilla populations were widely separated and may have differed culturally. MacKinnon [18] noticed a captive orang-utan make some sort of rope out of its bedding material and swing from it. Chimpanzees also interlace branches for their nests, though no definite knots have been reported; however, their technical culture is known to vary significantly in different local populations [15, pp. 15, 48]. There are anecdotes of captive chimpanzees untying and retying knots already tied in rope. All this at least suggests that the beginning of knot tying may well have preceded the evolution of the genus *Homo*.

Homo habilis

The 'handy man' *Homo habilis* who lived some 2.5 to 1.6 million years ago in eastern Africa (Koobi Fora, Olduvai) is credited with making the rather crude Oldowan stone tools and perhaps also bone tools. Tools similar to the Oldowan occur also in Asia, including even Siberia, though the age and the responsible species are unknown.

Homo habilis seems to have made caches of his stone tools, presumably for some future use, and also to have carried them over considerable distances [15, pp. 38, 48]. But there is no way of telling whether they were carried loose in the hands or bundled together in some way, which could have implied some use of cordage. A more or less circular accumulation of stones at a habitation site at a level of Olduvai Gorge dated at nearly 2 million years ago was at one time thought to represent a shelter of skin-covered branches held in place by the stones, which could have needed some cordage to stabilise it. But further examination showed that this was more likely to be a random association of rocks and stone tools, and few now believe in the shelter theory [14].

Thus the material evidence from this period provides little evidence about whether *Homo habilis* had either the cognitive status or the technological competence to tie knots, but it cannot be ruled out.

Homo erectus

Homo erectus ('upright man') has provided us with numerous skeletal finds from the three Old World continents, although some physical anthropologists tend to question his presence in Europe. Generally ranging from perhaps 2.0 to 0.4 million years or so ago, his remains have been found in various parts of Africa, in China, Java, and at a few sites in Europe. The period of his presence coincides with an increasingly differentiated tool industry, but in the later period it is far from certain which of these tools are attributable to *Homo erectus*, and which belong to the emerging archaic *Homo sapiens* of 500 000 to 300 000 years ago. The stone tool industry named Acheulian is often attributed to *Homo erectus*, and consisted of large, bifacial tools (called 'handaxes' or 'fist axes'), cleavers, scrapers and a repertoire of flake tools, all hand-held; prismatic blade tools appear in the late phases. The Asian and European sites of skeletal remains of *Homo erectus* are strikingly free of handaxes, even at localities where over a hundred thousand stone tools were recorded. It has been suggested that in Asia, bamboo tools might have replaced stone axes for some purposes.

Acheulian camp sites have been identified over much of Africa, across western, central and southern Europe, on the Black Sea, in the Levant and as far east as India. In Africa, the Acheulian begins perhaps with the appearance of *Homo erectus*, but continues long after his demise, to about 150 000 years ago. Even early Acheulian tools cannot, always and with certainty, be attributed to *Homo erectus*. Concurrent with much of this period, there are various handaxe-free tool industries, and there are no significant differences between the apparent methods of subsistence, cognitive development or technologies of late *Homo erectus* and contemporary archaic *Homo sapiens* populations. Essentially, these hominids began to use fire, erect simple shelters and stone walls, and lived in semi-permanent base camps. They hunted and scavenged, butchered large animals such as elephants and rhinos with their massive stone cleavers, and carried meat to their camps. Hence the two populations lived in very similar ways, and their tools do not identify the species responsible—only recognisable human bones could do that.

Objects such as small quartz crystals, presumably non-utilitarian, have been found at many sites of the Acheulian or contemporary industries, in all three continents, and from deposits up to 900 000 years old. Fossil casts on stone tools, too, seem to have been appreciated. At sites in the Czech Republic and in India, pieces of haematite have been found to bear abrasion markings indicating that they were rubbed against a rock surface. In India some rock engravings were discovered in a cave; their stratigraphical position proves their Acheulian age. They are the oldest rock art known in the world. Engravings on bone, elephant ivory and stone of similar age occur in Germany, and an Acheulian site in Israel has yielded a volcanic pebble with deep grooves held to be artificial, which emphasise the pebble's shape as of a female figure. These finds suggest that the people of the Acheulian engaged in non-utilitarian activities of various types and produced art-like objects and markings.

People dependent on making and using many tools would carry the tools with them when they shifted their home base, because of lack of time or resources to make new ones. A fully upright bipedal stance allows the hands to be used for carrying. But sooner or later it must have occurred to someone it is much easier to bind the objects together so that they can be carried at the waist, over a shoulder or round the neck, or collected together in a container such as a bag or basket. It could well be that one of the first uses of any sort of cordage and knots was for these purposes [15, p. 48], [18], though there is no evidence for such expedients until their depiction on portable art much later, less than 25 000 years ago.

The German site Bilzingsleben yielded skeletal fragments of late *Homo erectus*, together with fragments of a 65 cm-long, polished tapered point made of the split straight tusk of an elephant [8]. Splitting of elephant bones and tusks is often evident at this site, and obviously involved very considerable effort, in driving wedges into such objects with hammer stones. The ivory point, at about 350 000 years the oldest such object known to us, is too large to be used as a dagger, and too short to serve as a lance. Much longer ivory lances were fashioned just 20 000 years ago in Russia. The question then is, what was the Bilzingsleben point used for? These hunters focused on the largest animal species in their environment for their food. Possibly they used lances to allow the charging animals to impale themselves, just as modern humans did in Africa, in which case they had a need for hard points to penetrate the thick skins of their quarries. If the ivory point had been used in this fashion (and it was found broken), it would have had to be lashed to a staff, no doubt with cordage and knotting. Perhaps this is the first such indirect evidence we have.

About the same time, at Bilzingsleben and elsewhere, evidence of possible dwellings begins; although there is no proof that cordage was involved in their construction. At Terra Amata, near Nice, there are 10 sites with stones and post holes presumably outlining dwellings, with occupational litter and hearths inside, dated at around 380 000 years ago [33]. The poles may have been covered with skins or thatch; cordage may have been used to stabilise the structure or fasten the covering.

Cordage is also likely to have been used for suspending perforated objects. Beads or pendants begin around 300 000 years ago, the two oldest known specimens having been excavated in an Austrian cave, the Repolusthöhle. One is a nicely drilled wolf incisor, the other is a triangular, flaked bone point, perforated at the base (Fig. 1).

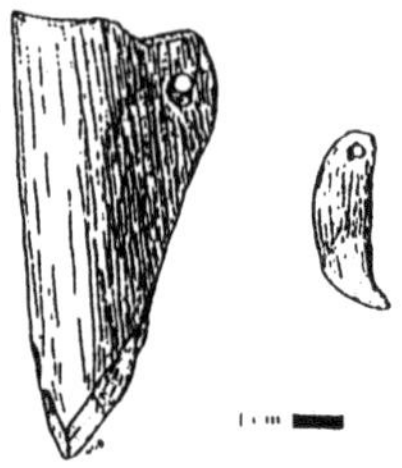

Fig. 1. The two oldest perforated objects known to man: a bone point and a wolf's incisor, from the Repolusthöhle in Styria, Austria. Thought to be almost 300 000 years old.

Spherical stones occur frequently in Lower Palaeolithic occupation sites up to at least 500 000 years old, in both Africa and China. They are usually from 6 to 12 cm in diameter, and it has been suggested that they may have been used as bola weights in hunting. If that were correct, it would undeniably prove the use of both cordage and knots, indeed probably some sort of bag or netting to hold the stones as well.

Other hunting tools, such as slings, snares, traps and nets for fish or birds, and fishing lines are also possible very early inventions, but there is no evidence at all for any of these until the late Palaeolithic.

Archaic *Homo sapiens*, including Neanderthals

The first signs of human species later than *Homo erectus* seem to have appeared about 500 000 years ago, but anatomically modern *Homo sapiens* did not become universal until about 35 000 years ago. The intervening period is a confusing one during which several groups or subspecies of archaic *Homo sapiens* developed, migrated and, with varying success, replaced previous groups, sometimes after a long coexistence. The details are still very unclear, and will probably remain so until much more evidence is found.

The best known of these groups comprises the Neanderthals, who are usually taken to be responsible for the Mousterian tool industry, lasting from about 200 000 to about 35 000 years ago. However, some non-Mousterian Neanderthals and some non-Neanderthal Mousterian tools are known.

During this period, there is copious evidence of fitting points to spears and the like from various sources and up to at least 80 000 years ago. It even includes moulded hafting resin with imprints of the wood shaft and the chipped stone tool, found at two German sites [19]; hafted points from Greece [26]; tools clearly intended for hafting due to their form, such as the tanged points of the Aterian industry of northern Africa; and stone tools with clear hafting wear [2]. Since resin setting by itself is not often adequate for effective hafting, the use of cordage was probably often involved.

The settlement of the temperate regions of Eurasia may have promoted the construction of substantial dwellings. In the loess plains of eastern Europe, this led to the construction of large shelters built of mammoth bones and tusks, perhaps covered with hides, thatch or turf, and possibly held down by cordage, no doubt fastened with knots. Although these tent-like structures are more typical of the later Upper Palaeolithic period, they first appear in the Mousterian of the Ukrainian sites Molodova I and V, over 40 000 years ago [10].

Other indirect evidence of the use of cordage comes from the proposition that long parallel cut marks on some bone cutting boards indicate where thongs of animal skins had been cut [20]. We have no knowledge of their use, but such evidence can be safely assumed to date back to beyond 100 000 years in Europe.

Early crossings of short stretches of water, such as rivers or small lakes, might have been made with the aid of driftwood logs or vegetation mats, but these are not suitable for extensive journeys. Almost any more-robust craft would need the use of cordage and knots. Whether bamboo rafts, inflated animal skins, coracles, reed boats, canoes or wooden rafts were used, they had to be tied, strapped or sewn together in some way. Dugout canoes might not need cordage, but they are very unstable unless provided with one or two outriggers, which do. Kefallinía Island, off the west coast of Greece, was visited by people of the Mousterian [30]. That island can be approached from the mainland by several routes, but all involve water crossings not less than 6 km, which implies that Neanderthals had the capacity of sea-going. Much earlier, humans crossed the Wallacean Line when they journeyed from Java to Flores at least 700 000 years ago [29] and later on to Timor [15].

There is also evidence from France of non-perforated pendants which were grooved, apparently for fastening to cordage (Fig. 2); knots must have been used, perhaps different knots from those used elsewhere. Some of these pendants were associated with Neanderthals of the Upper Palaeolithic [31].

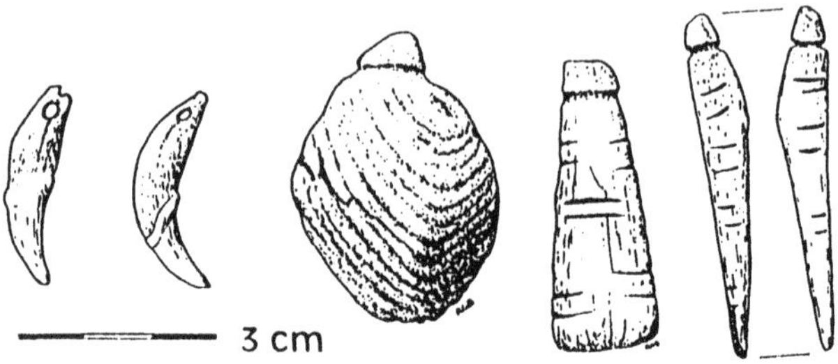

Fig. 2. Two perforated teeth and a grooved fossil shell, from the Châtelperronian of Neanderthals in Grotte du Renne, Arcy-sur-Cure, France (circa 33 000 years B.P.): and two pendants, grooved for attachment to strings, from the Aurignacian I of Abri Blanchard (left) and Abri Cellier (right, two views), both in France (probably about 33 000 years old).

Modern Humankind

There is still much controversy about the origin of anatomically truly modern humans, *Homo sapiens sapiens*, but many believe that they originated in sub-Saharan Africa, some 200 000 years ago. By 92 000 years ago, they had reached Israel, and then spread almost throughout the world. In western Europe, they became known as Cro-Magnon people, dating from 30 000 years ago. New tools of stone, bone and antler were produced, with clearer evidence of planned design; for the first time, both objects with plentiful decoration, and representational art have been found.

The remains of many Upper Palaeolithic tent-like and more complex dwellings have been found in western Europe, Poland, Russia, the Ukraine and Siberia, dating older than about 10 000 years ago.

We have no direct evidence of the use of snares, but it is extremely unlikely that Palaeolithic people would not have used such a reliable method of hunting. The cave bear seems to have been hunted some 30 000–35 000 years ago, in its hibernation haunts such as the deep caves of the European Alps. Arrangements of deep claw marks of this animal have been found on the walls of some caves [6]. They seem to indicate that an animal was restrained in its reach, while clawing the wall in different directions. Such locations seem to coincide with those most suitable for placing snares, so perhaps these hunters knew better than to take unnecessary risks with such a powerful quarry, and chose the safest option of harvesting the fattened bears. A cave bear was significantly larger than the modern bear, and no doubt a very powerful animal, so we would need to assume that the cordage used in these snares was of a breaking strength of perhaps several tons; though, of course, several slightly weaker cords might have been used.

The bones of many hundreds of small (10–20 cm long) fish were found at 19 000 year-old Ohalo II in Israel [24], in several heaps suggesting that they might have been confined in containers of some kind, perhaps fibre bags. No remains of fishing tools were found at the site, and it seems most likely that the small fish were caught in some sort of net or sieve. Some tiny fragments of twisted vegetable fibres were found at the site, and were thought to have come from either the nets or the containers.

The origin of bows and arrows is uncertain; tanged flints of a size and shape resembling arrow heads, have been found in Angola to date from about 30 000 years ago [15, p. 207]. Bows would require well-made cordage and secure knots.

Clothing would offer another potential application of knotting technology, but evidence concerning garments is notoriously hard to find. Prior to the Upper Palaeolithic, which begins 33 000–35 000 years ago, such evidence is almost entirely absent; even for the last phase of the Pleistocene, it remains rare, and almost entirely indirect. This shows once again how the archaeological record must be distorted, because the final period of the Palaeolithic is certainly marked by relatively great technological and artistic sophistication. Three burials at Sungir′, Russia, some 20 000–25 000 years ago, provide an inkling of the sophistication of these societies, as well as of apparel. Two juveniles and an adult were buried with rich grave goods. Among them, 13 113 ivory beads and well over 250 perforated fox teeth were apparently sewn to their apparel [31]. A few of the human figurines from Siberia bear a pattern that resembles quilt-like clothing, including a hood that conceals all but the person's face [1]. Some bone points, thought to be used in sewing skin clothing about 100 000 years ago, were found in China [11] and elsewhere; and at least 26 000 years ago in Tasmania [12].

Artefacts with perforated holes, whether for decorative or utilitarian use, were almost invariably used with cordage and knots. As we have seen, such items made of relatively durable materials occur back as far as 300 000 years ago, and we have no idea of how long before that perishable beads may have been perforated. More recent perforated objects, mostly animal teeth but still older than 35 000 years ago, come from Germany, Bulgaria, France and Russia. Between 33 000–35 000 years ago, beads become much more common, with large numbers found in France and Russia, and smaller numbers in numerous countries, from Japan to South Africa [7]. The earliest evidence of beads in

Australia is over 30 000 years old [22]. In a number of cases the perforations of pendants have been studied microscopically, and the wear from the supporting cord has been observed [4].

Stone axe heads grooved down the middle to take handles are known from the Huon Peninsula, New Guinea, from 40–45 000 years ago [13], from northern Australia from beyond 32 000 years [23], but from Europe at only much less than 10 000 years ago. It seems unlikely that resins alone could hold the head in place; cordage would have been needed.

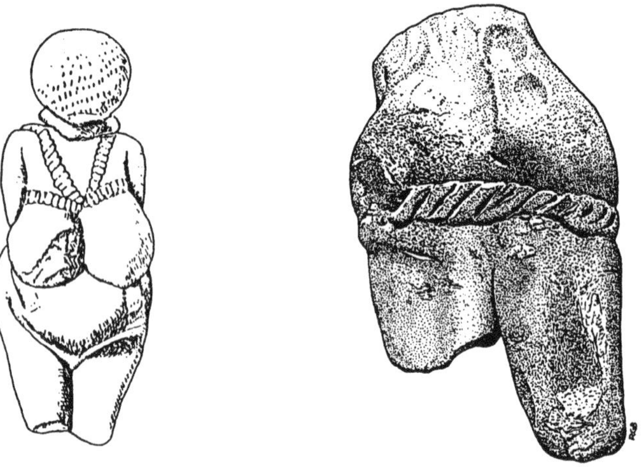

Fig. 3. (left) Female figurine of soft limestone. The head was separated and has been replaced in this reconstruction. Kostenki I, Russia, 21–24 000 years B.P. (right) Detail of hip belt of fragmentary female torso from Pavlov, Moravia, of fired clay. Multiple, carefully and evenly twined flexible elements are clearly recognisable. Similar belts on Russian stone figurines are usually modelled as zigzag patterns. Both after [17].

All of these possible uses of strings and ropes are based on indirect evidence, although much of it is quite convincing. But we also have some detailed and sometimes highly naturalistic depictions of decorative materials on human figurines from the Upper Palaeolithic. They are usually female and naked, except for belts, girdles or harness-like straps. Because of their consistent appearance it has been suggested that they have symbolic meanings, perhaps referring to the wearer's status [21]. Good examples of this occur on several Russian limestone or ivory figurines from Kostenki. In principle, these could represent decorated hide straps, linked beads of some sort or perhaps even braids, all of which would need cordage and knots (Fig. 3). But one belt from Pavlov, Moravia, on a fired clay figurine [17], has a much clearer structure, carefully twined triple (or perhaps double) flexible straps (or perhaps three, or two, passes of one such strap, a more sophisticated technique); the method of fastening is not clear on the figurine. Several depictions of females incised on stone plaquettes also include hip belts, including two from the French site La Marche, and one from the final Palaeolithic (about 11 000 years old) at the Dutch site Geldrop, in which some objects appear to be suspended from the belt. These probably non-utilitarian 'garments' are almost without exception accompanied by bracelets or armlets. There are no clear depictions of knots on any of these examples, but there can be no doubt that knotting technology not only existed then, but was already highly developed. The female figurines occur especially around 22 000–28 000 years ago, although earlier and more recent examples are also known.

Clothing and decorative attire can also be seen in some rock art, mostly more recent than the figurines. Some Magdalenian art (10 000–15 000 years old) from Europe seems to represent human figures with animal horns, tails and sometimes skins; the Trois Frères 'sorcerer' is the best known. These may be purely imaginative, but if they do represent actual happenings, then cordage and knots must have been used to fasten the objects to the body. Headdresses, armlets, anklets and other decorative clothing seem to have been represented in rock art from India and Australia; they would have certainly required some knots. They have not been securely dated, but are probably younger than the Magdalenian.

As yet, we have no actual remains that could possibly refer to Pleistocene sea-going, anywhere in the world. That is, until the sea level settled roughly where it is today, just under 10 000 years ago. This is not very likely to be a coincidence. It is much more likely the result of a particular taphonomic factor. Boats and any equipment used with them, such as paddles, oars, even tools for fishing, whaling or sealing are not usually transported far from any shore, which is why most of such evidence is now found in bogs and lowlands. The lowland bogs of the late Ice Age, however, are without exception below sea level today (due to the lower sea level of that entire period), hence any such remains from that time, if they could have survived, are effectively beyond our reach.

The first substantial sea voyage of which we have evidence is that undertaken by anatomically modern humans to the continent of Sahul (or Greater Australia, including New Guinea). It occurred over 60 000 years ago [32], at a time when the sea level was 60–140 m below that of today. The voyagers started from South East Asia; they could have used any of several routes, but all have several water crossings, with at least one exceeding 90 km [9]. Even with the northwest monsoon and the Leeuwin Current to aid them, they must have spent quite a while out of sight of land. The extensive coastlines of south-east Asia and the numerous nearby islands provided access to the abundant shellfish, the most concentrated source of protein in any habitat [15, p. 134].

The plentiful bamboo of the region would have provided material for bamboo rafts, excellent craft for the large stretches of warm and usually calm waters to the east of modern Indonesia. The greatest diversity of boat types can be found in southeast Asia, suggesting that this might have been the region where the use of watercraft first came to dominate any human societies [15, p. 188]. Sulawesi has been suggested as the centre of diffusion of the single-outrigger canoe [15, p. 196].

Later again, crossings of 30–50 km to New Ireland occurred, still before 30 000 years ago, followed a bit later by even longer voyages of about 180 km to Buka Island, though a roundabout route using some intermediate small islands and shorter distances is possible [32]. Bones of marine (not shoreline) fish were first found in New Ireland middens, dating from about 20 000 years ago. With open-sea fishing, the value of using the wind in some way would have become clear, and perhaps sailing was first practised extensively here [15, pp. 198–202]. By this time, it seems, the competence in systematically traversing stretches of ocean would not have been significantly different from that of modern hunting-foraging peoples.

All of this evidence is entirely indirect, however. The earliest direct evidence of any form of watercraft is from the early Holocene: the paddles recovered from peatbogs at Holmgaard, Denmark, and Star Carr, Yorkshire (about 9500 years old); and the earliest known boat, the canoe from the peat at Pesse, Netherlands, which is a little over 8000 years old.

Horses and other animals have provided transport on land for a much shorter time, and again their use would require cordage and knots. The earliest reliable evidence for horse use dates from only about 6000 years ago, and consists of some teeth worn as from a bit. Portable engravings of horses' heads marked as if wearing harnesses and dating from probably about 15 000 years ago have been found in France; the depictions are by no means clear and this interpretation is not generally agreed.

Some Upper Palaeolithic rock art is thought by some to show lassos, nets and pit traps in western Europe; if these highly dubious interpretations were correct, cordage and knots would have been used.

Post holes remain to suggest that ladders or scaffolding could possibly have been used during the production of some of the cave art in Europe from around 17 000 years ago. Some fossilised fragments of probably two-ply laid rope of about 7 mm diameter have been found in Lascaux Cave [16] from about 17 000 years ago. Perhaps the rope was used in conjunction with the extensive scaffolding that seems to have been built there, or perhaps to facilitate entrance to the cave. Access to several art sites and chert or ochre mines in caves is difficult and may well have involved the use of ladders or ropes. These include the entry down a sinkhole to Koonalda Cave in South Australia more than 20 000 years ago, and the 10 m vertical climb in Baume Latrone in France.

SUMMARY

TIME MARKERS						ARTEFACTS PROBABLY REQUIRING KNOTS			
TIME '000y ago	GEOL. EPOCH	HUMAN SPECIES	AGE	CULTURE	ART	DWELLINGS	BEADS & CLOTHING	TOOLS & WEAPONS	SEA VOYAGES
6		*Homo sapiens* Modern	Neolithic						Boats, Oars
8									
10	Holocene Pleistocene		Mesolithic Upper Palaeolithic	Magdalenian					
15							Costumes depicted	? Nets	
20				Magdalenian	Representational Figures	Many Shelters		Cordage	
25			Upper	Aterian			Awls (Aust.)	Many Lances	Solomons (180 km)
30		Cro-Magnons	Palaeolithic	Aurignacian			Blind Pendants	? Snares	
40		Neanderthals	Middle Palaeolithic	Châtelperronian Mousterian	Geometric Figures and Cupules	Mammoth Bone Shelters	Many Beads	Hafted Tools	
50									Australia (90 km)
60									
80		Modern		Aterian					
100							? Awls	Shafted Weapons	
150			Middle Palaeolithic	Mousterian					
200		Neanderthals ? Modern	Lower Palaeolithic	Acheulian	Markings		? Thongs		
250									
300		Archaic *Homo erectus*			Crayons	? Bilzingsleben	Beads & Pendants	? Lance	
400		*Homo sapiens*				Terra Amata			
500						?		? Bolas	
600									
800					Non-Utilitarian Collections				
1000									
1500		*Homo habilis*		Acheulian					
2000	Pleistocene	*Homo erectus*	Lower						
2500	Pliocene	*Homo habilis*	Palaeolithic	Oldowan					

Conclusion

Knots and cordage have been used by humans for a very long time. There is good evidence of them from non-perishable artefacts of up to more than 300 000 years ago (see the Summary Table, after the References). And there is good reason to believe that they might have been used with wholly perishable materials for a long time before that. But there was no evidence at all from artefacts or art of just what knots were used, until actual samples of knotted cord, preserved by some unusual circumstance, were found (for examples see Chapter 2).

The earliest evidence available suggests that binding knots and lashings of some kind were used, and evidence for bends is nearly as old. The early use of hitches would seem very likely; stopper knots, and fixed or (particularly) sliding loop knots, might also have been in very early use, but we would expect specific indications to be very rare and hard to recognise.

References

1. Z. A. Abramova, *Paleoliticheskoe iskussivo na territorii SSSR* (Akademiya Nauk SSSR, Moscow, 1962).
2. P. Anderson, 'A microwear analysis of selected flint artefacts from the Mousterian of Southwest France'. *Lithic Technology* **9(2)** (1980) 33.
3. C. W. Ashley, *The Ashley Book of Knots* (Doubleday, New York, 1944).
4. R. G. Bednarik, 'More to Palaeolithic females than meets the eye', *Rock Art Research* **7** (1990) 133.
5. R. G. Bednarik, 'Palaeoart and archaeological myths', *Cambridge Archaeological Journal* **2** (1992) 27.
6. R. G. Bednarik, 'Wall markings of the cave bear', *Studies in Speleology* **9** (1993) 51.
7. R. G. Bednarik, 'Art origins', *Anthropos* **89** (1994) 169.
8. R. G. Bednarik, 'Concept-mediated hominid marking behaviour in the Lower Palaeolithic', *Current Anthropology* **36** (in press).
9. J. Birdsell, 'The recalibration of a paradigm for the first peopling of Greater Australia', in *Sunda and Sahul*, eds. J. Allen, J. Golson and R. Jones (Academic Press, London, 1977) 113.
10. P. I. Boriskovski, *Paleolit SSSR* (Izdatelatvo Nauka, Moscow, 1984) 134.
11. L. Chen, *Chinese Knotting* (Echo, Taipei, 1982).
12. R. Cosgrove in G. Burenhult (ed.) *The First Humans* (University of Queensland Press, St Lucia, 1993) 167.

13. L. Groube, J. Chapell, J. Muke and D. Price, 'A 40,000 year old occupation site at Huon Peninsula, Papua New Guinea', *Nature* **324** (1986) 453.
14. D. Johanson and J. Shreeve, *Lucy's Child* (Viking, London, 1989) 152.
15. J. Kingdon, *Self-Made Man* (Simon & Schuster, London, 1993).
16. A. Leroi-Gourhan, 'The archaeology of Lascaux Cave', *Scientific American* **246** (#6) (1982), 104.
17. V. P. Lubine and N.D. Praslov, *Le paléolithique en URSS: découvertes récentes* (Academy of Sciences of USSR, Institute of Archaeology, Leningrad, 1987).
18. J. MacKinnon, *The Ape within us* (Collins, London, 1978) 170-1.
19. D. Mania and V. Toepfer, *Königsaue: Gliederung, Ökologie und mittelpaläolithische Funde der Letzten Eiszeit* (VEB Deutscher Verlag der Wissenschaften, Berlin, 1973).
20. A. Marshack, 'A reply to Davidson on Mania and Mania', *Rock Art Research* **8** (1991) 47.
21. A. Marshack, 'The female image: a "time-factored" symbol. A study in style and aspects of image use in the Upper Palaeolithic', *Proceedings of the Prehistoric Society* **57** (1991) 17.
22. K. Morse, 'Shell beads from Mandu Mandu Creek rock-shelter, Cape Range peninsula, Western Australia, dated before 30,000 b.p.' *Antiquity* **67** (1993) 877.
23. M. J. Morwood and P. J. Tresize, 'Edge-ground axes in Pleistocene Australia. New Evidence from S.E. Cape York Peninsula', *Queensland Archaeological Research* **6** (1989) 77.
24. D. Nadel, A. Danin, E. Werker, T. Schick, M. E. Kislev and K. Stewart. '19,000-year-old twisted fibers from Ohalo II.' *Current Anthropology* **35** (1994) 451.
25. R. G. Roberts, R. Jones and M. A. Smith, 'Optical dating at Deaf Adder Gorge, Northern Territory, indicates human occupation between 53,000 and 60,000 years ago', *Australian Archaeology* **37** (1993) 58.
26. C. Runnels, 'A prehistoric survey of Thessaly: new light on the Greek Middle Palaeolithic', *Journal of Field Archaeology* **15** (1988) 277.
27. I. T. Sanderson, *Animal Treasure* (Macmillan, London, 1937) 187.
28. G. B. Schaller, *The Mountain Gorilla* (University of Chicago Press 1963) 188.
29. P. Y. Sondaar, G. D. van den Bergh, B. Mubroto, F. Aziz, J. de Vos and U. L. Batu, 'Middle Pleistocene faunal turnover and colonization of Flores (Indonesia) by *Homo erectus*', *Comptes Rendus de L'Académie des Sciences Paris* **319** (1994) 1260.
30. T. H. Van Andel, 'Late Quaternary sea-level changes and archaeology', *Antiquity* **63** (1989) 733.

31. R. White, 'Technological and social dimensions of "Aurignacian-age" body ornaments across Europe' in *Before Lascaux: the complex record of the early Upper Palaeolithic* (CRC Press, Boca Raton, FL, 1993) 277.
32. S. Wickler and M. Spriggs, 'Pleistocene human occupation of the Solomon Islands, Melanesia', *Antiquity* **62** (1988) 703.
33. J. Wymer, *The Paleolithic Age* (Croom Helm, London, 1982) 128, 131.

CHAPTER 2

WHY KNOT? — SOME SPECULATIONS ON THE EARLIEST KNOTS

Charles Warner

Chapter 1 shows that the earliest knots we have any real evidence for date back some half a million years, and presumably comprised binding knots, round lashings, bends and probably hitches. There was no early evidence for the existence of loop knots of any kind or of stopper knots, though it seems likely that both must have been in some use. Indeed, it may well be that even the very earliest members of the genus *Homo* practised knotting of some kind some two and a half million years ago. In this chapter I speculate on just what knots were among the first tied, and how they might have been thought of in the first place.

My hypothesis is that largely random tucks of ends here and there played a major part in the development of different knots.

The Knotting Medium

We need a neutral term to describe the entire range of materials used to tie knots in ancient times. These include unmodified natural materials (such as creepers, grass, hemp, rattan, withies, or hair, sinews, intestines); modified natural materials (such as slit bark or large leaves, or thongs from skins); and manufactured materials (twisted, laid or braided fibres), which may have come into use much later than the others. I refer to all these things as *knotting media* or *media* [2] for short.

When the natural filaments were too weak for the purpose in mind, the obvious thing to do was to use several in parallel. This would often have been adequate for binding knots, but would not have allowed some other applications, nor extensions in length by mixing in new filaments. However, twisting the filaments together into yarns would have enabled gradual insertions of extra filaments to increase the length and made a satisfactory medium for many purposes (the filaments were mostly stiff and had rough surfaces, and would consequently tend to adhere together). In the course of twisting three filaments together, people might have come across the 3-ply Common Sennit ([1], #2965), which could also be used as a yarn; this would have been particularly likely for someone specially attracted to the decorative appearance. The making of laid rope by twisting several yarns together would have been a major invention. We do not know how old these techniques are; as seen in Chapter 1, samples of twisted filaments have been recovered from 19 000 years ago, and of two-ply laid rope from 17 000 years ago. Because these media are very much subject to decay, these figures can be taken as minimal ages.

First Attempts at Knots

We can now start speculating on how the knotting process might have started, what could have suggested to the early hominids that the process we call knotting could be useful.

Tropical forests have many plants that wrap around other plants, sometimes strangling them, splitting them, or pulling them over, and sometimes forming Overhand Knots or Half Hitches in the process. Spider webs can be large and thick, able to net and trap large insects. Some nesting birds and rodents can shred the fibres in palms and bark to weave into their nests. Entanglements of vines, brambles or saplings can be found in many jungles, obstructing movements.

All these things could have inspired a deliberate making of entanglements, ambushes, snares and traps, using at first vines and other media naturally present at the site, but eventually collecting the materials where they were common and moving them to a suitable site. At some times and places, traps and snares may have provided much more animal food than active hunting. Entanglements could also have been used to make fences and shelters, and media could have been used to make slings to carry objects and ornamental or symbolic necklaces, armlets, belts and the like.

Binding knots, lashings and perhaps hitches and bends would have been needed. The use of simple binding knots to help make entangling traps, or fences, or shelters could readily have suggested the use of similar bindings on hafted tools.

We postulate that the first knots consisted of random tucks of an end in somewhere, producing such partial knots as Overhand Knots and Half Hitches. A single such tuck or unit of structure is often insecure; the obvious remedy is to make one or more additional tucks, either similar or different, as necessary. We can distinguish two kinds of such compound knots:

(i) Conglomerations are haphazard collections of two or more units of structure. The units are often well separated, so that the knot can slip as the units work closer together under load. The normal tendency would then have been to add more units until the knot appeared secure. It is unlikely that the identical conglomeration would have been repeated when the need arose for a similar knot, even by the same person: the procedure would have been to simply add tucks, twists and hitches until the knot seemed secure. The average modern person with little training or interest in knotting is much more likely to make a conglomeration when tying up bundles or the like than anything else.

(ii) Composite Knots are deliberate sequences of two or more units of structure. Because the whole thing is deliberate, we may usefully talk of an *algorithm* or detailed method of tying the knot. Once a composite knot was found by someone to be suitable for a given purpose, that person was very likely to try to tie the identical knot whenever a similar need was recognised, and to deliberately teach the knot to others, particularly others in the same family group.

If the longer estimates of the period during which knots have been used are correct, it is possible that the first ordered composite knots were tied instead of conglomerates at about the time that the crude, randomly cracked stone tools called Oldowan began to be replaced by the much more deliberately regular Acheulian tools, perhaps as much as $1\frac{1}{2}$ million years ago. When stone tools became more and more finely made, their manufacture became more and more the job of specialist knappers, with ordinary people making only crude tools. Similarly, as composite knots became more and more finely made, they would probably have been made more and more by specialist knot tiers, with ordinary people content with conglomerates, as they still are today.

The design of stone tools changed only very slowly until late in the Stone Age, when creativity blossomed in many crafts, including rock art. Considerable manual dexterity was demonstrated, together with some originality of design and an obvious capacity to recognise patterns. It is possible that around this period, perhaps 30 to 20 thousand years ago, there was also a blossoming of creativity in the making of both working and decorative knots. The making of a wide variety of composite knots probably had to wait for the appearance of robust, flexible cordage; we know that twisted and laid cordage have been known for a period approaching 20 000 years and perhaps more.

Some Definitions

It is often useful to consider families of knots, differing only in the handedness or alignment of particular elements, and often dependent on the chance of tucking an end over or under, or to the right or left, of particular structures.

Many of these variants in any one family have similar properties; they also share the same shadow, the projection of the knot onto a plane surface, exemplified in Fig. 1. To identify knots I give a diagram, a common name and, if the knot is described in Ashley [1], the appropriate number (#xxx); #0 is used if the knot has not been found in [1].

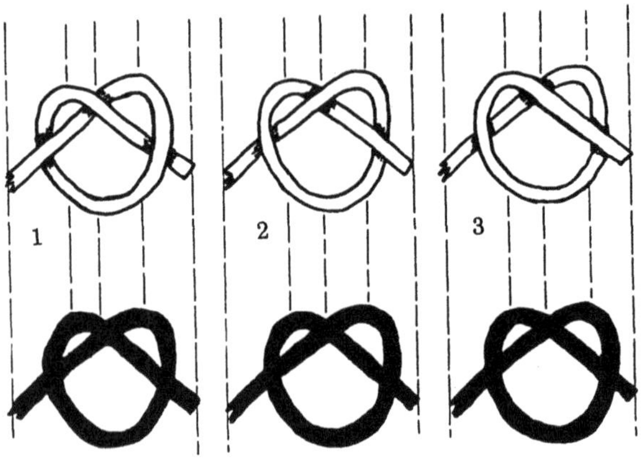

Fig. 1. The shadow or projection of an Overhand Knot (#46). Light shines from above; (1) is a **Z** or right-handed, (2) an **S** or left-handed Overhand Knot and (3) is a non-knot; all produce identical shadows

Both kinds of Overhand Knot, **Z** or **S**, have the same shadow, so that the shadow can be said to represent both knots; space and discussion can be saved by using a shadow rather than the various originals. The shadows of the non-knot (3) and other similar non-knots are also identical, but this kind of solution is generally ignored in this context.

We also need for this study some special technical terms to identify divisions of the standing part and directions along it. These are shown in Fig. 2 below.

The rest of this chapter discusses some possible routes to early knots via random tucks. No attempt is made to describe all possible moves; many others exist. Note that modern findings on the efficacy and security of knots tied in modern knotting media that are smooth, uniform and flexible have little relevance to the same knots tied in the rough, non-uniform, stiff media likely to have been used in really primitive times. It seems probable that in actuality more conglomerations would have been tied than composite knots, but only the composite knots will be discussed here.

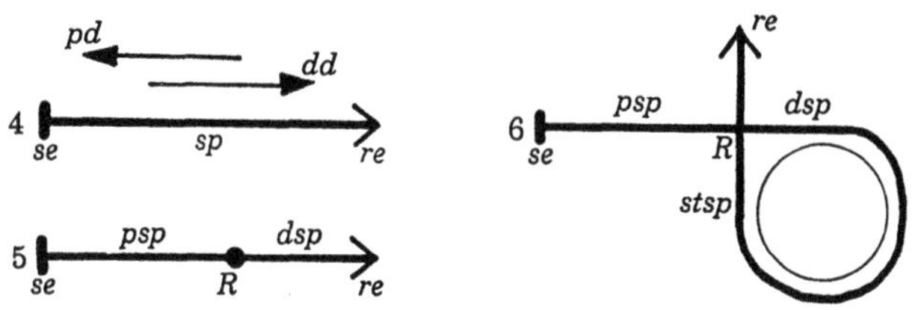

Fig. 2. Some special terms for the divisions of the standing part. (4) is a simple knotting medium. (5) has a reference point demarcated. In (6) the medium is taken round an object to cross its standing part at a reference point, as for the start of a binding knot or hitch; it is assumed that the standing end (*se*) is not in use, the proximal standing part (*psp*) is inactive. See below for further explanations.

Technical terms (see Fig. 2):

dd distal direction: towards the outer end; away from the point of attachment, the point of origin, the medial area, or the inner end.

dsp distal standing part: that part of the standing part distal (see *dd*) to the reference point *R*.

pd proximal direction: towards the point of attachment, the point of origin, the medial area, or the inner end; away from the outer end.

psp proximal standing part: that portion of the standing part proximal (see *pd*) to the reference point *R*.

R reference point, usually where a knot has been or is about to be tied etc.

re running end (this is a practical, not an abstract definition, so it includes e.g. a whipping or stopper knot at the end).

se standing end (again, a practical definition, including any attaching knot).

sp standing part: between the standing and running ends.

stsp subterminal standing part: that part of the standing part closest to the running end.

Binding Knots and Hitches, using a Single Turn

The first knots tied seem likely to have functioned as binding knots and perhaps hitches. An obvious first thing to do would have been to pass the running end of the knotting medium once around the object and then across the standing part at a reference point as in (6). This would have formed the start of the simplest binding knots (Fig. 3).

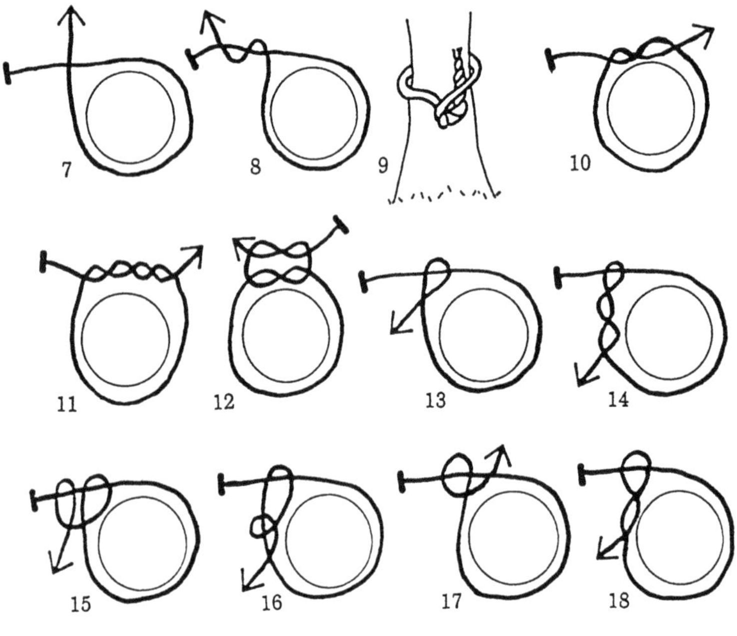

Fig. 3. Binding knots using a single turn round the object

This simple 'knot' (7) would have been completely ineffective in almost all natural media. The simplest and most obvious way to try to make the binding more secure was to twist the running end with the proximal standing part (8), even several times, but would have been little more effective. However, tucking the twisted ends between the object and the distal standing part (9), using a bight if the ends were long, would have been quite effective in rough media; this formation is the Binder Knot (#245). Probably the next most simple way to try to secure (7) was to tuck the running end under the distal standing part (10), forming an Overhand or Half Knot (#47). This would have held well with very rough materials, at least for a while. More security could be gained by making an extra tuck (11). The idea of turning the ends of (10) back and making another Half Knot on top of the first (12) to form a Granny Knot (#1206) or a Reef Knot (#1204) would probably only have occurred to someone using a flexible medium; the knots are neither obvious nor easy to tie with stiff media.

These last two binding knots could have been recognised as bends if the medium were slipped off the object; or if the distal standing part broke and the knot fell off; or by insightful observation. Of course, the use of extra tucks (11) to form such knots as the Surgeon's Knot (#1209) would probably also have occurred early. Modern knot tiers, accustomed to modern knotting media, know that the Granny Knot is less secure than the Reef as a binding knot, and that even the Reef Knot is insecure as a bend. It is doubtful that either property would be noticeable with most primitive media; both knots would have been found adequate as bends as well as binding knots.

The final simple way to try to secure (7) was to tuck the running end under the subterminal standing part (13). This forms a Half Hitch (#50); it could also have been formed by capsizing (10). An extra tuck could have been given to (13) in several ways. If taken round the subterminal standing part (14), the result is a Timber Hitch (#195). A second Half Hitch round the proximal standing part (15) forms Two Half Hitches (#54) if of the same alignment or handedness, or Reversed Half Hitches (#57) if of the opposite; these two knots could also have been obtained by capsizing (12). A second Half Hitch round the subterminal standing part (16) forms a Half Hitch Noose (#0); some of the possible knots with this shadow can capsize to the Overhand Noose (#1114).

Other two-tuck knots, starting with a turn round the proximal standing part, include (17) a sort of Noose (#0) and (18), a Figure-Eight Hitch (#1666) that, with an additional tuck or two, would form a variety of Timber Hitch.

Note that (13)–(18) would have functioned as secure binding knots only if they were drawn up tight and the load was kept steady; otherwise, the standing part would have had to be stabilised by some such means as a Half Hitch or an Overhand knot tied in the standing part round the running end. Note also that all these knots would have been effective hitches and that (15)–(18) would have functioned as nooses in most media if removed from the original object.

Thus with a single turn and no more than about two tucks, more or less at random starting from (7), a number of effective binding knots and hitches would be obtained and, if the object were removed, some effective bends and nooses. Any perceived remaining insecurity would quite likely have been met by making additional tucks.

Binding Knots and Hitches using Two Turns

With two turns round the object, a large number of binding knots or hitches could have been formed quite readily. I show here (Fig. 4) only those that require wrapping the medium twice round the object in a set pattern, followed by a single tuck of the running end. All these knots could have been obtained by chance tucking of the end.

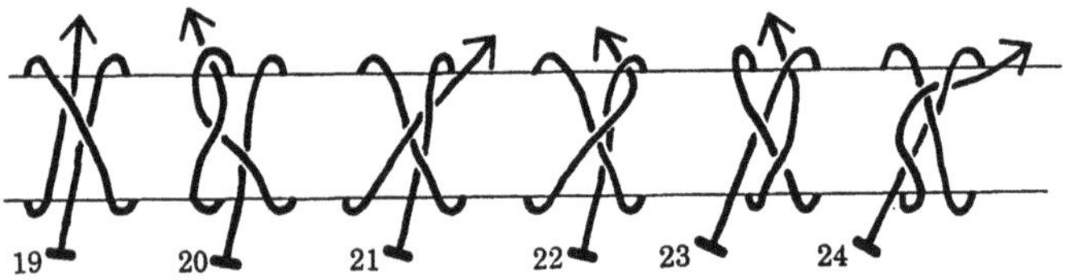

Fig. 4. Binding knots using two turns and a single tuck

If a simple Round Turn was used, with no riding turns, the only single-tuck knots possible were the analogues of (8), (10) or (13); these would have been very little more effective than the originals.

In (19), the running end was passed round the object, crossed over the proximal standing part and passed round again, tucking directly under the distal standing part. This forms the Clove Hitch (#1245). In (20), the final tuck was taken under the distal standing part in the opposite direction, forming a Half Hitch and Half Knot (#0). In (21), the medium was passed over the distal standing part and a tuck taken directly under the proximal standing part. This forms a Sack Hitch, shown in Ashley in various conformations and under various names in #277, 390, 1243, 1676. In (22), the final tuck was taken under the proximal standing part in the opposite direction to make a Snug Hitch (#1674).

In (23) and (24), the running end was passed round the object and crossed over the proximal standing part as before, but then brought back between the proximal and distal standing parts, tucking directly under the proximal standing part in (23), forming the Miller's Hitch (#389, 1242), or reversed in (24), forming the Bag Hitch (#388, 1241). Tucks under the distal standing part from this start would have given unstable knots, where the two ends could unlock and pass back round the object, leaving either a Round Turn or an Overhand Knot.

Additional knots could, of course, have been obtained by using more than one tuck or more than two turns, or by reversing the direction of the turns at some stage.

Stopper Knots, Bends and Loops

A third group of simple knots, almost certainly amongst the first knots tied, was based on the Overhand Knot (25) (#515). This knot (see Fig. 5) is sometimes formed naturally in vines and the like. Once knotting media of sufficient length and flexibility were available, it would have been formed as a wind knot in a flapping line, or, with seaweed or the like, a similar structure under water; it could also be formed as a tangle knot in an accumulation of cord. It could have been formed directly if the core object was removed from a Half Knot (10) or a Half Hitch (13) after capsizing. It could have readily been used as a stopper knot to provide a grip for pulling or a stop to hold in place such things as beads on a necklace or, as soon as any kind of manufactured cordage started to be made, to prevent yarns or strands from fraying or unlaying.

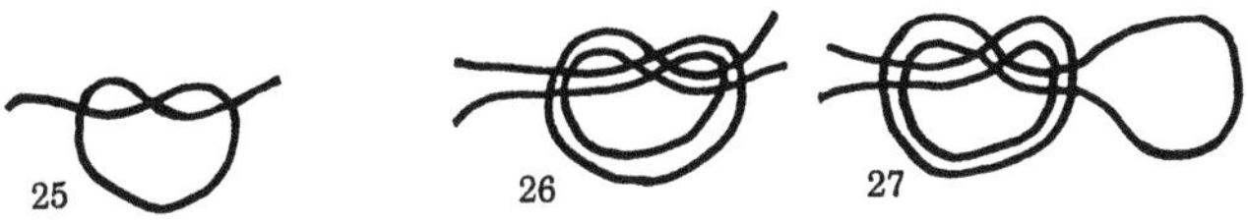

Fig. 5. Overhand knots

If two lengths of a medium or if a bight were used to make the knot, the result was the Overhand Bend (26) (#1410) or Overhand Loop (27) (#1009), perhaps the simplest bend or fixed loop that could be made.

With all these knots, once the knot had been obtained by random methods similar to those indicated, and once it had been recognised as useful, other algorithms would probably have been devised for tying it.

The use of binding knots, round lashings, bends and probably hitches can be inferred from the remains of many artefacts, as discussed in Chapter 1. It is difficult to imagine any use for a loop, whether sliding or fixed, that would leave any similar remains. However, sliding loops occur in many of the hitches already described and their use in making snares and the like would soon occur to someone, perhaps after noticing some naturally occurring ensnarement. But where would a fixed loop be first applied where no other knot would do? It is hard to imagine any application until animals started to be domesticated some 10 000 years ago.

Conclusions

The foregoing form only a sample of the large number of different effective knots that can be formed from very simple algorithms. While most of the early knots were almost certainly conglomerations, the insightful construction of deliberate composite knots has, again almost certainly, been taking place for at least tens of thousands of years. The existence of rock art over this sort of period in all major regions of the world suggests that at least the more gifted individuals had good pattern recognition, good manual dexterity and great skill in portraying their mental images—all qualities of experimental knotters.

Quite a number of these knots are not in the traditional mariners' repertoire (for example, they are not mentioned in Ashley [1]), though most could be expected to work adequately when tied in a suitable medium. The least satisfactory knots (whether because of their bad mechanical properties or their awkward algorithms) would no doubt be discarded soon after a more satisfactory one was discovered and recognised. However, as soon as a knot satisfactory for its purpose in the local circumstances was discovered, it would tend to be perpetuated, at least locally and for a while [3].

The propensity of individuals and groups (occupational, tribal or ethnic) to mark their individuality by using specific artefacts would tend to maintain repertoires of knots specific to that individual or group. Such a specific repertoire would tend to persist even when there was contact with other individuals or groups using other repertoires. A quick glance at other chapters in this book shows that the ancient Egyptians, the Inuit, deep sea fishers, climbers and heralds all made or make use of knots not mentioned in Ashley. In many instances, the knots in any list are not necessarily superior to those in other lists, they are just different, while equivalent in function.

The world population has probably numbered in the millions for hundreds of thousands of years. Just about every person probably tied (or had tied for them) several knots every day of their lives on clothing, tools, or weapons, or in the construction or repair of dwellings or the transport of materials and so on. A hunting-gathering life provides ample spare time for anyone interested to experiment with knots quite extensively, whether by random tucks or by using insight into knot structure. Many of the resulting knots would have been quite effective.

My guess would be that every simple knot possible has already been tied somewhere, by someone, at some time, very likely many times. There is nothing new under the sun! Modern claims to have been the first to have invented a simple knot are difficult to sustain. Just about all that anyone can hope to claim is that the knot has not been published in any common knotting publication. Since only a very small proportion of the world's knotting activities over the millennia has ever been written down, no other claim to 'originality' is likely to be valid.

It is impossible to say that the knots discussed here were necessarily the first composite knots tied, but it seems to me likely that the process of knot development must have been at least similar.

The preparation of this chapter was greatly stimulated by several days of intense discussion with Pieter van de Griend, who however is not to be held responsible for its final form.

References

1. C. W. Ashley, *The Ashley Book of Knots* (Doubleday Doran, New York, 1944).
2. P. van de Griend, *Knots and Rope Problems* (self-published, 1992).
3. P. van de Griend, 'Survival of the Simplest', *Knotting Matters* Issue 39 (1992) 23.

CHAPTER 3

ON KNOTS AND SWAMPS
Knots in European Prehistory

Gerre van der Kleij

The author of this article is an archaeologist, not a knotter. What will be presented in the article is therefore an archaeologist's view of knots and their problems, in a way which hopefully will be of help to those who approach knots from different perspectives.

In the following pages I will start by briefly discussing the conditions under which knots, from whatever period, can and cannot be expected to have been preserved; and the bias that this introduces in the sample of retrieved knots. I will then explain what barriers and hindrances await those who want to study the scarce knotting material that has survived the ages, problems that are mostly a result of the conditions under which the discipline of archaeology is forced to work at the moment. I will illustrate these conditions, or rather some of the problems they create, by discussing three examples, three archaeological contexts that are potentially rich sources of knotting-information. Finally I will make some suggestions as to what knotters themselves could do to increase the amount of information on prehistoric knots, and to make sure that this information is interpreted responsibly.

Knots are, by definition, made in a pliable material. Until the invention, archaeologically speaking quite recently, of synthetic fibres all such materials were organic, i.e. made from parts of plants and/or animals. Unfortunately for us, most organic materials are perishable. Some more so than others; and much depends on the conditions in which they are kept. Generally speaking, however, the materials traditionally used for knotting (plant fibres, inner bark of trees, sinew, leather and the like) will not survive for more than a few centuries in all but a few extreme circumstances. These extremes are, mainly:

1. A completely dry environment, accompanied by rapid desiccation of the material itself;
2. Permanent sub-zero temperatures;
3. Permanent saturation of the material and its environment with water, thus creating an oxygen-poor environment;
4. Chemical reactions, resulting in alteration and preservation of the material.

Even under these extreme conditions disintegration is usually only slowed down, not stopped altogether. The further we go back in time, the smaller our chances are of finding intact the fragile materials used for knotting.

Permanent conditions of extreme dryness or extreme cold are rarely found in the temperate zones of middle and northern Europe. These regions therefore have not, and will never, produce the impressive amounts of textiles, basketry and ropework found in, e.g., Egypt and the American Southwest—with, however, one recent exception, which will be dealt with later.

Waterlogged conditions, however, do occur quite frequently; and here our relatively cool climate is an advantage, for the lower the water temperature, the more the processes of disintegration will slow down.

Chemical preservation is also found fairly often. Thus, quite a few textile fragments have been preserved because they were in close contact with copper or copper-containing objects, such as brooches and beltplates. A combination of chemical preservation and saturation with water has preserved many bog finds from all periods, but especially impressive is the large number of rope- and textile-fragments, complete clothes, and finally well-preserved human bodies from the Iron Age.*

Our information about knotting in the European stone age comes almost exclusively from waterlogged sites, including bogs. Quite a few of these sites have been found, and more will be found and properly excavated, as the techniques and the knowledge necessary to excavate especially under-water sites are further developed. But at the same time, this introduces a bias in the sample of preserved knots. The vast majority of sites that were waterlogged when discovered were also originally located in or near water: bogs, the shores

*Iron Age: in northern Europe, the start of the Iron Age is conventionally placed at around 2600/2500 B.P. (B.P. = *Before Present*). All dates given in this article are calibrated ones: i.e. the dates obtained through C14 measurements have been corrected by allowing for fluctuations in C14 content in the atmosphere, as they are now known to have occurred (formerly the C14 content was assumed, wrongly, to have remained constant throughout the ages). The differences between calibrated and uncalibrated dates can vary from so small as to be negligible, to almost one thousand years. The latter is the case for a large part of the mesolithic and neolithic periods. Hence it is important to declare whether dates used are calibrated or uncalibrated ones.

of seas and lakes. This means that only that part of the total knot repertoire which was connected with the activities performed at those sites will be preserved. Thus one can, for example, expect an over-abundance of knots used for fishing-gear, lost or thrown out on or near the site. Any discussion of the repertoire of knots from stone-age Europe and their use must take this inherent bias into account.

Another problem facing the knotter interested in prehistoric knotting is the difficulty of collecting information about what little material has survived. This is due to three main factors: the enormous increase in the amount of archaeological material the present generation of archaeologists has to cope with; the gross underfunding of the archaeological discipline as a whole, and individual archaeological institutions in particular; the fact that electronic aids, such as computerised information exchange networks and open-access databases are only slowly beginning to be introduced into archaeology.

The first factor has forced archaeologists to become specialists in selected periods and regions only. Those who still possess a working knowledge of all of European prehistory are members of a dying race, and those who could claim to be familiar with world pre-history have long since died out, if ever they existed at all. Thus archaeological knowledge has become very fragmented. This situation is made worse by the second factor, underfunding. Most archaeological institutions cannot possibly cope with the wealth of information coming in, because both the number of trained staff members to process and publish the material, and the facilities to do so are lacking. The result is that the greatest part of all archaeological material at any given point in time is not published, and therefore is only known by those who excavated it, or who have free access to the storerooms and archives. The third factor, the lack of an easily accessible communication network, means that tracking down this fragmented unpublished information, if at all possible, becomes cumbersome and time-consuming in the extreme.

What this unfortunate situation means for the prehistory of knotting, I will illustrate through three examples, each taken from the Mesolithic or Neolithic period* of European prehistory. This choice was to some extent determined by the limitations described above, to which the author as an archaeologist is also subjected. These examples are the so-called Ice Man from Tyrol and his equipment, found in 1991; the Swiss Neolithic lake dwellings; and finally Danish and Neolithic sites with knots or would-be knots.

The Ice Man (or Similaun man, or even 'Oetzi' as he is affectionately

*Mesolithic—the period of nomadic to semi-sedentary gatherer-hunters/fishermen, from the end of the last ice age (around 12 000 B.P.) to the first known introduction of an agricultural component into the local economy. Neolithic—the period from that transition (southern-Europe, about 9000 B.P.; northern Europe, about 6000 B.P.) to the large-scale introduction of bronze (south, about 4500 B.P.; north, about 3800 B.P.)

called) was found on 19 September 1991 in a glacier, on a high mountain pass, in South-Tyrol, only about 90 m south of the Italian-Austrian border (much to the chagrin of the Austrians). His whole body and most of his substantial equipment were extremely well-preserved, because they had been frozen solid during the whole period following the man's death—almost 5400 years. The Ice Man is the only person from the European Neolithic who has been preserved in a better state than as a mere skeleton, and one of the very few from the whole of European prehistory. As for his equipment, the find is absolutely without parallel, at least for this period. Of interest for knotters is that this find shows how important was the role that knots played in the construction of every-day articles, thus giving a glimpse of all that is lost. Among the knot-bearing articles found on or near the body, there were: a sewn leather quiver containing fourteen arrow shafts, some of which had points and feathers attached; bundles of spare rope and yarn; bunches of spare arrowheads, kept together by a piece of string; a belt-purse; an axe-shaft with a copper axe, attached by leather thongs; a flint dagger with attached wooden handle; shoes with strings of leather and vegetable fibres; leather clothing, neatly sewn and coarsely repaired; a grass cloak; a bunch of tiny strings with a pendant/bead attached to it. The material used for the ropes and strings was leather, sinew, grass and other vegetable fibres.

The find is still under conservation and study, spread out over a number of different laboratories and institutions in Europe. So far, two professional publications on the Ice Man have appeared, both quite substantial.* But none of the authors is a knotter, and accordingly nothing is said about knots. The many published pictures do show that there are plenty of knots present, and from them some information can be gleaned. Among the yarns, three types could be identified†:

1. right-hand (Z) laid single-ply yarn used for weaving;
2. two-ply left-hand (S) laid cord made from two right-hand (Z) laid yarns;
3. two-ply right-hand (Z) laid cord made from two left-hand (S) yarns.

Among the knots, the following knots could be identified with some probability:

*Discounting all the more or less sensational publications, full bibliographies can be found in [11] and [19]. The former is written mainly for a specialist audience; articles are in German, English and Italian, all with short English summaries. The latter is accessible to non-specialists; the original edition is in German, but an English translation is available.

†The identification of the Ice-Man yarns and knots mentioned here was done by Charles Warner, to whom I am most grateful.

1. Single Hitches as parts of lashings;
2. Overhand and Half Knots;
3. a Reef Knot;
4. an Overhand Bend (Ashley, #1410);
5. a Strap Knot (Ashley, #1491);
6. some more complex knots, probably involving Overhand Knots in combination;
7. several kinds of plaiting and weaving;
8. some sort of net knot.

The Swiss lake dwellings form another context where large quantities of potential knotting material have been preserved. Throughout the Neolithic and bronze-age periods, large numbers of villages arose on the borders of the many cold lakes of Switzerland. The word 'village' is appropriate here, for these were groups of substantial houses of wood and/or wattle-and-daub, often, especially in the later periods, ordered in regular rows. It is still debatable whether these villages consisted of real pile-dwellings, i.e. actually built on the water, or whether the houses were built on the muddy, swampy shore. The important thing is, that any organic material cast out, or lost or fallen, quickly ended up in the cool mud and water of the lake, where they, thoroughly water-logged and cut off from most oxygen, were preserved until the day of their excavation, four to five thousand years later. Even pieces of fragile linen-fabric, fishnets, carrying bags, straw hats, shoes and the like were preserved, besides tons of wood, twigs, pottery and manure. Excavations of the different locations started in the last century, and many publications have appeared about them.* Unfortunately, only a few of those were available to me, and those few didn't contain any illustrations of identifiable knots. It is clear, however, that this material is a potential gold-mine for any dedicated knotter who is able to gain access to it.

The third and last example consists of all those Mesolithic and Neolithic sites known to me from Denmark where either the actual knots were preserved, or the character of the find was such that knots must originally have been present. Working at a Danish archaeological institute, for once the processes of fragmentation and limited access to information mentioned above were in my favour, and I will therefore go into more detail here.

---

*A long bibliography appeared in: Sakellaridis, Margaret 1979—*The Mesolithic and Neolithic of the Swiss area.* British Archaeological Reports International Series 67. Among those mentioned there is: Schwab, H. 1959/1960—Katalog der im Bernischen Historischen Museum aufbewahrten Faden, Geflecht und Gewebefragmente aus Neolothischen (evt. Bronzeitlichen) Seeufersiedlungen. In: *Jahrbuch des Bernischen Historischen Museums in Bern* 39/40, pp. 336–366. Unfortunately, attempts to obtain this book through interlibrary loan failed.

SITES WITH KNOTS

Sigersdal Mose (North Zealand). Bender Jørgensen 1986, p. 105–106.

A piece of string, probably made of vegetal fibres (closer identification not possible) and used as a noose. The string was wound around the neck of a skeleton, probably that of a juvenile person, found together with a second skeleton of a teenager in a bog near Veksøin northern Zealand. Both skeletons were dated around 3500 B.C. (calibrated), i.e. Early Neolithic. The rope was double twined, and remains of a knot were still present, now resembling a Granny, but probably partly undone (Fig. 1).

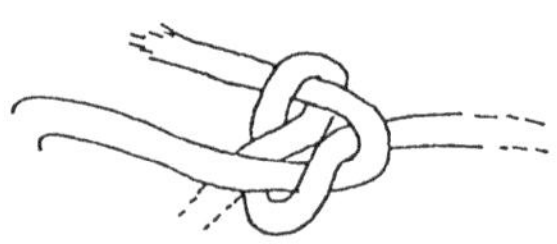

Fig. 1. The Sigersdal Mose knot
Source: Bender Jørgensen 1986, p. 105, fig. 29

Skjoldnæs (Ærø). Skaarup 1981, 1982.

A submerged site, part of the garbage dump of a late Mesolithic (Ertebølle) coastal site.*Among the many organic materials found was the lower part of a leister, almost intact, with substantial pieces of lashing still in place. The lashing was probably made of nettle, and was tightly wound around the leister and fastened by a row of half-hitches (Fig. 2).

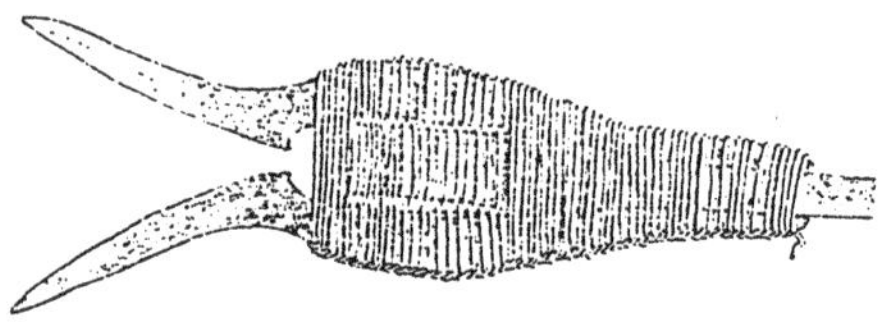

Fig. 2. Skjoldnæs knotting, about a leister
Source: Skaarup 1982, p. 166, fig. 1

*Ertebølle period—the last part of the Mesolithic in Denmark and N. Germany, from about 7200 until about 6000 B.P..

Tybrind Vig (West Fynen). Andersen 1980, 1985,1987.

A submerged settlement plus garbage dump, on the coast of west Fynen, now 3 meters under water, of late-Mesolithic (Ertebølle) date. A large number of objects of wood and bone were found, including pieces of textile (the oldest known so far in Europe) in needle-binding technique. Also found was a bone fishhook with a 5 mm long section of the line still attached, tied on the front of the hook by a clove hitch. The material of the line is sinew, gut or something similar (Andersen 1980) (Fig. 3).

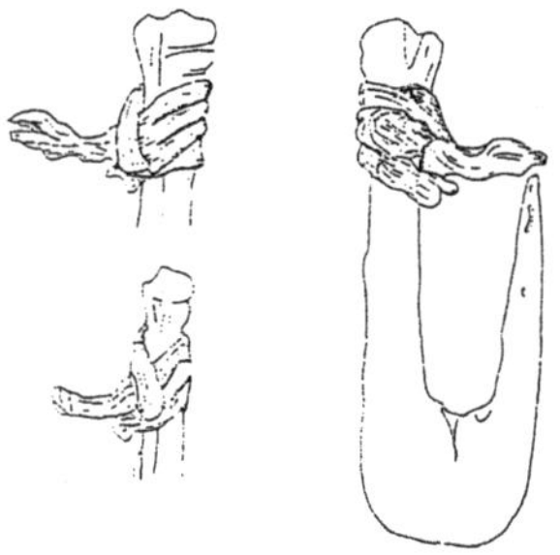

Fig. 3. Tybrind Vig knotting
Source: Andersen, S. H. 1980, p. 10, fig. 5

SITES WITHOUT KNOTS BUT WITH KNOTTING MEDIUM STILL PRESENT

Bokilde Mose (North Coast of Als) Bennike et al. 1986

A piece of rope was found around the neck of one of two Early Neolithic† skeletons, an adult man (the one with the rope) and an adolescent boy. Presumably the rope had been used to strangle/hang the man. It is braided with three braids, but no knot is preserved. The material is vegetable, but a closer identification is impossible.

Dejrø (Ærøskøbing) Skaarup 1980

A submerged site, the garbage dump of a Ertebølle-period coastal settlement. Among others, a wooden float was found, with part of the fishing line still attached which was apparently made of spun plant-fibre. No knot can be observed.

Kongsted Lyng (South Zealand) Mathiassen 1942; Becker, 1947

An Early Neolithic handlejar, found in a bog as an offering. A piece of rope still ran through two of the four ears, but no knot was preserved in

†Early Neolithic: in Denmark between about 6000 and 5500 B.P..

it (according to Mathiassen). The rope was double twined, and made of tree bast, but the exact species of tree could not be determined.

Sludegaard Mose (West Coast of Fynen). Albrectsen 1954.

A rope fragment of double twined lime-bast, found through the handles of a Early Neolithic handlejar which had been placed in a bog as an offering, together with other whole pots, parts of domesticated animals, and probably humans as well. No knot is mentioned in the description, and an illustration of the piece could not be obtained.

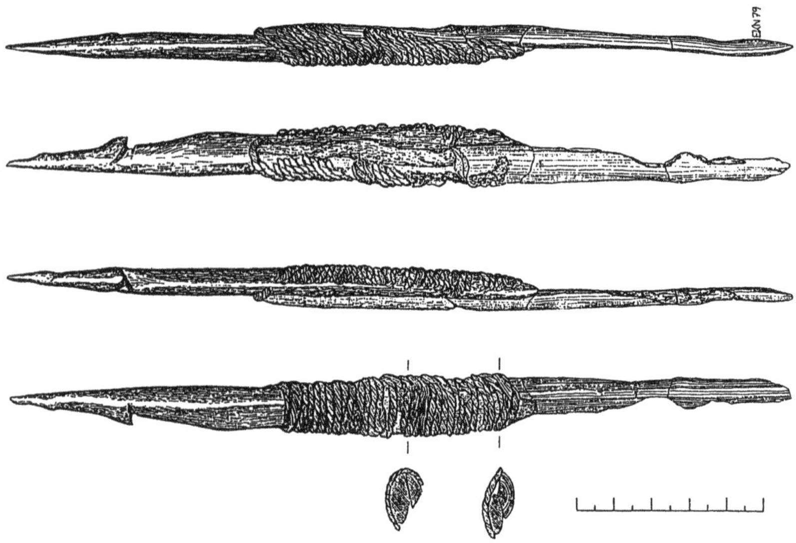

Fig. 4. Bone point, lashed to a wooden shaft Ulkersrtup Lyng. Source: Anderson, Sørgensen and Richter 1982, p. 76, fig. 68

Tulstrup Mose (North Zealand) Becker 1945; 1947.

Four fragments of rope, found, together with fragments of what perhaps was a carrying net, inside an Early Neolithic handlejar which had been placed in a bog as an offering. One of the rope fragments is said (Becker) to have a noose on one end, and therefore presumably also a knot. This is, however, not specified. All four pieces were made of lime bast.

Ulkestrup Lyng, Åmose (middle Zealand). Andersen, K. et al. 1982.

Two hut sites found at the edge of a former lake in the Åmose, dating to the Maglemose period (C14 date about 8200 B.P.) Among the finds was a bone point, barbed, still fastened to the remains of a wooden shaft by a lashing of turned string, probably made of lime bast. The string had been wound tightly and regularly around the shaft and point, but unfortunately, precisely the ends where the knots must have been had disappeared (Fig. 4).

Åmose, middle Zealand. Rønne, 1989.

Four natural stones found in a bog, three of which still had rope around them whereas the fourth had traces of rope. All four were lying on top of what were possibly the remains of a net, which was too fragile to be preserved. The rope was made of lime bast. No knots were preserved. The site was the garbage dump of a late Ertebølle settlement on a low island in the bog, but the large amount of pottery found makes the author suspect that the site might have been an offering place as well.

These ten sites are to my knowledge the only ones so far where artefacts have been found of which it can be assumed with reasonable certainty that they originally contained knots. Even so, only three knots have actually been preserved: a clove hitch, something resembling a granny, and a row of half hitches.

As was to be expected, all ten sites are water-logged sites. Five of them were coastal—or lake—settlements, and in all five the artefacts concerned were directly connected with fishing: a fishhook, a leister, a harpoon, netsinkers, a line float. The five other sites are all offerings in bogs, but here the contexts in which the knots were originally used are more diverse: two nooses used to hang or strangle people, two carrying slings for jars, one carrying bag (if that interpretation is correct).

In nine of the ten cases the material used for the rope was of vegetable origin. The one exception was the piece of fishline from Tybrind Vig, which had been made of some animal product. Five of the remaining nine artefacts had rope out of tree bast, and in four cases this could be more closely identified as lime bast. In three cases no more could be said than that the material was vegetable, but of one of these it could be established that it had not been tree bast. Finally, in one case (Skjoldnæs) nettle-fibres had been used.

The rope could be produced by braiding (Bokilde), but spinning and/or twining seems to have been more common (eight, perhaps nine cases).

It is clear that only the bog offerings offer a glimpse of the uses of knots and knotting in activities other than those immediately connected with water. It is therefore the more unfortunate that only one of those five sites has produced an actual knot (Sigersdal, Fig. 1), and a partly undone one at that. Of course, a source of potential bias with offerings is that they have strong religious/magical overtones, and that perhaps certain types of knots were reserved specifically for these contexts, and not used in others.

Indirectly, however, other find categories, such as the textile remains found at Tybrin Vig, can give us some impression of what we might expect to find. Those textile fragments had been constructed using the 'needle binding' technique [10], a difficult technique requiring skill and spatial insight, and clearly the result of a long tradition. The skill and the materials to construct many different and complicated knots were thus available by the late Mesolithic period at the least. This is not to say, of course, that such knots were therefore indeed produced. Technology and skill are by no means the only factors at work here. But it will not be too surprising if they do turn up some day.

As an extra, I will end by mentioning another find from outside Denmark, and early-Mesolithic fish-net fragment from Antrea (Vuoksenniska), on the Karelian isthmus, formerly east-Finland, which was recently brought to my attention. The fragment was found in 1913 by workers digging in, again, a bog which had formerly been part of a sea strait between the Baltic Sea and Lake Ladoga. The whole find consisted of the net itself, with its floats and weights, a number of bone, stone and antler artefacts, and other objects of stone and wood. It was interpreted as the remains of a fishing boat, with net and toolkit, which sank.* The net itself was originally quite long, about 27 to 30 metres, and between 1.3 and 1.5 metres high. It was made of willow, with its 31 stone-weights attached by willow bark. The meshes were about 6 cm apart, and knotted by what Pälsi [13, p. 16] calls a 'Russian knot', a type known from Estonia and the eastern regions of Finnish settlement (Fig. 5). There is one radiocarbon dating from one of the netfloats, which recorded 9280±210 B.P., a date which fits well with the geological date of the surrounding bog.

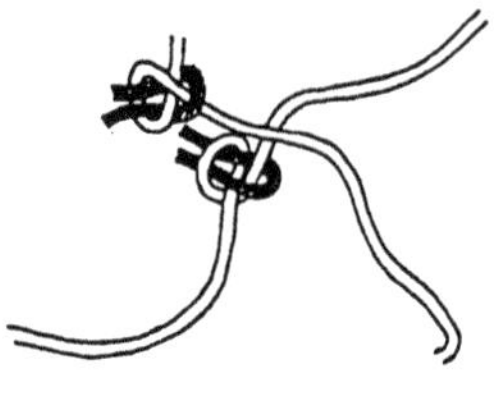

Fig. 5. The Antrea 'Russian Knot'
Antrea. Source: Pälsi 1920, 1–19

CONCLUSIONS

We can be certain that both the materials and the skill necessary to make complicated knots were present in Europe from a very early age, at least since

*Taavitsainen, unpublished mss.; thanks to Pieter van de Griend for providing a copy.

the end of the last ice age. This is evident from, among others, the Danish material, the textiles and ropes from Neolithic Switzerland, and the rich equipment of the Ice Man. But for knotters, interested in actual knots, the conclusion of this Chapter is bound to be disappointing, though not unexpectedly so, after what I tried to explain at its beginning. To put it bluntly, it is simply impossible at the moment to study the knots of the different periods and regions of prehistoric Europe, and the contexts in which these knots were used, thoroughly and responsibly. The dangers of using an *argumentum ex silentio*,[†] especially in a context such as this, where we know beforehand that the chances of our study material surviving the ages are slim, are well known. And unless the funding given to archaeological institutions increases manifold, and the possibilities for the sharing of archaeological information improve considerably, this situation is unlikely to change within the near future. Only when we can be fairly certain that we possess a *representative* sample of the total knot-repertoire and its contexts of the region and period under study, for example, can we begin to address such interesting problems as the symbolic meanings of the different knots, or their origins and spread.

But, to end on a more positive note, there are things that knotters themselves can do to improve the situation. Most archaeologists totally lack knotting expertise; but each of them usually knows fairly well what treasures are hidden in their Institute's storerooms.

Knotters, wanting to study the knots of their own region, can contact their local museum and discuss with the local archaeologist the conditions for getting access to those treasures. Then they can see the prehistoric knotting-material with their own eyes. Most archaeologists are only too happy when someone shows interest in their efforts, and will gladly guide a knotter in archaeological matters. If all knotters interested in the history of knots were to take this kind of action, it would then perhaps become possible to write a truly good 'Knots in European Prehistory'.

References

1. E. Albrectsen, *Et offerfund fra Sludegaards Mose* (Fynske Minder, 1954) 4–14.
2. K. Andersen, S. Sørgensen, and J. Richter, Maglemose Hytterne ved Ulkestrup Lyng, (1982).
3. S. H. Andersen, *1980—Tybrind Vig. Forløbig meddelelse om en undersøisk stenalder boplads ved Lillebælt* (In: Antikvariske Studier **4**, 1980) 7–18.

[†]litt. argument from silence—'We have not found it, therefore it did not exist'; conversely, 'We have no evidence that it did not exist, therefore it existed'.

4. S. H. Andersen, *1985—Tybrind Vig. A preliminary report on a submerged Ertebølle settlement on the west coast of Fyn* (In: Journal of Danish Archaeology **4**, 1985) 52–69.
5. S. H. Andersen, *1987—Tybrind Vig. A submerged Ertebølle settlement in Denmark* (In: European Wetlands in Prehistory [John M. Coles & Andrew J. Lawson, eds.] 1987) 253–280.
6. C. J. Becker, *1945—To Nordsjællandske Offerfund fra Yngre Stenalder* (In: Fra det Gamle Gilleleje 1945) 41–60.
7. C. J. Becker, *1947—Mosefundne lerkar fra yngre stenalder* (Aarbøger for nordisk Oldkyndighed og Historie 1947) 12–14; 42.
8. L. Bender Jørgensen, *The string from Sigersdal Mose* (In: Bennike, Pia & Ebbesen, Klavs: 'The Bogfind from Sigerdal. Human sacrifice in the Early Neolithic.' Journal of Danish Archaeology **5**, 1986) 105–106.
9. P. Bennike, K. Ebbesen, and L. Bender Jørgensen, *1986—Early Neolithic skeletons from Bokilde bog, Denmark* (In: Antiquity **60**) 199–209.
10. E. H. Hansen, *Nålebinding: definition and description* (In: 'Textile symposium in York, 6–9 May 1987', Textiles in Northern Archaeology. Nesat III, 1990) 21–28.
11. *Der Mann im Eis*, Band 1—Bericht über das Internazionale Symposium 1992 in Innsbruck. (F. Höpfel, W. Platzer and K. Spindler, eds. Eigenverlag der Universität Innsbruck, 1992).
12. T. Mathiassen, *Et udvalg af aarets mosefund* (Fra Nationalmuseets Arbejdsmark, 1942) 9.
13. S. Pälsi, *1920—Ein steinzeitlicher Moorfund* (Suomen Muinaismusitoyhdistyksen Aikakauskirja, XXVIII:2).
14. P. Rønne, *1989—Saenkesten* (Skalk, 1989) 21–24.
15. H. Schwab, *1959/1960—Katalog der im Bernischen Historischen Museum aufbewahrten Faden, Geflecht und Gewebefragmente aus Neolithischen (evt. Bronzezeitlichen) Seeufersiedlungen* (In: Jahrbuch des Bernischen Historischen Museums in Bern 39/40) 336-366.
16. J. Skaarup, *1980—Undersøisk Stenalder* (Skalk **1**, 1980) 3–8.
17. J. Skaarup, *1981—Lyster* (Skalk **6**, 1981) 10–11.
18. J. Skaarup, *1982—Ærø* (Journal of Danish Archaeology **1**, Recent discoveries, 1982) 166.
19. K. Spinrad, *Der Mann im Eis. Die Ötztaler Mumie verrät die Geheimnisse der Steinzeit* (C. Bertelsmann Verlag GmbH, München, 1993).
20. J. P. Taavitsainen, *The Antrea Net* (unpublished manuscript, 1993).

CHAPTER 4

ANCIENT EGYPTIAN ROPE AND KNOTS

Willemina Wendrich

Introduction

Rope making and knotting were perhaps the first techniques available to early man. Before man knew how to make pots or stone knives, he would have used organic materials, such as wood, twigs, grasses and leaves. Knotting would be necessary to connect wooden posts for making shelters. In many areas of the world those shelters were probably made of matting, plant stems or twigs tied with strings to form the sides and roof. For basket making, which was a human invention pre-dating pottery, the same materials were used as for rope making and knotting. Because stone and pottery survives to a much greater extent in the archaeological context than organic materials, our image of ancient cultures is undoubtedly lopsided: stone-age man did of course use stone tools, but in daily life the use of organic materials must have been even more important than the use of stone objects. 'Basket-age' would probably be a better modifier than stone-age, to indicate the most important artefacts of that period. On the other hand, the indication 'basket-age' is valid for the most part of human history, with the exception of the last 40 years of our history, in which the use of plastic is taking over rapidly the niche of basketry.

In Egypt, archaeologists are in a favourable position to catch a glimpse of the importance of organic materials in daily life. Because rainfall is extremely limited in large parts of the country, the preservation of rope, basketry, textiles and leather is extremely good. Until a decade ago, excavators have failed to pay systematic attention to these materials, probably because their attention was focussed on pottery and written material. Furthermore, Egyptology traditionally concentrated on palaces, temples, tombs and (religious) inscriptions. Not until the last twenty years has there been an increasing interest in settlement archaeology and a focus on the life of ordinary people in ancient Egypt. The objects of daily life, such as rope and knots, are used as a source of information on the social context of the ancient Egyptian inhabitants. An analysis of the artefacts is used as the basis for an interpretation of the social context of their producers and users.

The function of knots is many-sided, from purely functional to symbolic and religious. Information on the use of rope and knots can be found in excavations, but also on wall paintings, reliefs and in texts. In two of the following sections the Egyptian sources of our knowledge on knots will be surveyed, first the archaeology, secondly the information from texts. These sections do not pretend to be exhaustive or conclusive. Many more years of fundamental research will be needed to come to conclusions on a more firm basis. They merely form an introduction into ancient Egyptian knotting, by presenting a selection of the information available at the moment.

Rope and knots from archaeological excavations

Because most excavation reports lack detailed information on rope and knots, it is not possible to make a chronological survey of their use in Egypt. The information referred to in this chapter has been recorded by the author during ten expeditions to four different archaeological sites. These sites represent several periods, types and regions (Fig. 1).

Tell el-Amarna is the earliest site, dating from 1350 B.C. It was built in Middle Egypt as the new Capital of Pharao Amenhotep IV / Akhenaton. The main city was built in a desolated area near the River Nile, and abandoned less than a century after its foundation. In the main city, hardly any organic material survived. The rope and knots were excavated at a workmen's village which was located further inland, in the foothills of the Eastern desert. The village was built for the workmen who made the tombs in the nearby mountains for the Pharaoh and the Nobles of the city. The workmen and their families lived there for a relatively short period of only one or two generations. Because of the brief period of occupation of the village, the rope and knots can be precisely dated.

Berenike is much later than Amarna. It was a harbour town on the Red Sea coast, founded in 275 B.C. and abandoned in the late 5th century A.D. Berenike was the most important harbour for trade between India, Africa south of the Sahara, the Arabian peninsula and Rome. The merchandise had to be transported on camel back from the Red Sea coast to the Nile, following a route through the Eastern Desert which took 12 days. The rope and knots referred to below, were found in a trash deposit dating from the 4th and 5th centuries A.D.

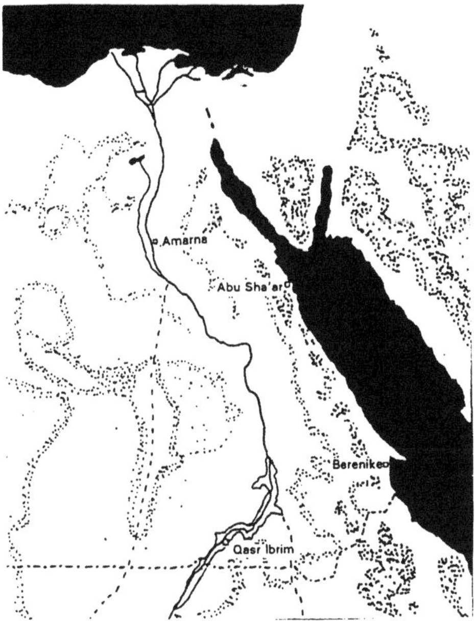

Fig. 1. Map of Egypt with Tell el-Amarna, Abu Sha'ar, Berenike and Qasr Ibrim

Abu Sha'ar is also located at the Red Sea coast, 30 km north of the modern town Hurghada. Abu Sha'ar was founded by the Romans as a military installation in the 3rd century A.D. After a brief period of abandonment, part of the fort was used from the 4th to the 7th century A.D. by an early Christian community. These were probably heremites, living a life of contemplation in the deserted fort. Rope and knots were found in the trash dumps, barracks, street areas and in the main building, which was in a later phase modified into a church. The rope and knots were dated to all periods of occupation.

Qasr Ibrim was a town built as an eagle's nest high above the River Nile in Egyptian Nubia. In periods of unrest, the population left the hamlets down by the river and built their houses in the town, protected by strong defences. Qasr Ibrim has a long history, from the New Kingdom period (1500 B.C.) to the 19th century. The rope and knots from Qasr Ibrim are all from one house-area. The house is dated to the Meroitic and the X-group period (3rd century B.C. to 6th century A.D.).

Before continuing with a survey of the rope making and knotting techniques occurring at these sites, the terminology used in this section should be briefly reviewed. Any terminology is derived from a classification, however implicit and inconsistent the criteria for such a classification may be. The criteria of the ancient Egyptians are difficult to retrieve, but they certainly will have been different from those of twentieth century England. Therefore, the composition of rope and string is indicated by a descriptive formula, rather than implementing present-day English terminology. The same hesitation should be felt in applying English names for knots to those found in Egypt. Since the complexity of knotting formulas will entangle most readers, the knots are recorded by two drawings: one conveying the appearance, the other showing the construction. An exception is made for the five most commonly occurring knots: the half knot, overhand knot, reef knot, granny knot and mesh knot. For practical purposes the names of these well known knots are used below. The orientation of these knots is indicated in the same manner as that of rope.

The rope formula indicates the direction of twist and the number of strands used. The direction of twist is indicated by a letter S, Z or I, the middle stroke of the letter indicating the three variations (Fig. 2). The direction of twist varies at the different levels: usually spinning, plying and cabling are orientated in opposite directions. Thus zS_2Z_3 rope consists of fibres spun in the z-direction(z). Two of those strands are plied in the S-direction (S_2). Three of those S-plied, z-spun strings are cabled in the Z-direction (Z_3 see also: Wendrich 1991: 32). Throughout the entire Egyptian history, the ropemaking orientation encountered by far the most is zS_2.

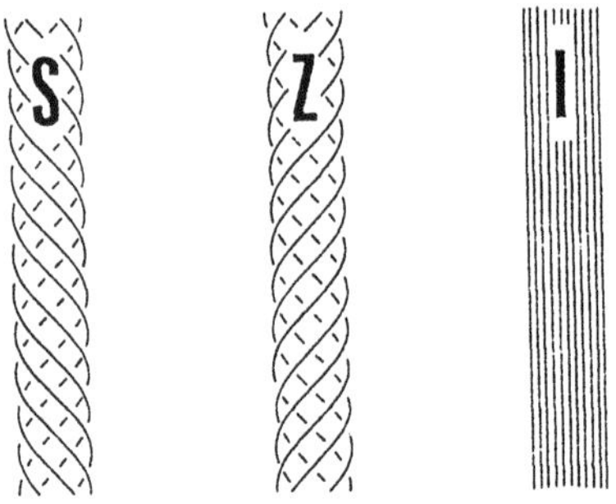

Fig. 2. The direction of twist for spinning, plying or cabling is indicated with a capital letter S, Z or I, the middle strokes of the letters indicating the orientation

The same indications, S and Z, are used to indicate crossings and orientations of knots. Crossing strands can be orientated *overS* or *underS*; *overZ* or *underZ* (Fig. 3). Overhand knots (Fig. 4) or mesh knots (Fig. 5) are $S-$ or $Z-$orientated. Reef knots are made of two half knots in opposite orientation: they are either SZ or ZS orientated. Granny knots are made of two half knots in the same orientation: SS or ZZ (Wendrich 1991: 42-3).

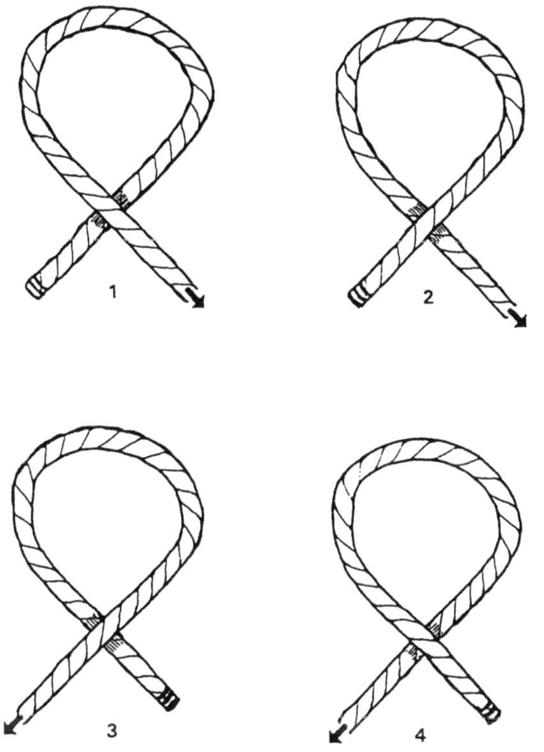

Fig. 3. Orientation of crossing strands: 1 = over-S crossing; 2 = under-S crossing; 3 = over-Z crossing; 4 = under-Z crossing

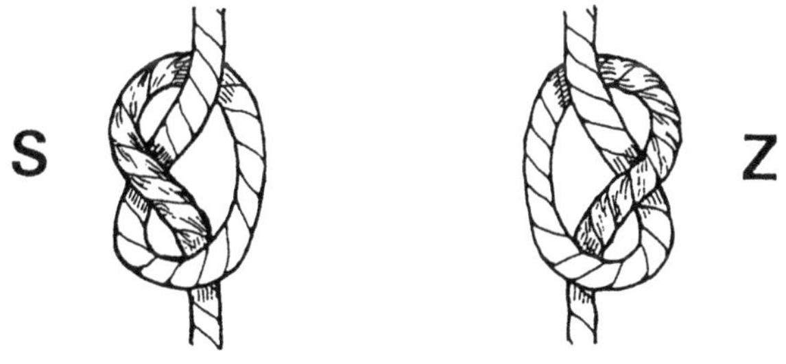

Fig. 4. Overhand knot (#46)*in S and Z direction

*The symbol # refers to the numbering system of *The Ashley Book of Knots* [1]

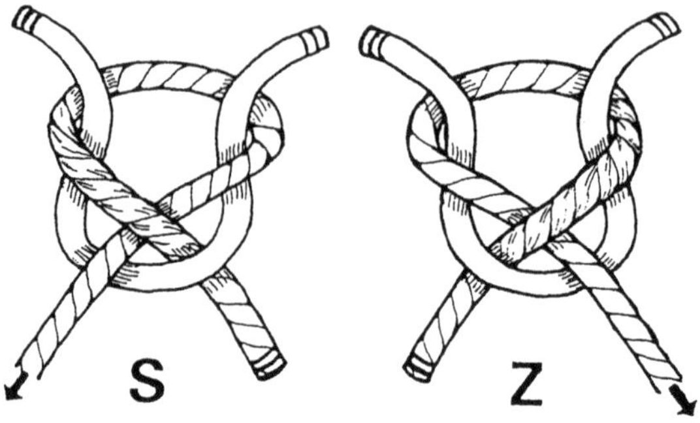

Fig. 5. Mesh knot (#402) in S and Z direction

Two formulas are used to describe plait patterns found in Egypt: \1/1\\1 indicates the plait pattern 'under 1, over 1, with a shift of 1'; \2/2\\1 stands for 'under 2, over 2, with a shift of 1' (see below, Fig. 28).

Amarna

In Amarna most string is made of tall grasses (*Desmostachya bipinnata* and *Imperata cylindrica*). In large ropes complete plants, roots and all, are used. The string and rope is made in one production phase: two bundles of grass are spun by rolling them between the hands. After rolling, the bundles are plied in the same movement by picking up the first bundle and laying it behind the second. In the next rolling movement the second bundle is lifted and put behind the first. The opposite orientation of spin and ply locks the string. This method or ropemaking is still in use in Egypt today. Farmers and fishermen do not buy their string, but make the required length of rope on the spot. With artificial fibres becoming generally available in the Egyptian country side, the art of rope making might well disappear in the next few decades.

In the ancient material there are clear indications that zS_3 string (three z-spun yarns plied in the S-direction) is first made as zS_2 string, and that the third yarn was spun and added in a second production phase. From tomb paintings it is also known that for larger ropes, used in shipping, the fibres were twisted with the help of weights (Teeter 1987: 72). Apart from grass, the Amarna rope was also made from the leaves of the dom palm (*Hyphaene thebaica*) and from the rind of papyrus stems (*Cyperus papyrus*), which is strong and purple brown. The rind had to be removed before the white spongy inner part of the culm could be cut into thin slices for making papyrus for writing. Almost all grass string was zS_2 string. Rope was not cabled, but was made of thicker bundles of grass, made into zS_2 or zS_3 rope*. The palm leaf

**String* is defined as having a diameter less than 10 mm. *Rope* has a diameter of 10 mm or more.

and papyrus string was often made in an opposite orientation to that used for the grass string: sZ_2 and sZ_3. Very thin yarn was made of flax (*Linum usitatissimum*). None of the raw materials was available to the inhabitants in the direct vicinity of the village, but all plants could be found at an hour's walking distance, near the Banks of the Nile.

The workmen at Amarna were not specialised in making knots. The knots used are very basic and surprisingly similar to the standard repertoire of knots available to ordinary people in the modern Western world. Most frequently present is the *overhand knot*, in both orientations, but the Z-overhand knot is encountered most. The *half knot*, *reef knot* and *granny knot* are also found often. Because the finds are fragmentary, it is often not possible to decide what the exact use of the knots must have been, but it is apparent that the people in Amarna were using knots without much expertise. The 'knotters' curse' was found a number of times: using a reef knot to connect two strings with different diameters and often made of two different materials. Many knot constructions gave the impression of being unplanned: windings and half knots; loops through which ends have been pulled unsystematically. These constructions have no modern or ancient parallels and seem to have been 'invented', or rather muddled up on the spot. Many stopper knots were used, mostly to keep strings from unravelling, but also to keep a string from slipping through a loop. The most frequently used stopper knot is the overhand knot, but occasionally other knots were used: combinations of overhand knots, or one of the unsystematic newly invented constructions. None of the more complex knots occurred and neither did splices. Loops were made by pulling a string through the start of the ply (see below, Figs. 14, 22).

Peculiar to Amarna are large grommets made of string, or bundles of grass, which seem to have functioned as stoppers or pads. The pad of Fig. 6 is made with sZ_2 string, which was wound in Z-direction in four stages (see Fig. 7). Lashings did occur also: the handles of brushes are lashed, and in one case a hitched lashing is used (Fig. 8). This particular brush had been used as a paint brush for whitewashing walls.

Nets were made with sZ_3 flax yarns and mesh knots. They had a mesh size of 30 x 32 mm. A carrying net for amphoras was made of sZ_2 papyrus string and overhand knots. From the 70 fragments found, the construction of the net could be inferred: the base of the carrying net was formed by a ring (Fig. 9), made of a bundle of grass, wrapped with strings. At six points, divided equally over the ring, two strings were fastened, which formed the end of the net (Fig. 10). The net was started at the handles and made with Z-overhand knots in two rows of six knots. The two handles were made of sZ_2S_3 rope, fastened with a knot (Fig. 11). Two men were needed to carry a pole, which was put through the handles, with its heavy load of net and amphora in between them.

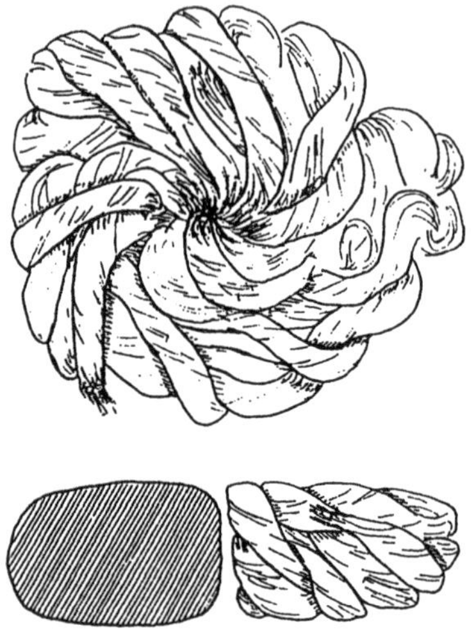

Fig. 6. Grass pad, made as a grommet

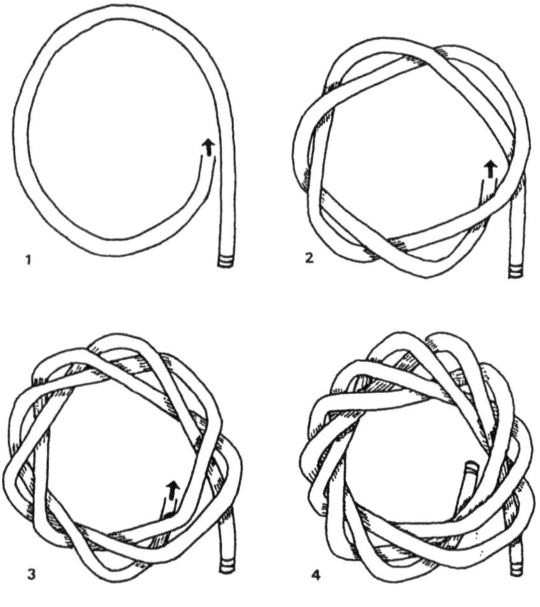

Fig. 7. Schematic drawing of the pad of Fig. 6

Fig. 8. Paint brush, made of a bundle of dom palm-leaf strips, tied just above the middle (Reef Knot, #74). The leaf strips are folded back and fastened with hitched lashings (Half Hitches, #49, #53, #56)

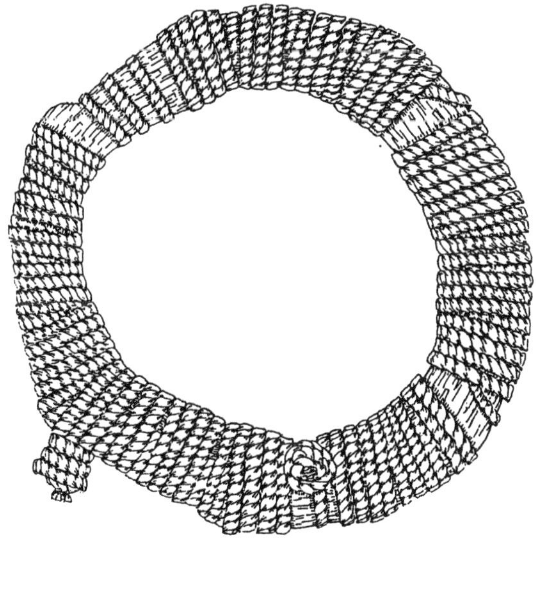

Fig. 9. Bottom ring of a carrying net for amphoras

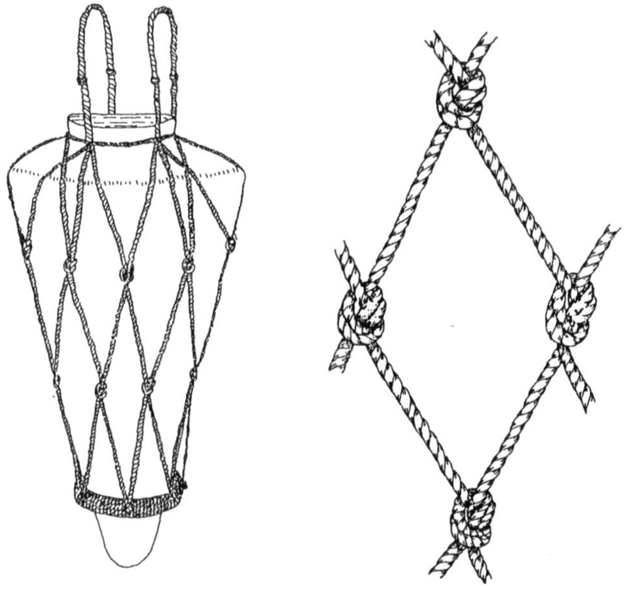

Fig. 10. Reconstruction in drawing of the carrying net and detail of the knotted net. The pointed base of the amphora was held by the ring. A net, made with overhand knots in double strings, held the body (Overhand Net Knot, #406)

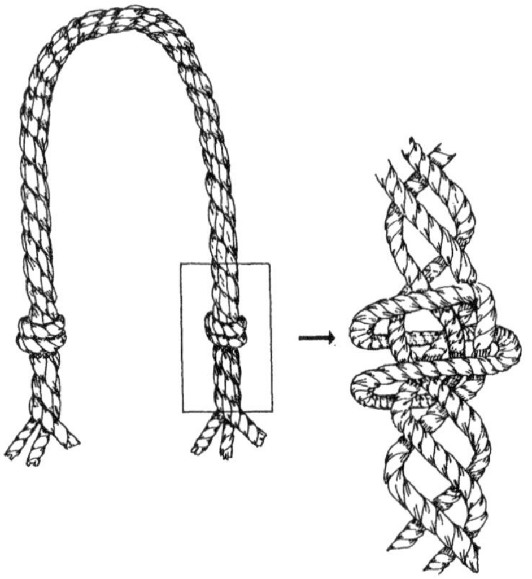

Fig. 11. One of the handles of the carrying net for amphoras, with a reconstruction of the knot (Double Overhand Knot, #516)

Except for simple three- and four-strand plaits, plaiting is basically unknown in Pharaonic Egypt. Complicated knots, which can be considered related to plaiting do not occur either.

Berenike

Berenike, a harbour town at the Red Sea coast, differs from Amarna both in date and region. Both sites have in common that they were located in a more or less isolated area. The workmen's village near the tombs in the desert, Berenike in the Eastern desert at the edge of the sea. The plant material in the vicinity of Berenike was limited to desert shrubs. The ropes used in seafaring were mainly made of the leafsheath fibres of the date palm (*Phoenix dactylifera*). This material must have been imported from the Nile Vally, a journey which took 12 days by camel. Other materials used were dom palm (*Hyphaene thebaica*), grass (*Desmostachya bipinnata*), flax (*Linum usitatissumum*) and cotton (*Gossypium arboreum*). Cotton was not introduced into Egypt until the fourth century A.D. (Germer 1985: 123).

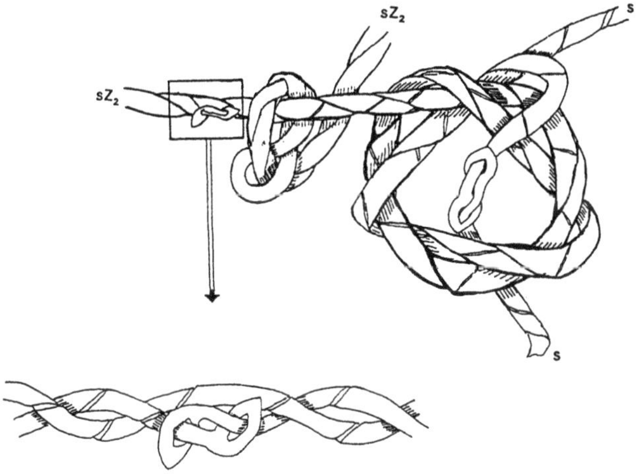

Fig. 12. A rope made of untanned camel hide, with a detail of the method of connecting strips of hide (Strap Bend, #1491)

The rope found at Berenike is well made. Unlike the rope from the Pharaonic period, which consisted mainly of grass (thin bundles for string, thick bundles for rope), the rope from Berenike is made of cabled date palm fibre. Another method of making strong rope was by wrapping a bundle of grass with zS_2 grass string. This type of rope must have been particularly resistant against pulling forces on the width of the rope and might have been part of a Roman *ballista*. Apart from vegetable materials, grasses and palm leaf, a large part of the string was made of animal material. Fine string was made of goat hair. This material was always spun, plied and cabled in the opposite direction as was usually the case: sZ_2S_3. Apart from goat hair, untanned camel hide was used occasionally. Of course camel hide cannot be spun. The solution for connecting strips of rawhide was to cut the ends in the form of slitted triangles, which could be hooked one in the other (Fig. 12). Two S-twisted lengths of connected camel hide strips were Z-plied. This use of animal material does not only point to a possible lack of vegetable material, but also indicates that there were contacts between the inhabitants of the town and the local Bedouin population.

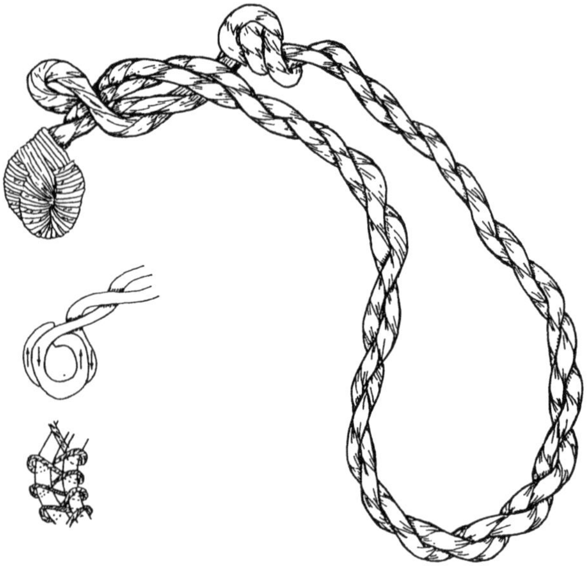

Fig. 13. Wrapped stopper knot, loop and Z-overhand knot in a zS_2Z_2 rope, made of date palm fibre

Not only the raw materials are different at Berenike, but also the knots and other applications. From a 5th century A.D. trash dump is a beautifully made stopper in a zS_2Z_2 rope, made from date palm fibre (Fig. 13). Loops were made by pulling a string through an opening, made by twisting open the end where the plying was started. Fig. 14 shows a zS_4 string*, made of very fine date palm fibre, with a loop and an S-overhand knot to prevent the loop from sliding. The construction is very similar to that in the rope made of camel hide shown in Fig. 12. In general the most frequently found knots are

*This string can also be recorded as zS_2S_2.

again overhand knots, looped overhand knots, double overhand knots and reef knots. Fig. 15 shows a fragment of date palm leaf with four overhand knots, three in Z-direction, one in S-direction. This knotted strand and two other fragments from the same context formed a strand with eight overhand knots: four in S-direction and four in Z-direction.

Fig. 14. Loop in zS_4 string, of fine date palm fibre. The loop is prevented from sliding by an S-overhand knot

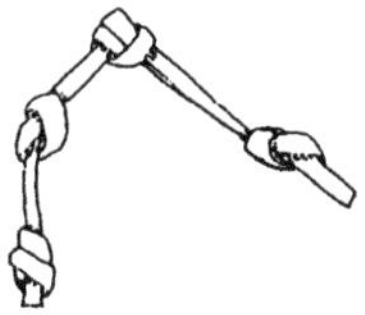

Fig. 15. Strand of palm leaf with three Z-overhand knots and one S-overhand knot (Overhand Knots, #46)

Fig. 16. Four-strand plait, made with strips of date palm leaf

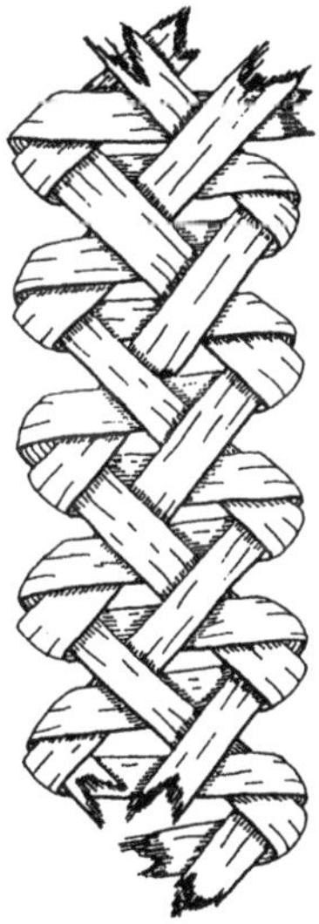

Fig. 17. Schematic drawing of the plait of Fig. 16

Plaiting does occur at Berenike and is used for making a type of basketry which is very common in Egypt in the post-Pharaonic period. More will be said about plaited basketry when discussing the cordage of Qasr Ibrim. Apart from these plaits some decorative plaiting occurs at Berenike, for instance a four-strand plait, made with strips of date palm leaf (Figs. 16 and 17).

Abu Sha'ar

The Roman Fort at Abu Sha'ar was in use between the 3rd and the 7th century A.D. Its situation was very similar to that of Berenike: at the Red Sea coast, isolated by the mountainous Eastern desert which lies between the Red Sea coast and the Nile valley. The nature of the finds at Abu Sha'ar differs considerably from those at Berenike. In Abu Sha'ar no luxury items were found. Even the cordage from the Roman fort was very basic and utilitarian. Vegetable materials used were grass (*Desmostachya bipinnata*), the leaf and leafsheath fibre of the date palm (*Phoenix dactylifera*) and to a lesser extent rushes (*Juncus arabicus*) and flax (*Linum usitatissimum*). A large number of grass string ties were found, with a diameter ranging between 100 and 120 mm, tied with an S-half knot (Fig. 18).

Fig. 18. Schematic drawing of two stages of tying a zS_2 grass string around bundles of rushes

A number of these were found in situ, tied around bundles of rushes (*Juncus arabicus*). These bundles, imported from the Nile Valley, were used for several purposes. Complete bundles were found in the roofing layers, indicating that they were used on the roofs, perhaps to weigh down other roofing materials such as twigs and matting. Rushes were also used to make fish traps and baskets. Considering the large number of grass string ties that were retrieved, the rushes must have been imported in large quantities. A plausible explanation is that the rushes were mainly used as fodder for the camels and goats present at or near the fort. At Abu Sha'ar a large amount of goat hair string was found. This string was largely made in an orientation opposite to that of grass and datepalm fibre string: sZ_2 and sZ_2S_2. This is consistent with the occurrence and orienation of the goat hair string from Berenike, also located both at the Red Sea coast and near to the Eastern Desert. The knots found at Abu Sha'ar show an awareness and practical training in the use of knots. Apart from the half knots used around bundles, the reef knot is used frequently for the same purpose. The reef knots are not mis-used for tying two ropes together and there are very few granny knots. Fig. 19 is an example of two grass ropes of different sizes, connected with a hitch.

Fig. 19. Connection of ropes of different size: the string is hitched to the larger rope with a Clove Hitch, #1670. There are two Half Hitches at the end (#54)

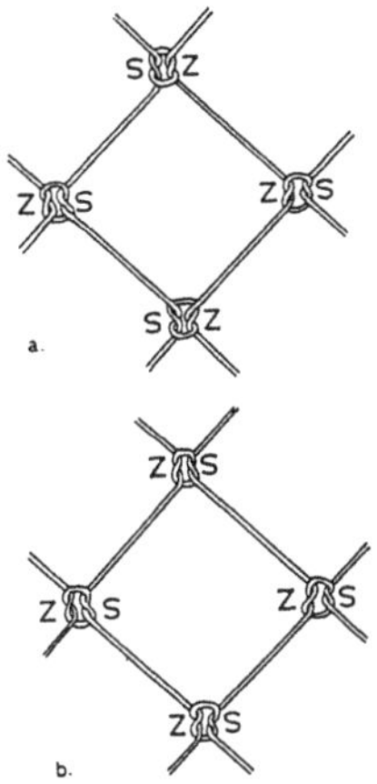

Fig. 20. Schematic drawing of netting with reef knots. a: orientation differs per row: SZ and ZS; b: orientation is the same for all reef knots (ZS)

Reef knots were used for making carrying nets. For a well-knotted fabric the reef knots should be orientated differently in each row of knotting: first a row of *SZ* reef knots, then a row of *ZS* reef knots, in order to prevent the net from curling up (Fig. 20a). In Abu Sha'ar, however, all reef knots are orientated the same: only *ZS* reefknots are used (Fig. 20b). Fishing nets were not made with reef knots, but with *Z*-mesh knots. A large quantity of fishing nets were found at Abu Sha'ar, made of *s*-spun and *Z*-plied flax yarn. The mesh size differed for the two main periods of occupation (Wendrich & Van Neer, in press).

Qasr Ibrim

The materials used for cordage at Qasr Ibrim are mainly leaves of the dom palm (*Hyphaene thebaica*); leaves and leaf sheath fibre from the date palm (*Phoenix dactylifera*), grass (*Desmostachya bipinnata*), flax (*Linum usitatissimum*) and flax (*Gossypium arboreum*). Cabled rope does occur frequently in Qasr Ibrim and is usually made of date palm fibre. Here, as in Berenike, large ropes occurred, consisting of grass bundles wound with zS_2 string. Plaits with three or four strands, usually made of palm leaf or bundles of grass, occur quite often. Plaiting around a core was found a few times, with six and eight strands, made of flax yarns or strands of date palm leaf. Fig. 21 shows eight-strands plaited around a core in a \2/2\\1 pattern.

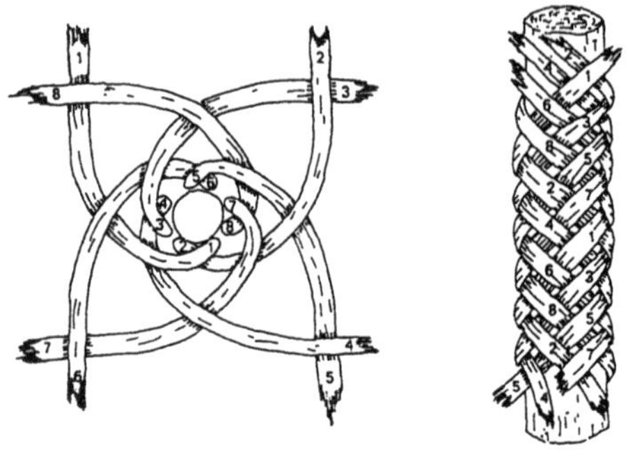

Fig. 21. Eight strands plaited around a core in a \2/2\\1 pattern; left: schematic cross section; right: overall appearance

In the X-group house at Qasr Ibrim knotting was used both functionally and decoratively. Loops were made by pulling a string through an opening at one end, a stopper knot preventing the loop from slipping (Fig. 22).

Fig. 22. Loops found at all Egyptian sites were made by pulling the string through the part where the plying was started. In this example from Qas Ibram, the loop is prevented from slipping by a stopper knot consisting of a Z overhand knot on top of an S-overhand knot

Other examples of functional knotting are found in roofing, bags, baskets and sandals. In Qasr Ibrim many roofs and floors covering cellars, consist of wooden beams, covered with a lattice made of the midribs of date palm leafs, tied with *S*- twisted dom palm leaf (Fig. 23). These light constructions were covered with old pieces of matting and finally with a layer of mud. Knotting was used often in combination with plaiting, which was the most frequently occurring basketry technique in Qasr Ibrim. Plaiting did not occur in Egypt before the Third Intermediate Period (1000 B.C.). From that period onwards plaiting was increasingly important, becoming the dominant basketry technique in Egypt by the ninth century A.D.† Small palm leaf bags were plaited with date palm leaflets.

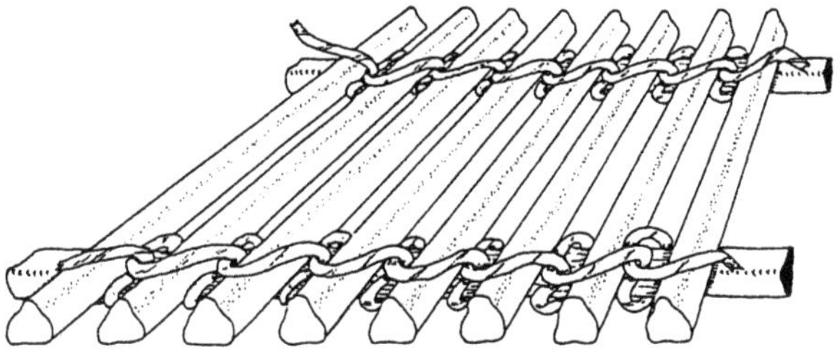

Fig. 23. Schematic drawing of a roof made of date palm midribs, tied with half knots

The date palm is a feather palm, of which each leaf consists of a large mid rib, with small leaflets at both sides. These leaflets can be split along the vein, with both halves still connected at the base (Fig. 24). Thus prepared, these leaflets were cleverly interlaced, so that the closed ends formed the top side of a bag, and the tips of the leaflets were fastened at the bottom side, forming a strong circular edge. Separate strands were pulled through the fabric and knotted at the base with reef knots, or grannies to close the bags. Many of these little bags have been found and they all had been opened at the top side, the natural connection being torn. This aspect clarifies the function of the bags: they were disposables, since opening them at the top made them unsuitable for re-use. Just as we mistreat plastic bags by tearing the plastic, rather than opening the knot, the people of Qasr Ibrim tore the palmleaves and left the knots in place. These bags must have been a standard packaging for commodities, but the archaeological context thus far has not yielded information on the original contents of the bags.

†Before the introduction of plaiting, all baskets were made with variations of coiling and twining (see: Wendrich 1991 and in press).

Fig. 24. The date palm is a feather palm, with small leaflets branching off a tall midrib. The leaflets are split in the length and used for the production of rope and basketry

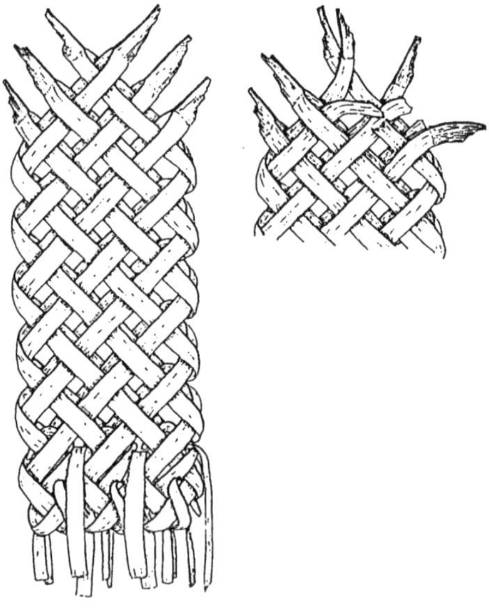

Fig. 25. Schematic drawing of plaited 'disposable' bags from Qasr Ibram

Apart from these disposable bags, plaiting was also used for making baskets and mats. These were made of plaited strips, laid out parallel and sewn by pulling a string alternately through the edges of both plaits (Fig. 26). The finishing of the edge of the mat was made by sewing on a separate plait. Baskets were made by sewing a long plait spirally from centre to rim (Fig. 27).

Fig. 26. Mat made of plaited strips which were sewn in the edges. Because the edges are pulled inside each other, the strands seem to form continuously plaited fabric. The plaits were made with 17 strands in a \2/2\\1 pattern

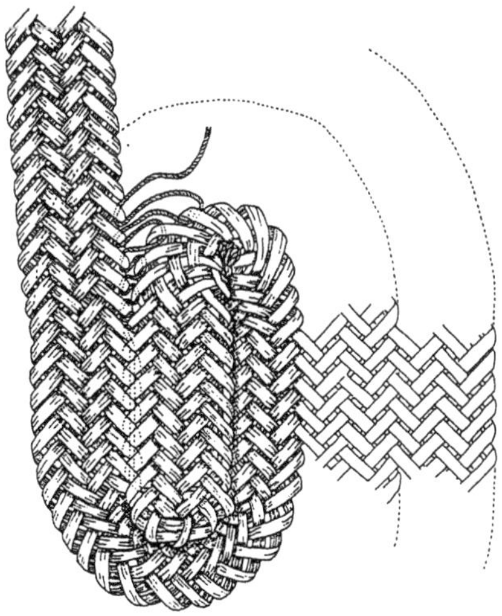

Fig. 27. One long plait is sewn spirally to form a basket. The plait is made with 9 strands in a \2/2\\1 pattern

The sewing was sometimes done with zS_2 string, made of palm leaf. Most often, however, unspun strands of dom palm leaf of 2 mm wide were used for sewing. Nine meters of plaited strip were sewn with only 500 mm long strands. The short sewing strands were connected with overhand knots, generally in Z-orientation. The use of string to sew the strips seems to have been an Egyptian tradition, while the use of unspun strands seems to have African origins. In Qasr Ibrim, which is located in the influence sphere of both North and South, Egypt and Nubia, both techniques occurred. The plaits used for making basketry all have in common that their edges are S and Z orientated (Wendrich 1991: 60-64). Of the five plaits shown in Fig. 28, three are suitable for making basketry. The first plait in Fig. 28 is made with four strands, which results in two S-orientated edges. The plaits with five and seven strands can be linked in such a way that the plait pattern gives the appearance that there is one continuing fabric (cf. Figs. 26 and 27), but the four strand plait cannot be linked in that manner. This implies that plaits in a \1/1\\1 pattern can only be used for making basketry, if they have been made with an odd number of strands. A seven strand plait in a \1/1\\1 pattern is suitable, but a seven strand plait in a \2/2\\1 pattern, is not. For a \2/2\\1 pattern only plaits made with 9, 13, 17 or 21 strands can be sewn into a seemingly ongoing fabric. The edges of 7–, 11–, 15– and 19-strand plaits have a wrong orientation.

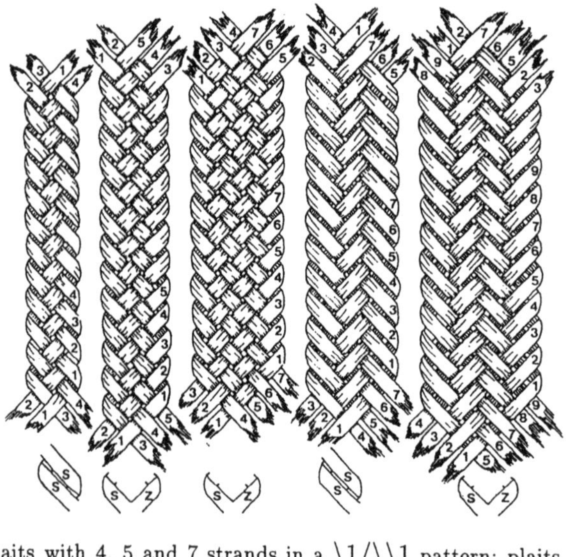

Fig. 28. Plaits with 4, 5 and 7 strands in a \1/\\1 pattern; plaits with 7 and 9 strands in a \2/2\\1 pattern (not to scale). Because of the orientation of the edges, the four-strand plait in \1/1\\1 pattern and the seven-strand plait in \2/2\\1 pattern cannot be sewn into a basket or mat

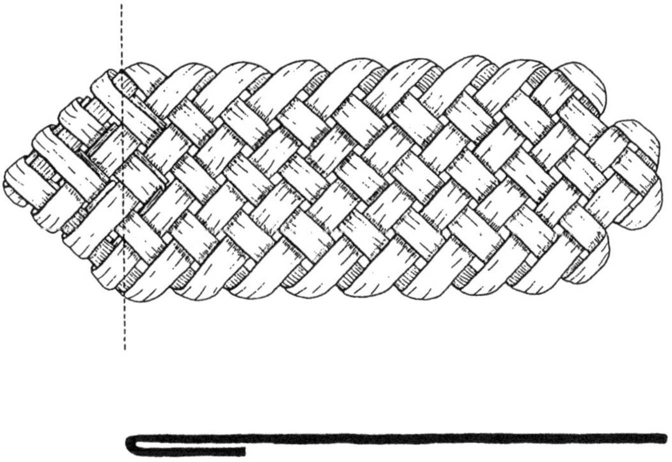

Fig. 29. Schematic drawing of a sandal plaited with eight strands (not to scale). One end of the sandal is folded over to give extra strength to the heel

The most simple sandals found at Qasr Ibrim are flip-flops made by plaiting strips of dom palm leaf into a sandal shaped mat (Fig. 29). The pointed end was folded to make a double layer underneath the heel. The mat was fastened to the foot by a strap, tied at three points: the two sides of the foot and between the toes. The strap was pulled through the fabric of the sandal and held in place under the sole by a *Z*- overhand stopper knot. The coarser the sandal, the larger the stopper knot had to be, in order to prevent the strand from slipping through the holes between the plaited strands. In some cases the size of the stopper knot must have been a hindrance for the wearer while walking. In the more complex sandals, which consisted of an inner sole, outer sole, and sides covering the toes, the strap was still fastened with a stopper knot underneath the foot.

In Qasr Ibrim, knots were occasionally used for decoration. Two examples of this are a palm leaf wreath and a small finger ring. The wreath was made of a midrib of the date palm split in half (Fig. 30).

The two halves were both tied into a circle, both sides being re-connected by knotting the leaflets with *ZS*-reef knots. Perhaps the most moving find from Qasr Ibrim is the small finger ring (Fig. 31). Made of a strip of straw, which has retained its golden gloss, the ring is a monument of simplicity and beauty. In small objects such as these the people of the past are suddenly very near to us.

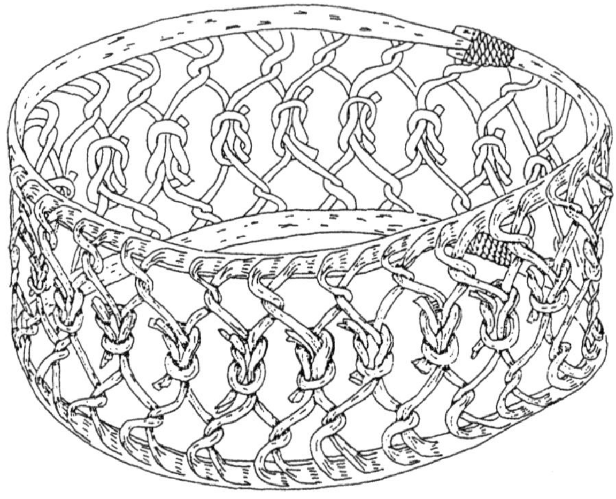

Fig. 30. Schematic drawing of the wreath found at Qasr Ibrim. The wreath is made from one palm leaf, which is split in the middle. The two halves have been tied in two circles, and connected by knotting the the leaflets with ZS-reef knots

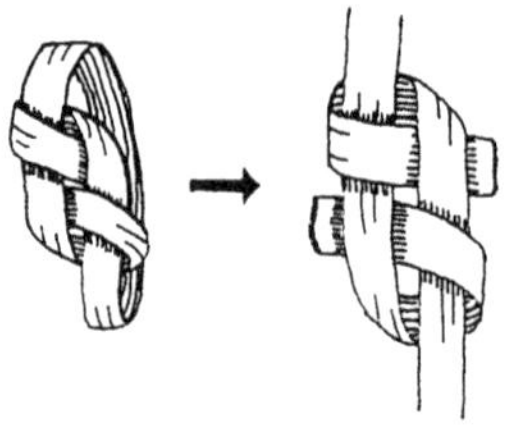

Fig. 31. Small knotted ring, and schematic drawing of the knot (Grass Knot, #1490)

Knots in texts and images

References to knotting occur regularly in ancient hieroglyphic texts. Usually the context is religious or magical, a division which would have been meaningless to the people of ancient Egypt. The world of the gods was a world like ours, and magic was a means to communicate with the other world. Similarly there was no difference between depictions of knots as a symbol and the magic use of knots, because to the ancient Egyptians all pictures had a magical reality.

The best known example of—what we would consider—a symbolical knot, is the *uniting of the two countries* (Fig. 32). Two Nile-gods, wealthy fat-bellied gods with river plants on their heads, are knotting a lotus and a papyrus, the plants symbolic for lower and upper Egypt. The plants are knotted around a depiction of the lungs and the trachea. This is the hieroglyph *semaa*, which means 'to unite'. This symbol refers to mythical times in which the south of Egypt and the Nile delta (lower Egypt) were united by the gods to be ruled by the first Pharaoh. Alternatively the two gods Seth and Horus are depicted in the same action. Horus, with the head of a falcon, is holding the papyrus of Upper Egypt, while Seth is knotting the Lotus of Lower Egypt. Throughout Egyptian history, the land was always thought of as consisting of a united twosome. The reef knot does not only symbolise the historical bond, but is also causing it. The depiction of the gods in their knotting action was an eternally repeated fastening of the existing order and thus a magical action, rather than a mere symbol.

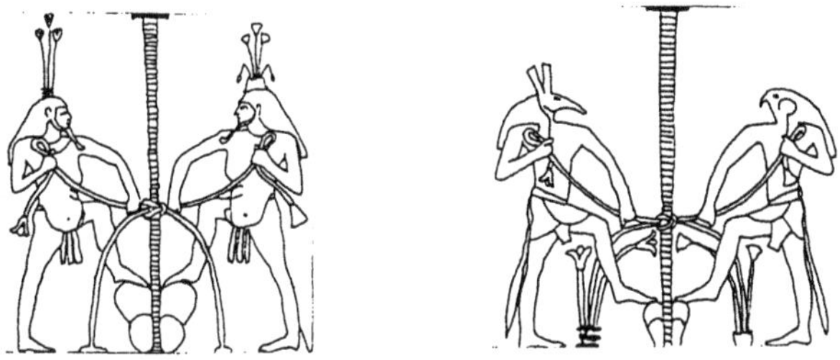

Fig. 32. 'Uniting the two countries'. To the left two Nile gods; to the right the gods Seth and Horus are making reef knots in plants symbolizing Upper and Lower Egypt

In the same manner, the reef knot was used to symbolise, and at the same time ensure, the coherency of the human body. On the mummy of Tut-ankh-Amun two gold amulets were found, shaped like reef knots. They were put on the thorax of the body, parallel to the arms (Carter 1933, III: plate 83). The use of the reef knot amulet can be explained with another Egyptian myth. In the dawn of time there were four important gods, two brothers and two sisters. These were Osiris who was married to his sister Isis, their brother Seth and their sister Nephtys. Seth murdered his brother and went after the son of Isis and Osiris, the child-god Horus. Isis escaped with Horus into the dense marshes of the Nile delta. Seth cut up his brother's body and hid the parts all over Egypt. Isis went out to bring together all the limbs and she brought Osiris back to life by knotting together all the parts. From that time onwards, Osiris was the god of the underworld and the dead. In the Egyptian religion there was a strong belief in the afterlife. The only requirement was that the body of the deceased remained intact. Therefore, it was important that bodies were mummified; but that was not enough. To ensure the integrity of the body, knot amulets which refer to the re-uniting of the body of Osiris, were placed on the mummy.

The same notion can be found in texts included in burials to protect the dead. Spell 50 from the Book of the Dead is a spell which the deceased should use to protect him from 'going inside the slaughterplace of the god'. The slaughterplace is the nightmare of all dead, because here the body is cut up, just like that of Osiris, and the deceased meets his 'second death', which is the final end of the afterlife. In the text of spell 50, the deceased is protected by three knots, knotted around him by three gods.

Fig. 33. The ancient Egyptian hieroglyph for the word *tjeset*, 'knot'

The ancient Egyptian word for knot is tjeset (Fig. 33). This word is used for the reef knot, but also for the half knot, as is clear from magical texts used by the living. Seven knots in a piece of string, tied around the neck, is used to cure headaches. This is done in analogy of the cure found in the myth of Isis, hiding the child Horus in the marshes. When Horus is ill, Isis cures him by transferring the illness to a young swallow. She makes seven knots in seven strings and ties this amulet around the neck of the child (Grapow, Westendorff 1958: 293). In another myth Seth cures a headache of Horus by tying a string with seven knots around Horus's left foot (Borghouts 1971: 18). One explanation of the use of the half knot for medicinal purposes is that the knot is thought to hold the power of spells recited whilst making the knot. Parallels for knots which are able to 'store' power are the windknots, brought on ships to make certain that there is enough wind for sailing (Day 1967: 44). Another explanation is that the knots are used to block the way of evil. This seems to be the case in a number of texts which contain spells for the protection of mother and child. A number of spells prescribe that a specific number of knots should be made to ward of dangers. One spell describes even the time at which the knots have to be made: 'One has to knot seven knots, one knot in the morning, another one in the evening, untill seven knots are knotted' (Erman 1901: 40-41)* Thus the last knot is made on the morning of the fourth day. This might indicate that knots in Egypt to some extent also had a mnemonic function.

*The spells in connection to which the making of knots is prescribed are M, N, O, P, Q and U.

Conclusion

The most widely known and used knots in ancient Egypt, the reef knot and the half knot, also had a religious, magical meaning. The reef knot was used to unite, the half knot was used to hold or withhold power. In the archaeological context it is often difficult to decide on the precise function of a knot. A string with three, four or seven half knots could be an amulet, a means of keeping count, or the result of boredom. The difference between a protective knot and an ordinary knot is determined by the way the knot was made, used and discarded. The archaeological context can only give an indication of a special meaning of knots, when they are used and discarded in a special way. The most important context, however, is that in which the knot was made. The meaning of knots depends on the people who were present, the words that were spoken, the things that were thought. Archaeologically speaking these vital, fleeting moments are thinner than air, never to be retrieved, always to be borne in mind.

Acknowledgements

The excavations at the four sites, as briefly described on pages 30 and 31, are directed by the persons named below. Details of the sponsoring institutions are also given below. We are grateful to all these persons and institutions for permission to use information gained from the excavations.
(1) The excavations at *Tell el-Amarna* are directed by B. J. Kemp (Cambridge University) and financed by the Egypt Exploration Society, London. The excavations in the Workmens' Village took place from 1979 to 1986, the basketry and cordage was recorded by the author in 1987, 1989, 1992 and 1994.
(2) The Dutch-American expedition to *Berenike* is co-directed S. E. Sidebotham (University of Delaware, USA) and the author (Leiden University), who is also responsible for analysing the basketry and cordage. The project is financed mainly by the National Geographic Society and the Netherlands Foundation of Historical Research (SHO-NWO). The first excavation season took place in the winter of 1994.
(3) The excavations at *Abu Sha'ar* were directed by S. E. Sidebotham (University of Delaware). The basketry and cordage were recorded by the author during five excavation seasons, from 1989 to 1993.
(4) The excavations at *Qasr Ibrim* were organised by the Egypt Exploration Society under directorship of M.C. Horton. The basketry and cordage was recorded by the author during excavation seasons in 1990 and 1992.

I wish to thank Charles Warner for editing this essay and supplying the names and identifying numbers for knots, from [1]. His many suggestions have helped to improve the chapter.

References

1. C. W. Ashley, *The Ashley Book of Knots* (Doubleday, New York, 1944).
2. J. F. Borghouts, *The magical texts of papyrus Leiden 1348* ((OMROLI) Leiden: Rijksmuseum van Oudheden, 1971).
3. H. Carter and A.C. Mace, *The Tomb of Tut.Ankh.Amen*, 3 Volumes (1923–33).
4. C. L. Day, *Quipus and witches' knots* (Lawrence, 1967).
5. O. Erman, *Zaubersprüche für Mutter und Kind* (1901).
6. R. Germer, *Flora des pharaonischen Ägypten* (Philipp von Zabern: Mainz am Rhein, 1985).
7. H. Grapow, and W. Westendorff, *Grundriss der Medizin der alten Ägypter IV* (Berlin, 1958).
8. S. E. Sidebotham and W.Z. Wendrich, *Berenike 1994* (Leiden: Centre of Non-Western Studies, 1994).
9. E. Teeter, 'Techniques and Terminology of Rope-making in Ancient Egypt', *Journal of Egyptian Archaeology* **73**, (1987) 71–77.
10. W. Z. Wendrich, 'Preliminary Report on the Amarna basketry and Cordage', *B.J. Kemp et al., Amarna Reports V*, 169–201, (London: Egypt Exploration Society, 1989).
11. W. Z. Wendrich, *Who is Afraid of Basketry?; A guide to recording basketry and cordage for archaeologists and ethnographers* (Leiden: Centre of Non-Western Studies, 1991).
12. W. Z. Wendrich, *The World According to Basketry*; analysis and interpretation of Egyptian archaeological and ethnographic basketry (with a 60 minutes video of present day Egyptian basket makers) (Leiden: Centre of Non-Western Studies, in press).
13. W. Z. Wendrich and W. J. Van Neer, 'Fish and fishing gear at the Roman fort of Abu Sha'ar (Red Sea coast, Egypt)' *Proceedings of the Workgroup for Fish Remains, Leuven 1993* (in press).

PART II. NON–EUROPEAN TRADITIONS

This Part contains three chapters, each dealing with the history of knots as used by a particular non-European civilization.

The first chapter describes the knotted cords, or *quipus*, of the Incas of South America. The Inca culture existed from about A.D. 1400 to A.D. 1560, in a region that is now Peru and parts of Ecuador, Bolivia, Chile and Argentina. The quipus were used to store data—to keep detailed accounts of the various activities involved in governing the kingdom. How the knotted cords were constructed, and used to represent and store numbers, is explained by the author.

The second chapter traces the exceptionally rich history of knots in China, beginning with the first hints of their existence in the late paleolithic period, some eighteen thousand years ago. The main theme is to describe how knots have been put to decorative use in ancient China.

Inuits are Eskimos of Greenland or North America. The third chapter describes something of their history, from 4500 B.P. to the present day. It examines a wide variety of evidence of their ability to manufacture ropes and to form and use knots. The evidence is drawn from two major sources, ethnographical observations in the literature, and archaeological records or museum examples of knotted artifacts found at many different excavation sites.

CHAPTER 5

THE PERUVIAN QUIPU

Antje Christensen

Introduction

In many different cultures throughout the world and in different epochs, knots served as symbols to denote numbers. In the most simple systems, a number was symbolized by tying the appropriate number of simple knots into a string. Several cultures of the New World, such as the Zuñi of New Mexico and the Peruvian Incas, developed more sophisticated systems. The system of the Peruvian Incas is the main subject of this text. It not only allowed them to symbolize high numbers in a way that made them easily recognizable, but also it provided the structure to put these numbers in a meaningful order. Enclosing information into knotted strings became one of the most important cultural techniques in the highly developed Incan culture, playing a role comparable to that of writing in other cultures. We do not know much about the Incas, as their culture was completely destroyed by the European conquerors within a few decades after the first contact. Yet we can deduce from European writings contemporary to the conquest, and from archeological finds, that science in general and mathematics and astronomy in particular were quite advanced in the Incan culture. While we do not have much direct information about Incan science, we share with the Incas the ability of mapping reality into the abstract world of numbers and structure. Within this world we have a chance to reconstruct their ways of thinking. Hence their knotted strings could be a rich source of information on their culture, if we were able to read and understand them profoundly.

'One knot–one object' Systems

In the most simple systems which use knots to denote numbers, one knot, for example a simple overhand knot, corresponds to one object. To denote the number 'two', two knots are tied next to each other, and so on. One of the most widely known examples where this system is used is the mariner's logline, a device for measuring the speed of a ship. It consists of a rope with a piece of wood, called the chip, tied to its end. The chip is thrown overboard and stays more or less stationary in the water. As the ship moves ahead, the rope runs out astern. It is divided into lengths of 15 m, marked by pieces of string which are inserted between the strands of the rope. The strings carry as many overhand knots as there are lengths between the string and the chip. The number of lengths that run out astern in 28 seconds is the speed of the ship in nautical miles an hour, or 'knots' (see [9], p. 14).

Another example is the rosary of the Greek Orthodox monks of Mount Athos, where knots instead of beads are used in order to count the prayers ([9], p. 14). There are many more examples, as this simple system has been invented many times independently.

The Zuñi System

The Zuñi of New Mexico used a base-ten positional system, which allowed them to denote high numbers without tying many knots. It worked much like the Roman system. There were distinct knots for one, five, and ten. Two 'one'- knots meant 'two', three 'one'-knots 'three'. A 'one'-knot above a 'five'-knot had to be subtracted, hence both together meant 'four'. If the 'one'-knot was under the 'five'-knot, it had to be added, giving 'six' (see Fig. 1, and [7]).

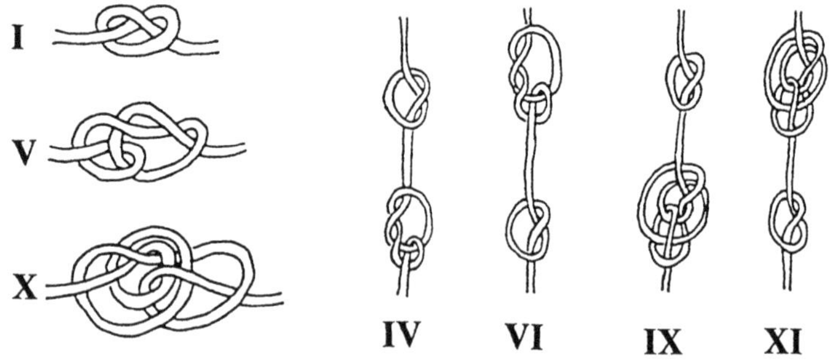

Fig. 1. Zuñi Knots (adapted from [7], p. 301)

Sources on Incan Quipus

When Francisco Pizarro reached the western coast of South America in 1533, he met a complex and flourishing culture named the Incas. The Spaniards did not come as guests, but as conquerors, and they managed to destroy the native culture within a few decades. Only a few of them, now known as the chroniclers, noted down their observations on the Incas. But as they came from a completely different cultural background, there were many things they did not understand, and as they usually felt superior to the Incas, their evaluations were often not very careful. An important source of information are authors who stood between the cultures, such as Garcilasso de la Vega. He was born in Cuzco, the capital of the Inca state, in 1539, as the son of a Spanish cavalier and an Inca princess. Raised within Inca culture until the age of twenty, he later travelled to Spain, learned Spanish and wrote about the culture of his mother's people.

Another source is Felipe Guaman Poma de Alaya, also the son of a Spanish father and an Incan mother. In the 1610s, he wrote a 1179-page letter to the Spanish king in which he describes the Inca culture and begs for a better treatment of the Incas by the conquistadores. This letter is of special interest because it contains almost four hundred drawings depicting scenes from Inca life before and after the conquest. In fact he wrote that letter 80 years after the Spaniards had entered the land, so he did not know life before the conquest from personal experience, and a strong Spanish influence can be found in his argumentation.

More direct information can be gained from the physical remains of the Inca culture, that is from archeological finds. These are mostly either buildings or jewellery taken away by the conquerors for the value of the material. Some, however, are rather plain pieces of knotted cords, called 'quipus'. The Spaniards, who did not know what was laid down on the quipus, either regarded them as witches' knots and burned them, or simply threw them away as the material was not valuable to them. So there is nothing left of the large quipu 'library' that existed in Cuzco before the Spaniards arrived. The comparatively few ancient quipus that we do have come from graves in the dry coastal areas, where the material did not rot. It is not always clear how they found their way into the museums, so we often do not know precisely where they were found.

The first scientist who wrote about the quipu code was the US-American anthropologist L. Leland Locke. In an article which appeared in 1912 he described the number system and the values of the knots, following a description he found in Garcilasso's writings. He also showed that a special kind of cord, the top cords, contain the sums of the numbers on other cords associated with them. So he could confirm Garcilasso's description and deduced the reading direction. In 1923 he published a book with a detailed description of several quipus.

The Swedish ethnographer Erland Nordenskiöld was the first who tried to interpret the content of the quipus. In 1925, shortly after Locke had uncovered the code, he published two booklets, 'The Secret of the Peruvian Quipus' [11] and 'Calculations with Years and Months in the Peruvian Quipus' [12]. He assumed that the numbers indicate days and that the quipus contain astronomical numbers. Unfortunately, his analysis lacks structure, and his calculation methods are rather far-fetched. A first evaluation of his calculations was made as late as in 1967 by Cyrus Day [9], a professor of English literature at Kansas University.

In 1931 Henry Wassén translated parts of Guaman Poma's letter into English, and interpreted an object on one of the drawings as a calculation device [13].

The analysis of the contents was taken up by an American couple, the mathematician Marcia Ascher and the anthropologist Robert Ascher. They stressed the structure of the quipu as a whole and deduced statements about Incan mathematics from what they found. Due to them is a detailed uniform description of 215 quipus from museums and private collections all over the world [5], and a beautiful, well-understandable book about quipus and their cultural and historical context, 'Code of the quipu' [6].

The Inca Culture

The Inca culture existed from about A.D. 1400 to 1560 in a region that is now Peru and parts of Ecuador, Bolivia, Chile and Argentina (see Fig. 2). The landscape varied widely from the deserts on the Pacific coast to tropical forests and mountanious areas up in the Andes. Its three to five million inhabitants formed a state under a king, the Sapa Inca, who reigned at the capital Cuzco. They had a common language, Quechua, a common religion, an effective system of irrigation, an extended road system, a system of taxation and storehouses where they stored the agricultural and finished products they had gathered as taxes in order to distribute them again in times of need. Therefore they needed an extended bureaucracy. Yet they did not have any writing in the sense we use the word, that is a direct transcription of spoken language. They kept their accounts by tying knots into cords using an elaborated system of number representation. These knotted cords were called 'quipus', the Quechua word for 'knot'.

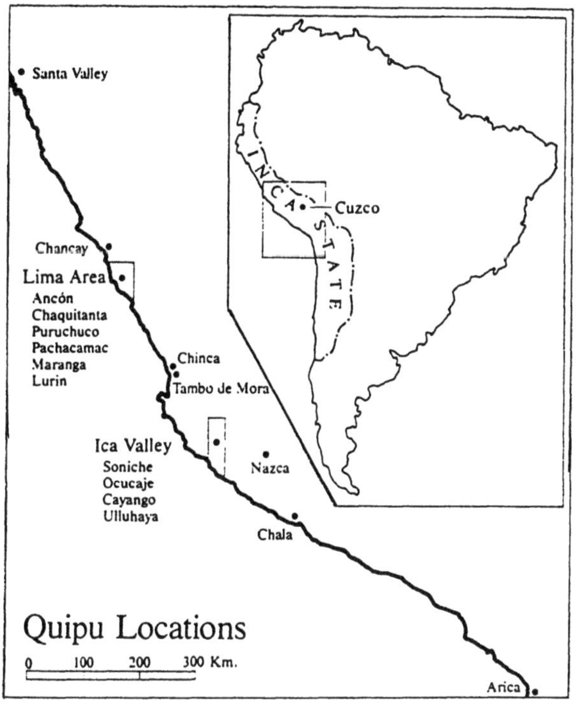

Fig. 2. Map of the Inca State and Locations of Quipu Finds ([6], p. 68)

Structure and Design of the Quipus

This section gives details of the materials used to make quipus, their structural layout, and the manner by which numbers were represented upon them.

The Materials

Quipus were made from coloured cotton or wool cords which were spun together so that there was an eye at one end. A knot was tied into the other end to prevent it from unravelling. The colours were achieved by dying the material before it was spun, and colour combinations were produced by spinning cords of different colours together.

The light material made quick transportation possible. There was a runners' post, with post stations along the main roads at intervals of a few kilometers, where the quipu was handed over to a fresh runner. By this method, a message could reach the capital Cuzco from the outermost edge of the country within a week.

The Spacial Layout

The backbone of a quipu is the main cord, which is usually thicker than the other cords. Attached to it are the pendant cords. There can be as few as three or as many as two thousand of these. Some cords fall in the opposite direction, they are called top cords. Top and pendant cords can have subsidiaries, which can again have subsidiaries, and so on, so that a tree structure occurs.

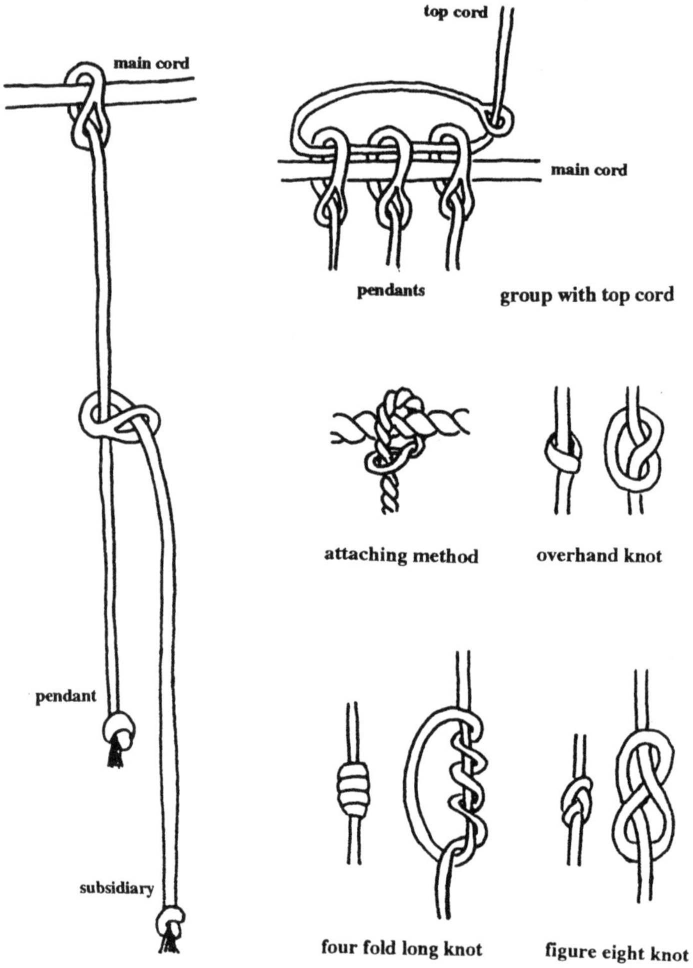

Fig. 3. Quipu Knots (attaching method, overhand knot, four fold long knot and figure eight knot adapted from [9], p. 16)

There can be up to six different levels and up to ten subsidiaries per level. The pendants, top cords and subsidiaries are the cords which carry the knots.

They are between 20 cm and 50 cm long, while the length of the main cord depends on the number of pendants. It varies from a few centimeters to more than a meter. Sometimes an extra cord dangles down from the eye of the main cord, it is called dangle end cord and can also carry knots. The top cords can either be attached to the main cord in the same way as the pendant cords or passed through the loops formed by several pendant cords, thus uniting these to a group (see Fig. 3 above).

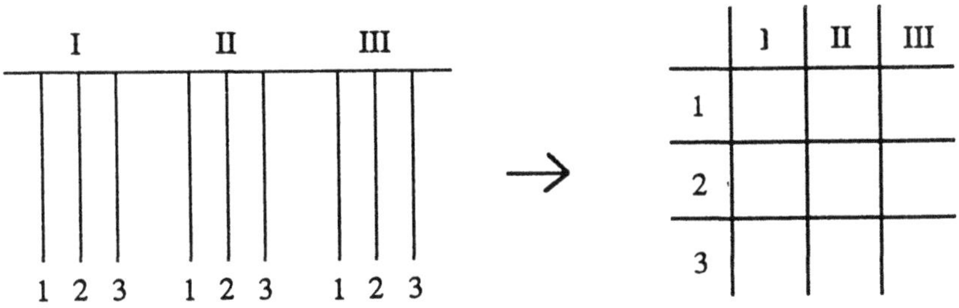

Fig. 4. Tabulation

Top cords of this kind are used for summing up the numbers on the pendant cords of the group. Groups of pendant cords can also be formed by inserting spaces between them or by colouring. Hence, tables of several dimensions can be formed, e.g. with the group indicating the column, and the position within the group, or the colour, indicating the row (Fig. 4 above).

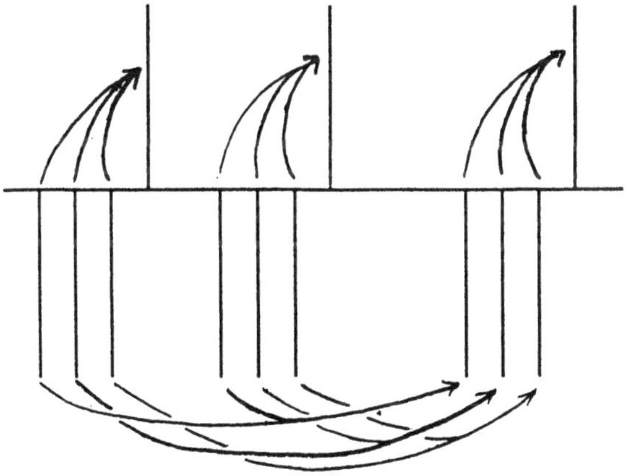

Fig. 5. Arrangement of Sums

While the sums within the columns are indicated on the top cords, the sums within the rows are recorded in an extra group where the first cord contains the sum of the numbers on the cords which hold the first position in the other groups, the second the sum of those in the second position and so on (see Fig. 5).

We can assume that the whole quipu was completed before the knots were tied into it, because several quipus without any knot were found, and all the knotted single cords that were found were obviously broken.

The Number System

The Incas used a base-ten positional system. Hence they had developed independently the same number system that we use today. The digits were represented by clusters of one to nine single overhand knots, and the positions were separated by spacing. The clusters closer to the main cord represented the higher positions, while the unit position was close to the end. The digits in the unit position were represented by special knots, namely by multiple overhand knots whose number of turns indicated the digit (see Fig. 3). They are referred to as 'long knots'. As an overhand knot with only one turn is the same as the simple overhand knot used for the other positions, a 1 in the unit position was represented by a figure eight knot. Using different knots for the unit position allowed the placing of several numbers on one cord without ambiguity in where one number ends and the next one begins. Several numbers on one cord were actually found on ancient specimens.

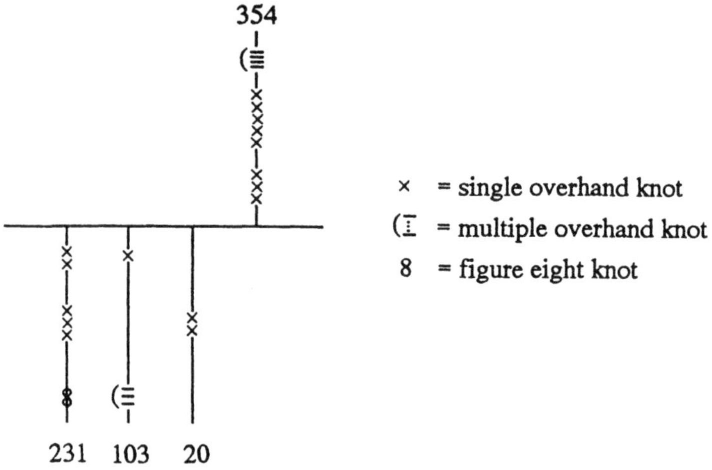

Fig. 6. Example of Numbers and Sum Recording

There are two ways to pull a long knot tight. Either hold the knot in one hand and pull the end which in Fig. 3 points upward with the other hand; then the belly of the knot stays straight. Or pull both ends, then the belly will curl around the spine, and the windings that are counted to determine which number the knot denotes are in fact the windings of the belly. Knots of both kinds were found on ancient specimens. Those of the first kind were found only in the North of the country; those of the second kind only in the South. Hence it was probably a matter of tradition which method was used, and every quipu keeper used only one of them.

There was no special sign to indicate an 'empty' digit, like our 0 in 101 for example. The positions were carefully aligned from cord to cord, so that a space without knots could be associated unambiguously with its position. There remains ambiguity in the cases of small quipus where a position is empty on every pendant cord. In these cases, which probably occurred rather seldom, the length of the spaces might have been considered. Fig. 6 shows a schematic example.

The Number Zero

Although there was no special sign for zero, we can assume that the Incas had the concept of nothingness in their number system. They represented zero by a cord without knots. We can conclude this because the colour code allowed meaningless cords to be omitted, as the following example illustrates. Imagine a quipu designed for recording the number of sandals owned by the people of a village. Each pendant group represents a family, with the number of the men's sandals on the first cord, the women's on the second and the children's on the third. The second categorization can be reinforced by using colours, for example green cords for men, yellow cords for women and red cords for children. If a family does not have any children, the red cord can be omitted without ambiguity. A red cord without knots would mean that there are children in the family, but they do not own any sandals.

As cords without knots, and also repeated colour patterns with single cords omitted, occur frequently on ancient quipus, we have to assume that it was on purpose; and this is the most likely explanation. We also know that another ancient American culture, the Mayas, had the number zero, and it is possible that the Incas took it over from them.

The Contents

This section deals with the types of information that Incas stored on their quipus. These include statistical data and astronomical observations; possibly non-numerical data were also recorded, using number codes.

Statistics

The chroniclers describe the use of quipus for recording statistical numbers. There was a regular census, as we know from Garcilasso. Every year the population figures were recorded, divided into provinces, villages, sex, age in decade intervals, and groups of unmarried, married, and widowers. There were also regular statistical reports on resources like agricultural produce, herds of animals, and weapons; and of course accounts were kept of the goods held in the storehouses. These data being available made it possible to organize the state's economy as a whole. When a harvest failed, crops were taken from the well-filled storehouses in other parts of the country, and during the conquest the provinces which had to pay most to the Spaniards were resupplied by the wealthier ones. This could only be organized if the bureaucrats in Cuzco were able to compare the supplies in the different provinces.

Astronomy

Astronomy played an important role in Incan religion, as the Incas claimed descent from the sun and the moon. They had observation towers, and we know from the chroniclers that they knew astronomic data like approximations for the solar year, the lunar month, and the synodic revolutions of Jupiter, Mercury, and Venus. It is reasonable to expect to find these data on quipus. This idea was examined by Erland Nordenskiöld in [11] and [12]. He conjectures that quipus which were put into graves '... were intended above all as puzzles for the spirits. The dead man was supplied in the grave with a quipu to occupy himself with, and to prevent him ... from walking after death.' ([11], p. 36). After intricate calculations, he finds all the astronomical numbers mentioned above. Unfortunately, his calculations are not at all suggested by the structure of the quipu. Many other methods of calculation would be just as probable or unprobable and would lead to a completely different set of numbers. Even worse, he does not give any argument for his assumption that the Incas knew the arithmetical operations he applies. This question will be investigated in the next section.

Here is a more satisfactory explanation of the significance of grave quipus, as suggested by Marcia Ascher ([6], p. 68): Only a few of the graves that were opened by archeologists contained quipus, and usually some of these were incomplete. These graves may have been the graves of quipu keepers, who were buried together with the quipus they owned at the time of their death, and not with quipus produced especially for the burial.

Non-numerical Data

The single data on the quipus are natural numbers. As far as we know, there is no direct transcription of speech on them. Yet the chroniclers claim that the quipus were used for writing down laws, peace negotiations, and history. The quipus designed for this purpose may merely have been mnemonic knots, such that every knot recalls associations to the quipu's owner which, taken together, make up the whole history that the owner has learned before. The technique is the same as we use when we tie a knot into a handkerchief in order to remember something. In this case the knots do not have a universal significance, but can only be read by the person who has tied them or was told about their meaning. There are several ancient specimens with irregularities on them that might indicate that these knots did not represent numerals. For example, Nordenskiöld describes ten- to fifteen-fold long knots ([11], p. 18) and long knots which appear in the tens' position ([12], p. 16).

A second, more refined possibility for noting down non-numerical information on a quipu is a label code. There might have been a system of translating spoken words into numbers, like we translate letters into numbers by using the ASCII code. In this case it would be possible to read a quipu when just knowing the code, but not its content. Yet there was no possibility of writing down a translation table. Reading this kind of quipu would have to be learned from oral instruction, and so the skill, if it existed, vanished with the Inca people under the Spanish conquest.

Incan Mathematics

The collection of numbers on the ancient quipus allow us to reconstruct some of the mathematics behind them. In this section, some of the studies in this field are reviewed. Possible arithmetic operations and geometric interpretations are discussed. Finally, a calculation device is described which the Incas might have used.

Arithmetics

From their analysis of the structure of a vast number of quipus, Ascher and Ascher deduced ([6], pp. 133–155) that the arithmetics used by the Incas must have included at least:

(1) addition
(2) division into equal parts
(3) division into unequal parts
(4) multiplication of integers by integers and fractions.

That means that the Incas dealt with fractional values in the form of division into parts and common ratios, though fractions cannot be encoded on the quipus.

The following examples are to demonstrate the occurrence of the various arithmetical operations. They are all taken from *Code of the Quipu*. The quipus are referred to by the labels by which they are catalogued in [5]; the letters denote the author who first described the quipu, and the number indicates the order of publication.

Addition: We already saw in the section about the spatial layout that top cords or extra groups carry the sums of the numbers on other cords. This summation appears on about 25% of the quipus examined by Ascher and Ascher.

Division into equal parts: AS161 The quipu consists of two groups with two respectively six pendant cords. It has a total value of 200 divided up as shown in Fig. 7.

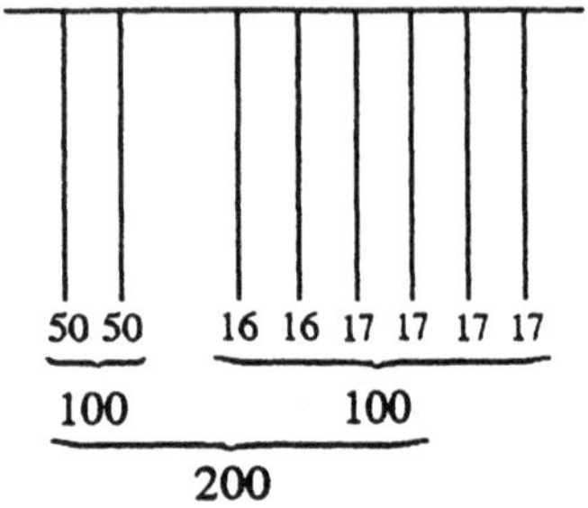

Fig. 7. Division into Equal Parts

Division into unequal parts: AS120 The quipu consists of three groups of eight pendants each and a summation group. In each group the third cord has a subsidiary. We denote the value on the jth cord of the ith group by p_{ij}. The jth cord of the summation group is denoted by $p_{\Sigma j}$, and the subsidiary in the ith group by p_{is}. The arrangement is shown in Fig. 8.

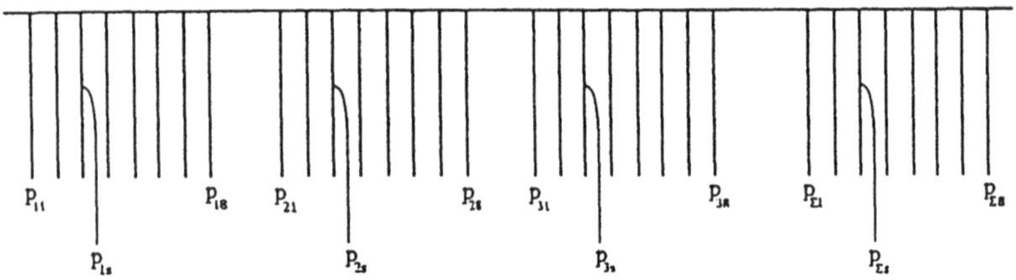

Fig. 8. Division into Unequal Parts

Thus we have $p_{1j} + p_{2j} + p_{3j} = p_{\Sigma j}$ for $j = 1, ...8, s$. The numbers on the quipu range widely, namely from 102 to 43,372. We find that

$$\begin{aligned} p_{1j} &= 0.340 \cdot p_{\Sigma j} \\ p_{2j} &= 0.425 \cdot p_{\Sigma j} \qquad \text{for } j = 1, ..., 8, s \\ p_{3j} &= 0.235 \cdot p_{\Sigma j} \end{aligned}$$

That is, the values in the first group are $340/1000 = 17/50$, the ones in the second group $425/1000 = 17/40$ and the ones in the third group $235/1000 = 47/200$ of the values in the summation group. Hence this quipu can be interpreted as a division table: Given the values in the summation group, these were divided into three unequal parts which then were encoded in the first three groups. As the exact result of the calculation is not always an integer, the numbers had to be rounded in order to be encoded on the quipu. Therefore we get an error which we state in percent of the number on the quipu. For the first group this error is about 0.6% for $j \neq 3$ (i.e. $0.33796 \leq (p_{1j}/p_{\Sigma j}) \leq 0.34204$) and 1.4% for $j = 3$, in the second group 0.7% for $j \neq 1$, and in the third group 0.9%, again for $j \neq 1$. That means that the actual numbers are close to those we must expect when the quipu really is a division table.

Multiplication of integers by integers and fractions: AS55 and AS56 These are two small quipus with seven and three pendants respectively which were found together. Though the pendants are not grouped by spacing or colour, an implicit grouping appears when examining the values. We name the pendants on the first quipu $p, q_1, q_2, q_3, r_1, r_2, r_3$ and those on the second quipu s_1, s_2, s_3. With this notation, we find several attractive relations:

$$\begin{aligned} q_1q_2q_3 \cdot r_1r_2r_3 &= s_1^2s_2^2s_3^2 \\ q_1r_2s_3 &= r_1s_2q_3 = s_1q_2r_3 \\ s_1s_2s_3 &= q_1r_2s_3 \\ q_j \cdot r_{j+1} &= s_j \cdot s_{j+1} \qquad \text{for} \quad j = 1, 2, 3 \end{aligned}$$

where the addition in the indices is mod 3, e.g. 3+1=1. We even find a more general pattern, namely

$$\begin{array}{lllllllll} q_1 &=& B^3Cx & r_1 &=& B^6C^2x & s_1 &=& B^4Cx \\ q_2 &=& x & r_2 &=& BCx & s_2 &=& Cx \\ q_3 &=& y & r_3 &=& B^2C^2y & s_3 &=& B^2Cy \end{array}$$

where $B = 7/8$ and $C = 34/33$. Hence the numbers are multiples of the values on the third and fourth cord on the larger quipu *by fractions*. Of course, we cannot be sure that this is indeed the pattern behind the quipu. But it fits very well, as the error we make is remarkably small, namely in all cases smaller than 0.4 %, in ten out of the twelve cases even smaller than 0.2 %. For a further examination of these quipus including the first cord and the three subsidiaries on the larger quipu, see [6], pp. 149–151.

Geometric interpretation: AS120, AS143 and AS149 The quipus AS143 and AS149 have the same underlying structure as AS120, which we considered when dealing with division into unequal parts. AS143 has four and AS149 five pendant groups. The ratios $p_{ij}/p_{\Sigma j}$ are the numbers b_1, b_2, c, a and b, c_1, c_2, c_3, a in the following table, thus they stand in the same order as on the quipu.

	AS120		AS143		AS149
		$b_1 =$	0.110		
		$b_2 =$	0.228		
b	0.340		0.338		0.222
				$c_1 =$	0.105
				$c_2 =$	0.534
				$c_3 =$	0.017
c	0.425		0.437		0.656
a	0.235		0.225		0.122

Naming the ratios from quipu AS120 a, b and c in the order of increasing value, we find that they solve the equation

$$\frac{c-a}{b-a} = \frac{c}{a} .$$

This relation was also studied in ancient Greece. But don't be tempted—we cannot assume that there was any contact. If we add the first two values of AS143, i.e. b_1 and b_2, and the middle three values of AS149, i.e. c_1, c_2 and c_3, we get three values per quipu which, named in the same order as in AS120, also solve this equation. They can be illustrated as in Fig. 9, where they appear as areas. To get a standard for all the three quipus, we take c as a new unit; and $X = 1 - a/c$.

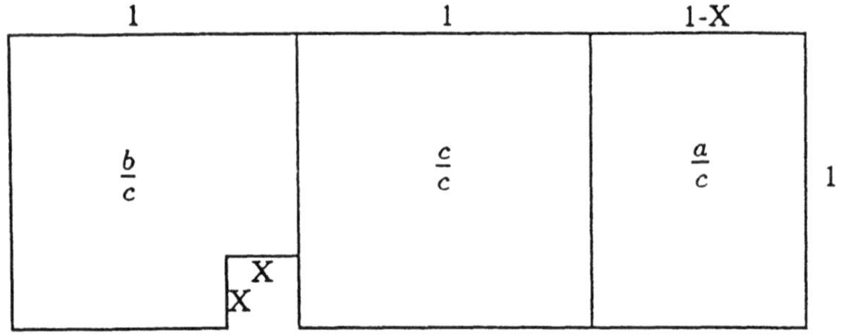

Fig. 9.

Adding the b_i and c_i was not arbitrary, as we can also find them in the figure, see Fig. 10 below. There, c_1/c is the area of the unit square minus the rectangle with sides 1 and X and the circle with diameter $1 - X$. Ascher and Ascher assert that this figure is 'quite similar to a geometric form thought to be important and persistent in the cosmology of western South America' ([6], p. 146). If the Inca who knotted these quipus really thought of this figure, he

must have had a geometric visualization for arithmetics, and he also knew the area of the circle in the figure.
As this is only one single case, we do not know if the Inca had a formula for the relationship of a circle's diameter to its area, and hence an approximation for π. If he did have, we can calculate from this case a correctness of 4.61% for π:

$$\frac{c_3}{c} = \left(\frac{1-X}{2}\right)^2 \pi \iff \pi = 2,997 .$$

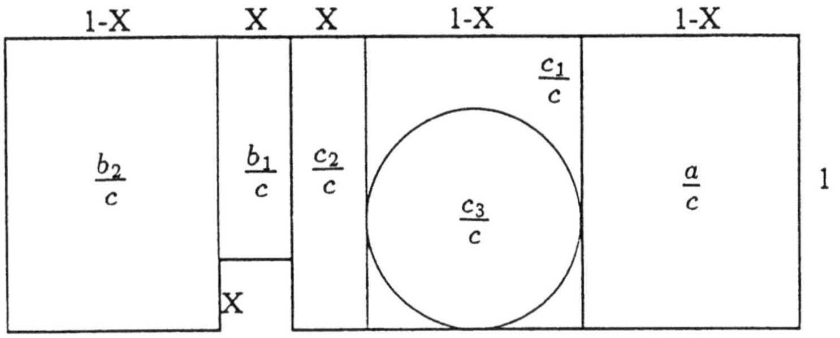

Fig. 10.

We still have to take into account an error already introduced when calculating the values in the table. A further error can come from the calculation or encoding of this special case. For the values of the various errors, see [6], p. 146.

Guaman Poma's Calculation Device

A quipu is well adapted for storing numbers in a permanent and transportable form, but it can hardly be used as a tool for calculating, as it takes too much time to tie and untie the knots. But as the Incas handled large numbers (of up to five digits), they probably had a calculation device. Father José de Acosta, who had lived in Peru from 1571 to 1586, wrote in his *Historia Natural y Moral de las Indias:*

> 'To see them use another kind of quipu with maize kernels is a perfect joy. In order to effect a very difficult computation for which an able calculator would require pen and ink for the various methods of calculation, these Indians make use of their kernels. They place one here, three somewhere else and eight I know not where. They move one kernel here and three there and the fact is that they are able to complete their computation without making the smallest mistake. As a matter of fact, they are better at calculating what each one is due to pay or give than we should be with pen and

ink. Whether this is not ingenious and whether these people are wild animals let those judge who will! What I consider as certain is that in what they undertake to do they are superior to us.' ([9], pp. 32/36).

Fig. 11. Guaman Poma's Picture of a Quipu Keeper and the Supposed Counting Board ([6], p. 66)

One of the drawings in Guaman Poma's letter (see Fig. 11) shows a man with a quipu and an object which may be the device described by Father José. It might depict a wooden board with regularly ordered holes in it. The white circles may be empty holes, and the black ones holes containing a maize grain. As we do not have a more detailed description, we can only guess how this abacus worked. There have been several different suggestions (see [9]). None of them is really satisfying. But we do not know if Guaman Poma himself knew how to use the device and therefore was familiar with the system, or if he simply drew a random pattern. Moreover, we cannot be sure that the object really is a calculation device, and, as Marcia Ascher correctly remarks: '... it might not even depict an object used by the Incas'([1], pp. 265–266).

Conclusion

The Incas met their vast bureaucratic tasks with the help of the quipus, cords which were tied together to form a meaningful structure and in which knots were tied that represented integers. These were not simply in one-to-one correspondence to the objects, but a base ten positional system was used. From the chroniclers who came to Peru during the Spanish conquest we know for what and how the quipus were used. Analysis of surviving specimens allows reconstruction of Incan mathematical ideas, which are far from primitive. Unfortunately for us, the Incas had no writing in the usual sense of the word. Their culture was completely destroyed by the Spanish conquerors before they could make contact with writing cultures, which was intensive enough to provide us with the equivalent of a rosetta stone. Therefore the only things we know for sure is what the single signs mean, and that there is in fact some profound meaning behind the Incan knots. We can reconstruct some of the ideas which stand behind the quipus, but not with any certainties. We cannot be sure that our interpretations are right; and we cannot say anything about the concrete meanings of the ancient quipus which we find in the museums.

References

1. M. Ascher, 'Mathematical Ideas of the Incas', in: Michael P. Closs, ed., *Native American Mathematics.* (University of Texas Press, Austin, 1986) 261–289.
2. M. Ascher, *Ethnomathematics. A Multicultural View of Mathematical Ideas.* (Brooks/Cole Publishing Company, Pacific Grove, 1991).
3. M. Ascher and R. Ascher, 'Code of Ancient Peruvian Knotted Cords (Quipus)', *Nature* **222** (1969) 529–533.
4. M. Ascher and R. Ascher, 'Numbers and Relations from Ancient Andean Quipus', *Archive for History of Exact Sciences* **8** (1971–72) 288–328.
5. M. Ascher and R. Ascher, *Code of the Quipu: Databook.* (1978) and *Code of the Quipu: Databook II.* (1988): available on microfiche from Cornell University Archivist, Ithaca, NY, USA.
6. M. Ascher and R. Ascher, *Code of the Quipu. A Study in Media, Mathematics and Culture.* (University of Michigan Press, Ann Arbor, 1981).
7. F. H. Cushing, 'Manual Concepts: A Study of the Influence of Hand-Usage on Culture-Growth', *The American Anthropologist* **V** No. 4 (1892) 289–316.
8. C. L. Day, 'Knots and Knot Lore: Quipus and Other Mnemonic Knots', *Western Folklore* (1957) 8–26.
9. C. L. Day, *Quipus and Witches' Knots.* (University of Kansas Press,

Lawrence, 1967).
10. L. L. Locke, 'The Ancient Quipu, a Peruvian Knot Record', *American Anthropologist* **14** (1912) 325–332.
11. E. Nordenskiöld, 'The Secret of the Peruvian Quipus', *Comparative Ethnographical Studies* No. 6 part 1, Oxford, (1925).
12. E. Nordenskiöld, 'Calculations with Years and Months in the Peruvian Quipus', *Comparative Ethnographical Studies* No. 6 part 2, Oxford, (1925).
13. H. Wassén, 'The Ancient Peruvian Abacus,' *Comparative Ethnographical Studies* No. 9, Oxford, (1931) 191–205.

CHAPTER 6

THE ART OF CHINESE KNOTWORK: A SHORT HISTORY

Lydia H. S. Chen

Introduction

Ancient Chinese mythology has it that 'when Heaven and Earth had been separated, there was still no human race. The goddess Nu-wa then shaped mankind out of yellow earth. However, as this task was too fatiguing and time-consuming, she trailed a rope in the mud, removed it and created men. The noble and the rich were made out of the yellow earth, while poor and lowly people were created from the mud-covered rope'[15]. Mythical though this legend may be, it nonetheless signifies the inextricable ties between rope and men, and this connection is even more significant when we realize the important role that rope has played in the real life of mankind.

By the time that primitive man had learned to cover his body with tree leaves and animal skins to ward off cold, he would be aware that he could make knots by interlacing ropes. While we are not certain when the craft of tying knots first came about, we do know from scientific investigations of archaeological finds that approximately one million years ago fire was used in the cooking of food by the inhabitants of China [3]. It is therefore not illogical to assume that men might have discovered the technique of knotting around the same time. The first hint of the earliest knots in China dates back to the late paleolithic period, some 18 000 years ago. Cultural relics from that era found in a cave at Chou-k'ou-tien in Hopei Province include bone needles, pierced shells, and dyed stone beads [16]. Archaeologists maintain that these instruments were used for sewing, and that the inhabitants possessed some rudimentary ideas of aesthetic appreciation. The presence of these artifacts also indicates that knots and splices must have already been in existence at that time.

Knots and splices are useful because they bind things; yet, in ancient China knotted ropes were employed to keep records as well. Such practice as exercised by Hsuan-yuan, Fu-hsi, Shen-nung, and other notable mythological figures is well documented in a chapter in the classic *Chuang-tzu* [7]. It is also set forth in a commentary in the *I Ching* that 'in prehistoric times, events were recorded by tying knots, and in later ages, writing was used for the purpose instead' [6]. In his classic *I Chu*, the Han scholar Cheng Hsuan (127–200) further expanded on this passage to point out that 'great events were recorded with large knots, while smaller knots signified events of less importance' [6]. Some fine examples would be that the numbers 10, 20, and 30 were symbolized with the knots ,  , and  respectively, and that the word 'end' was represented by the  knot [8].

As the civilization of China progressed, existing knot types and their variations could no longer meet the increasingly sophisticated requirements in the recording of events. While men had been able to replace the knot device with more advanced systems, such as pictography and writing [15], knots were still widely used in daily life and were depended upon by men. Some of these primitive knots were even decorative in structure.

The Characteristics of Chinese Knotwork

Chinese knotwork has primary utility as decoration, and its intrinsic aesthetical value is truly beyond compare. Not only does it occupy an important position in the decorative arts of China, it has also assumed a catalytic role in the development of the art of knotting in both Japan and Korea. That the Chinese ways of making knots could have exerted such a widespread influence in East Asia is by no means a coincidence; the author believes that the causes must be found in the structural diversity of the knotwork and its versatility in application.

The Chinese knots of concern to us in this article are those of both decorative and practical value. In general, they have the following characteristics:

First, Chinese knots are very compact in structure. The strain which pulls against the knots would draw the constituent parts tightly together, allowing them to hold. At the same time, Chinese knots are highly decorative, making them suitable for a variety of applications. The practice in T'ang China (618–907) of tying presents with red and white strings or cords serves as a good example; and as a matter of fact it is in this Chinese tradition that the Japanese mizuhiki finds its origin [12]. While the first mizuhiki emerged in the 7th century, the Japanese have through the ages shown a stern loyalty to 'tradition' and confined the application of the mizuhiki exclusively to gift-wrapping, and, as a result, no significant or innovative breakthrough has so far been witnessed in the development of Japanese knotwork. This may explain why the decorative knots of Japan are still quite simple in form and relatively loose in structure.

Second, Chinese knots are complex insofar as interlacement is concerned, and in this structural complexity exist many pattern variations. This characteristic, to be sure, clearly distinguishes Chinese knotwork from its Euro-American sibling, macramé tatting. While the square knots and half hitches are frequently used in the making of macramé, they are nonetheless rather plain, aesthetically speaking. Further, only when a group of square knots and half hitches are well put together may the beauty of a piece of macramé be realized. On the other hand, each of the knots in a Chinese knotwork is in itself a delicate piece of art, with its own cosmos of looping, weaving, hitching, and cording. When a group of these knots are assembled, countless combinations of decorative patterns emerge.

Similar to Chinese knots in terms of technical maturity are the knots of Korea. While tracing the history and development of knotting in Korea is by no means possible, as source materials conducive for such an endeavor are scarce, it is believed that Korean knotwork also has its origin in China. What is even more interesting to note here, though, is that the Koreans, having acquired their sophisticated knotting techniques from the Middle Kingdom, have been able to retain their own traditions and designs and create an art that is truly unique [12, p. 32].

Third, Chinese knots, for the most part, are symmetrical in form. While this design is very much in line with the underlying philosophy of Chinese decorative arts, it has nevertheless curtailed the artists' options in the selection of themes.

Fourth, Chinese knots are three-dimensional; they consist of two layers of cord, with an empty space in between. This type of interlacement can strengthen the structure of the knot without having to alter its shape, rendering it tenacious enough for hanging. At the same time, unwanted loose ends can be tucked into the space inside. Further, beads, jewels, and other auspicious objects may also be sewn in, adding to the overall beauty of the knot.

The Development of Chinese Knotwork and Its Application

Despite the fact that Chinese knotwork is in its own right a very exquisite form of artistic expression, documented references to its making or evolution have been scant. As Dr. Ch'en Ch'i-lu, former Chairman of the ROC's Council for Cultural Planning and Development, so succinctly pointed out in 1986, "Historians have always been confined by the concept that 'those with high forms constitute the Way, while those with low forms constitute the Tools,' and none has deigned to record for posterity the techniques of handicrafts. Consequently, the artistic formulas of master folk craftsmen have been completely left out of historical texts" [1]. The situation is even worse in the realm of knotwork. However well-tied, knots have traditionally played a rather secondary, supporting role in the decorative arts of China. As they have been employed primarily to enhance the beauty of other dominant art pieces, their significance has easily become unrecognized. Compounding the dilemma of scarce documentation further is that historically the preservation of knots has been given inadequate attention; hence we are left with virtually nothing from days gone by, except for a few pieces from the Ch'ing Dynasty (1644–1911).

Fig. 1. 1st Century Knotting: Intertwining Tails

Lacking historical source materials pertinent to the evolution of knotwork in China, one has to resort to other areas for possible references to this delicate form of art. Literature is one area of interest, as such knot terms as T'ung-hsin Chieh, Ho-huan Chieh, Shuang-shuang Chieh, and Hui-wen Chieh occasionally appear in the poetry of ancient China [9]–[11]; yet, we have no way of finding out how they actually looked. In addition, a good many works of sculpture, carving, pottery, and painting also feature decorations of knotwork. The study of these works may therefore yield some insights into the techniques employed in the making of certain knots, the historical period when a particular knot-form first emerged and how it was used, as well as a host of other relevant topics.

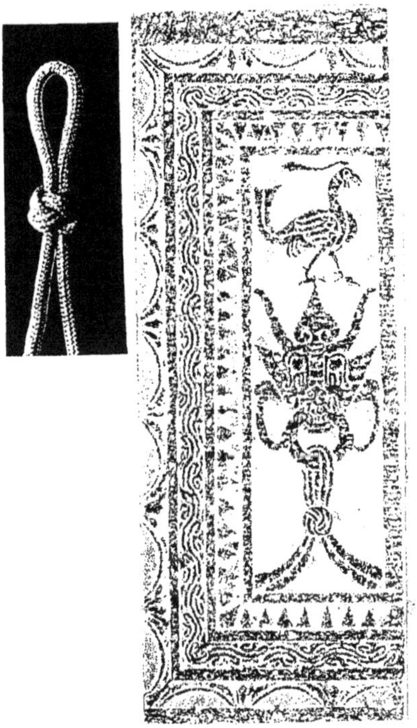

Fig. 2. A Knot from the Han Dynasty

The development and application of knotwork in China may best be viewed from the following two perspectives:

1. Knotwork as Symbolic Icon in Art

Rich in symbolic content, the Shuang-ch'ien Chieh (Double Coin Knot) is the knot that has the longest application history, insofar as our knowledge goes. The earliest artifact upon which the Shuang-ch'ien design can be found is a piece from the Warring States era (403–221 B.C.).

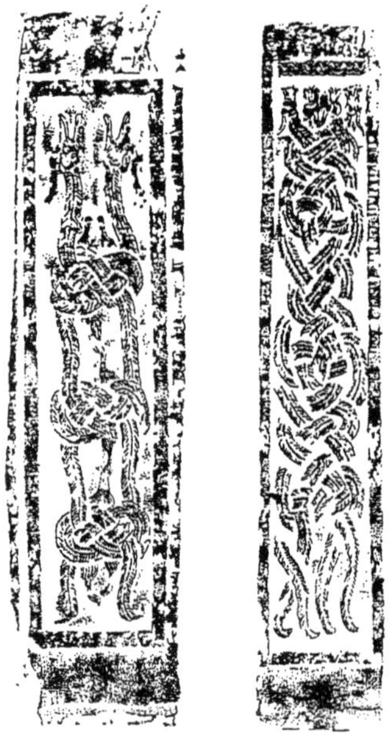

Fig. 3. Intertwining of Four Dragons

Yet, in those days this design was simply a decorative pattern, and it was not related in any way to the knot itself [14]. The emergence of the Shuang-ch'ien motif in the form of a knot is believed to be in the Han Dynasty (202 B.C. to A.D. 220), as many stone sculptures and brick inscriptions from the 1st century show the intertwining of either the tails of two dragons or the bottom halves of the semi-human, semi-bestial deities Fu-hsi and Nu-wa (Fig. 1) [4]. Fu-hsi was the god who 'tied knots in rope to make nets in order to hunt and farm' [6], and Nu-wa, as noted earlier, was the creator of humanity.

As soon as the goddess had established the system of marriage and was depicted joined to Fu-hsi, she was clearly his wife. It is in this connection that the author believes the Shuang-ch'ien Chieh is in fact the T'ung-hsin Chieh (True lover's knot) to which ancient Chinese poets had referred in their works. On the other hand, while the dragon is the earliest mythological beast in China, symbolizing the fertility of the earth and imperial power, the Shuang-ch'ien motif nevertheless signifies the fable that the Chinese are its descendants. The Shuang-ch'ien Chieh and the Niu-k'ou Chieh (Button Knot) both appear in stone sculptures and brick inscriptions of the Han Dynasty (Fig. 2).

An examination of its structure indicates that this particular form is essentially a variation of the Shuang-ch'ien motif. The Northern and Southern Dynasties period (386–589) witnessed the rise of yet another variant based on the Shuang-ch'ien design, with the intertwining of four dragons creating a beautiful, albeit structurally more complex, pattern (Fig. 3) [13].

Fig. 4. Wan-tzu Chieh (Sauvastika Knot)

2. Knotwork as Decoration

With artistic beauty and delicate structure embedded, the knotwork of China lends itself perfectly for purposes of decoration, and the earliest example the author is aware of is, once again, the Shuang-ch'ien Chieh. For instance, in the jade collection of the National Palace Museum, Taipei, Taiwan, is found a Han pendant carved in the style of the Shuang-ch'ien design, and upon the pendant itself is inscribed a simplified P'ing Chieh (Flat Knot) decor.

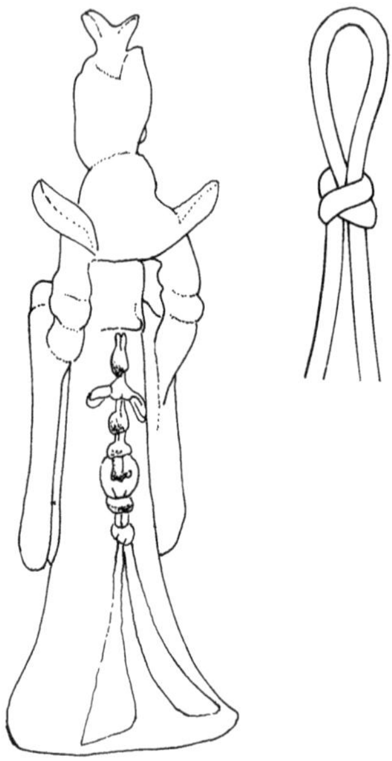

Fig. 5. Shuang-lien Chieh (Connection Knot)

Apart from being a source of inspiration for designers and carvers of personal adornments, knots of different types were actually put to use in real life. What follows is a brief review of three major areas in which the application of knotwork has been noted.

– Knots as costume accessories

Four types of knots have to date been identified, and they are listed here along with the earliest artifacts upon which they are found.

> Wan-tzu Chieh (Sauvastika Knot, Fig. 4): the sash on a statue of the Buddhist deity Kuan-yin (Avalokitesvara) from the Sui Dynasty (581–618).

Fig. 6. Chi-hsiang Chieh (Good Luck Knot)

Shuang-lien Chieh (Connection Knot, Fig. 5): the back of the sash on a san-ts'ai (three-color) pottery figurine of a T'ang palace maid. A Wan-tzu Chieh is also featured on the sash.

Chi-hsiang Chieh (Good Luck Knot, Fig. 6): the sash on a T'ang carving of the Buddhist deity Kuan-yin.

Shih-tzu Chieh (Cross Knot, Fig. 7): the holder on a T'ang or T'ang- style belt now in the collection of the Shoso-in in Nara-shi, Japan [5].

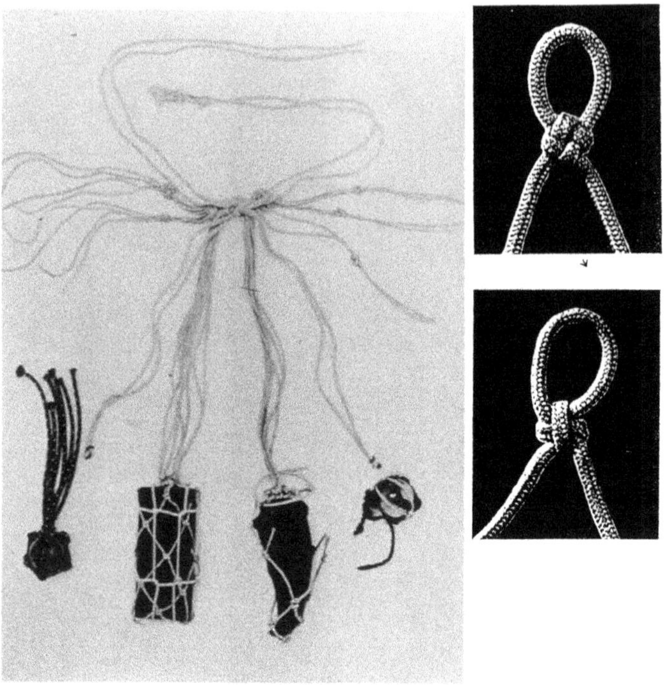

Fig. 7. Shih-tzu Chieh (Cross Knot)

– Knots as furniture ornaments

As representative pieces of mid- and late-T'ang painting, such as *Wan-shan Shih-nu T'u* and *Kung-yueh T'u* shows, knots of the Wan-tzu pattern were attached to chairs as ornaments (Fig. 8).

In the portrait *Sung Chen-tsung Chang-i Li-huang-hou Tso-hsiang*, Empress Chang-i (wife of the Sung emperor Chen-tsung, reigned 995–1003) is seen sitting on a chair decorated with several sets of pendants strung with knots (Fig. 9). During the Ming Dynasty (1368–1644) knotted pendants were gradually replaced by tassels as furniture ornaments.

Fig. 8. Wan-tzu pattern knots, attached to chair

Fig. 9. Pendants of knots, decorating a chair

Fig. 10. Screen, with P'anch'ang pattern knots

Fig. 11. Knots on 5th-century Stone Inscription

In the portrait *Ming Hsiao-tsung Tso-hsiang*, the folding screen situated right behind the Ming emperor Hsiao-tsung (reigned 1488–1505) is decorated with pendants made of knots of the mystic P'an- ch'ang pattern (Fig. 10). P'an-ch'ang, by the way, is one of the eight treasures or implements used by Buddha.

– **Knots as household ornaments**

In ancient China knots were tied to a multitude of household items besides furniture. They were in effect ornaments to other decorations, lending elegance and subtlety to objects that would otherwise seem commonplace.

Sunshades or canopies were often decorated with knots of different types, as depicted in the images from the scroll *Lo-shen Fu T'u-chuan* (painted by the renowned artist Ku K'ai-chih in the 4th century) and the 5th-century stone inscription of imperial pilgrimage unearthed in Honan Province (Fig. 11).

Fig. 12. T'uan-ching Chieh (Round Brocade Knot)

Fig. 13. Shou-tai Chieh (Cordon Knot)

Fig. 14. Ts'o-chiang-ts'ao Chieh (Cloverleaf Knot)

The 3rd- or 4th-century brick inscription of the legendary Seven Sages of the Bamboo Grove excavated in Nanking shows that the musical intrument held by the sage Juan Hsien (n.d.) is ornamented with a knotted pendant.

From time to time makers of household objects in ancient China would turn to knotwork for design inspiration. Two examples from the T'ang Dynasty are pertinent here. A silver kettle unearthed in Hsian (known today as Ch'angan) exhibits a pattern that is entirely based on the T'uan-ching Chieh (Round Brocade Knot, Fig. 12), while the decoration of a bronze mirror is modeled after the Shou-tai Chieh (Cordon knot, Fig. 13). The auspicious Ts'o-chiang-ts'ao Chieh (Cloverleaf Knot, Fig. 14) had even laid the groundwork for the pattern design of a Sung porcelain box.

Decorative knots in use during the Ch'ing period were many, indeed; they were no doubt the culmination of a tradition dating back to ancient times. Knots were attached to a wide variety of objects, such as fans, scepters, pouches, sachets, and eyeglass cases; in fact, most of the knot patterns that we are familiar with today were pretty much in vogue then.

Fig. 15. Intricate Knotwork

While we do not know how extensively knots and splices were used in the Shang and Chou periods (1766–221 B.C.), we are certain that they were crucial to many personal and household objects. For one thing, bronze mirrors were forged with rings, so that they could be tied to walls by knotted cords. For another, prized ornaments, such as those composed of several small pieces of carved jade, warranted equally attractive cord mounting, and putting the pieces together in a string certainly called for intricate knotwork (Fig. 15).

Other evidence also leads to the conclusion that knots were cherished as an essential part of everyday life. As a matter of fact, Chinese gentlemen of the Chou Dynasty (1122–221 B.C.) carried a special tool tied to their waist sashes, known as hsi. Made of ivory, jade, or bone, the hsi was a device to untie knots.

By the end of the 1st century, Shuang-ch'ien Chieh had met with wide popularity, as had the Niu-ko Chieh and Ping Chieh. In the early part of the 5th century, a multitude of structurally sophisticated variants of the Shuang-ch'ien design emerged. The development of Chinese knotwork reached its first peak during the Sui and T'ang Dynasties (581–907), during which many innovative forms emerged. A survey of some of the paintings that depict court life of the period and the practice of decorating the tombs of palace maids with knots indicates that the art was very much favored by members of the imperial family. The Ch'ing Dynasty (1644–1911) witnessed yet another state of prominence in the history of Chinese knotwork. The forms that are basic to us today, along with their more structurally complex variants, were widely employed. In fact, the art of knotwork was so popular that even the author of the *Hung-lou Meng*, or *The Dream of the Red Chamber*, made references to such knots as I chu-hsiang (Incense), Hsiang-yen-k'uai (Elephant's Eye), Fang-sheng (Twin Diamonds), Ch'ao-t'ien-teng (Sunflower), Mei-hua (Plum Blossom), and Liu-hs (Willow Leaf).

Some of the finest pieces in the collection of the National Palace Museum are furnished with decorative knots of the Ch'ing era.

Fig. 16. Author's Knotwork

The development of knotwork in late imperial and early republican China did not proceed as smoothly as it had previously, as the political chaos and social unrest so characteristic of the time were by no means conducive for the continued growth of the art. It was not until 1976 when the knotwork of China gradually regained popularity, thanks to a series of articles published in the Echo Magazine. A few years later, with the offering of relevant courses at the Shih-chien College of Home Economics in Taipei, the publication of the *Chinese Knotting* (1981) and its sequel *Chinese Knotting 2* (1983), and the founding of the Center for Chinese Knotwork (1982), more and more creative spirits and cultural aficionados began to try their hands at tying knots. Some practitioners have even gone so far as to invent new designs and patterns and experiment with a broad range of materials (Figs. 16–17). This promises to add a new and exciting dimension to the delicate art of Chinese knotwork.

Fig. 17. Author's Knotwork

Acknowledgements

The author wishes to thank Chao-Ling Sung, for help with translation.

Figure 5 is a line drawing from a photograph which appeared in *Ancient China*, published by Time Incorporated, 1967.

References

1. Ch'en, Ch'i-lu, 'Hsu,' in *Chung-hua Min-kuo Kung-i-chan = Arts and Crafts from the Republic of China,* ed. Huang Ts'ai-lang. (Taipei: Council for Cultural Planning and Development, Executive Yuan, 1986) 5.
2. Chieh, Hsi-ssu, 'Chieh-yang-ch'ang T'zu,' in *Ching-yin Li-tsao-t'ang Ssu-k'u Ch'uan-shu Hui-yao,* Chi Pu, **99** (Taipei: Shih-chieh, 1988) 446.

3. Chinese Academy of Social Sciences. Archaeology Institute. *Hsin-chung-kuo Ti K'ao-ku Fa-hsien He Yen-chiu,* 1st ed. (Beijing: Wen-wu, 1984) 4–6.
4. Chung-kuo Mei-shu Ch'uan-chi Pien-chi Wei-yuan-hui. *Chung-kuo Mei-shu Ch'uan-chi, 18: Hui-hua Pien: Hua-hsiang-shih Hua-hsiang-chuan* (Shanghai: Jen-min, 1988) 33.
5. Fu, Yun-tzu. Cheng-ts'ang-yuan *K'ao-ku-chi* (Tokyo: Bunkyudo, 1941) 4.
6. Han, K'ang-po (annot.), 'Chou-i Hsi-tz'u-hsia Ti-pa [Shu],' in *Shih-san-ching Chu-shu 1: Chou-i* (Taipei: I-wen, 1955) 168.
7. Kao, Ming (comp.). *Hsien-Ch'in Wen-hui,* **1** (Taipei: Chung-hua Ts'ung-shu Wei-yuan-hui, 1963) 614.
8. Kao, Hung-chin, 'Fu-shih,' in his *Chung-kuo Tzu-li = The Origin and Development of Chinese Characters,* 3rd ed., **1** (Taipei: Kwang-wen, 1964) 201.
9. Liang Wu-ti, 'Yu-so-ssu,' in *Hsien-Ch'in Han-Wei-Chin Nan-Pei-Ch'ao Shih* (Taipei: Mu-to, 1983) 1514.
10. Liang Wu-ti, 'Ch'iu-ke,' in *Hsien-Ch'in Han-Wei-Chin Nan-Pei-Ch'ao Shih* (Taipei: Mu-to, 1983) 1517.
11. Li, Pai, 'Tai Tseng-yuan,' in *Ch'uan T'ang Shih,* comp. Ch'ing Sheng-tzu, **3**, Chuan 184 (Taipei: Ming-luen, 1971) 1880.
12. National Museum of History. *2nd Joint Exhibition of Decorative Knots from the Republic of China, Japan, and the Republic of Korea* (Taipai: Center for Chinese Knotwork, 1985) 48.
13. Shan-tung Sheng Po-wu-kuan, 'Shan-tung Ts'ang-shan Yuan-chia Yuan-nien Hua-hsiang Shih-mu,' *K'ao-ku,* no. 137 (February 1975) 124–134.
14. Yang, Hua, 'Hu-nan Te-shan Ch'u-t'u Ch'u-wen-hua,' *K'ao-ku,* no. 34 (April 1959) 207.
15. Ying, Shao, 'I-wen Chuan I,' in his *Feng-su T'ung-i,* reprinted ed. (Taipei: Cheng-ta, 1968) 83.
16. Yuan, Te-hsing, 'Shih Ch'ien,' in his *Chung-hua Li-shih Wen-wu,* **1** (Taipei: Ho-luo, 1976) 1–46.

CHAPTER 7

INUIT KNOTS

Pieter van de Griend

A Little About the Arctic and the Inuits

The Inuits* occupy nearly all of the coastline from Greenland and Labrador in the east to the Bering Sea in the west, together with a short stretch of Siberian shore of the Chũkchi Peninsula. However, the area, with which this chapter primarily will be concerned, is roughly situated around Davis Strait and Baffin Bay (see Fig. 1). This is a region the Inuits have inhabited relatively undisturbed by foreign influences for an estimated four millennia. Generally accepted theory places the first of several migration waves of Inuit tribes in the Thule district around 4500 B.P. One may wonder what drove these people to settle in this harsh environment where vegetable foods are unprocurable, iron extremely scarce and trees exist in only one or two marginal districts. Large areas lack wood of any kind, including driftwood. Yet these people came from the west over Ellesmere Island, crossed Nares Strait and wandered right into the rich food supplies of western Greenland. Obviously only a people of great ingenuity and endurance could have survived in a region that lies rigid under snow and ice for 6–9 months of the year. Certain groups along the Arctic coast of Canada remained practically untouched by the outside world until the 20th century and even by the beginning of the 1960s a few Inuit were still maintaining the life of their forefathers almost unaltered, though none remain as such any longer.

Knowledge about Inuit history is in many ways sparse. What we know about their prehistory is due to their own orally transmitted legends and the picture emerging from a huge archaeological puzzle. From a Western view-

*An *Inuit* is an Eskimo of Greenland or North America.

point the recorded history about this part of the Arctic is closely linked to that of navigation, polar exploration and whaling. In crude outlines the first two started around A.D. 800 if, as is claimed, Irish monks in hide boats visited Greenland and continued their way over to Newfoundland. This impressive feat was followed by Vikings settling in south Greenland a century or two later. Their interaction however did not remain restricted to that area. Archaeological evidence has shown Viking influence on Inuits living as far north as Smith Sound. Due to climatic changes that area was left derelict during the 17th century. The Inuit had travelled southwards towards renewed meetings with Europeans, the Vikings having vanished by this time. The earliest recording of a re-encounter appears to be with Sebastian Cabot, who claimed to have had contact with Inuits in Hudson Bay in 1498 [11, pp. 185–186].

Fig. 1. Map of the Arctic, from Hudson Bay to Greenland

The British expeditions to find gold, riches and the northerly seaway to the Orient date from the last half of the 16th century. A few of these pursuits reached far north along the Greenlandic west coast. Around the 17th century the Spitsbergen whaling industry had followed the whales which were fleeing the climatic change. They had come south along the Greenlandic east coast, rounded Kap Farvel and were on their way north. The British expeditions aided a local increase in whaling and fishing activity representing many nations. Notably, the Dutch were early traders and whalers in the Davis Strait area. They were already in these locations during the last two decades of the 17th century with up to 150 ships during peak years of their activity [19, p. 15]. The whalers were very industrious in Davis Strait. The species they hunted were available in the open sea. There was thus no need to penetrate the ice pack to the north. The strong Dutch influence declined considerably after 1721 when Denmark colonized Greenland and lit the path for their missionary Hans Egede. His ambition was to convert all Inuit tribes, but he achieved this only from Kap Farvel to central west Greenland.

Meanwhile the quest to find the Northwest Passage to the Orient continued and yielded not only geographical results, but also increased ethnographical knowledge. In August 1818 Sir John Ross's expedition, well up north in the Baffin Bay ice, came across Inuits who had come south from the Thule district to hunt narwhals. He was led to believe that they belonged to a hitherto unknown tribe. Later, away up in the high north, Robert Peary's 1891 expedition was to re-establish contact with them. Ancestors of these peoples had been known to the Vikings, but that knowledge had vanished together with those early settlers. As the Inuits are semi-nomadic people a fair amount of travel, in particular along the west coast of Greenland, can be postulated; which means that the Polar Inuits which Sir John Ross met most certainly had been influenced by other Inuit tribes. It may be safely assumed that all the tribes in the area knew of each other's existence; but contacts between them may have been as infrequent as once a generation or so [26]. Perhaps the encounter was that specific tribe's first with Europeans, but not necessarily their first with elements from the latter's culture. It should not be forgotten that, at that time, the region's periphery had already been exposed to over 100 years of the whaling industry. Moreover, until then the Inuit culture, and especially their technical heritage, had not only been influenced from the east, i.e. by the Hudson Bay Company and other whaling ships, but most of all by what they brought with them from early times and possible encounters with Indians during their eastbound trek over the Arctic.

Nowadays the Inuits are integrated into the Canadian and Danish societies, but in the Greenlandic case they enjoy a considerable degree of home rule. Compared to their ancestors they are more adapted to, and have taken over larger parts of, modern life; all of this at the cost of the disappearance of their aboriginal culture.

Why study Knots used by the Inuit?

From a reductionist view, I prefer to see knots as solutions to rope problems.

This means that, given a medium in which to realise a knot, such as a rope, and a problem, such as tethering a cow, one holds the basic ingredients for what I term: a rope problem [13]. Whilst tangling the rope, tentative solutions to that problem, i.e. knots, are generated. Those structures which result as solutions can count on being able to enjoy a spell of being known, which could suitably be called their life-span. When rope users are involved in the bringing forth of a knotted structure I have observed that there is a complex play going on. The knots most frequently encountered empirically are usually structurally relatively simple. However, there are typically also similar solutions, which moreover are subject to parallel evolution. In this light consider von Brandt's attempts [5, p. 63] to outline evolutionary classes of some specific knots used by fishermen. During their life-span there is a continuous battle between all these solutions; and all possible means offered by the environment in which they are to survive are exploited. Survival is influenced by a number of factors, and as the situations in which the knot is used may change, they eventually result in what I call 'Survival of the Simplest' [14]—simplest not solely in a structural sense, but rather a combination of tying method and structure. As a rule the simplest is found to triumph over its competitors sooner or later.

The two most prominent exterior factors taking the stage are what I call the propagation of technological knowledge and the ethnographical concept known as cultural identity. The former speaks for itself and relates to the question *How does the knowledge of knots spread amongst peoples?* The latter denotes the fact that individuals display characteristic behaviour, like using certain knots, because their social environments have taught them to do so or perhaps force them by tradition. This, moreover, is often purposely maintained as a mechanism for uniting the social group. The concept answers part of the question *Why do certain groups of people employ specific knots?* I will use these ideas for showing that in a sense the first factor provides the required dynamics whereas the second factor usually displays a tendency towards arresting a solution's evolution within a cultural group. They occasionally work together and effectively prevent a knot from becoming locally extinct in time and place.

These ideas are particularly easy to apply in the case of the knots used by Inuits, but their knots are interesting for a number of other reasons too. For primitive people to survive in the Arctic, knowledge of knots and the media in which they can be realised is absolutely essential. Since we wish to subject these solutions of rope problems to further investigation we must determine which aspects of them we want to study.

Knots are medium dependent; different media require different knots. In the Inuit case there is a range of materials which have biased their knotting techniques. For example, Baffin Land Inuits used slipped knots in thong lashings to facilitate untying by means of a thrust of the foot, without having to take off their mittens. On photographs of the kamiks (boots) of the Kilaqitsok mummies, knots similar to the Codline Knot appear to have been used [27]. It is interesting to know to what extent Inuit materials have influenced their knots, i.e. which classes of solutions were achieved? As a rule they have chosen the simplest of solutions expressing an understanding of physical and spatial phenomena such as friction and, often, symmetry.

Structurally Inuit knots are of interest since they are as a rule either single-stranded or formed as rawhide splices, using slit materials, though combinations of both occur. The leather material Inuits have used does not readily permit multistrand structures, such as splices, to complicate affairs, as found in other peoples' knot repertoires. In general, techniques using slit materials are preferred. These techniques are related to ordinary knots also, since they are based on recognising and appreciating some topological properties of a loop. However, slitting techniques were a solution to only a portion of the problems they met. Frequently Inuits were driven to employ knots instead of rawhide splices, since the slit material could give way under tension. Thin material, found in for instance lashings, may therefore be knotted. Moreover, the making of a rawhide splice involves using the non-working end of the material, which does not facilitate the tying process.

The Inuits were among some of the last so-called primitive peoples; and their knotting techniques may indicate how they have developed them. Knots are very rudimentary examples of technological knowledge, and therefore analysis of their use places one in a position to determine how people reason in the learning process, along with obtaining indications of what influences they have been subjected to. Common sense does not always explain the use of particular knots. Peoples' reasoning need not adhere to any immediately apparent logical consistency. The Inuits are a pantheistic people; I would not expect pantheistic people to assign symbolic functions to knots, and do not know of any explicit example. However, Bourke [4, p. 561] mentions the selling of wind knots by 'the inhabitants of Greenland' and references Grimm: *Teutonic Mythology* [**2**, p. 640]. Though the Inuit had decorative knots, we will mainly focus on practically applied tangles in this chapter.

In the following discussion there will be recurrent reference to the social group called Mariners. I would like to define what I mean by that term in the context of this paper. By 'Mariners' I denote western seafarers from the early 16th until the second half of this century. To many people the sailor profession of the clipper heyday has the supreme claim to knots, but in my eyes the romantic powers of the sea have caused that claim to become an exaggerated common misconception. Mariners, by nature of their trade, have had a considerable exposure to rope problems along with the necessity of having had to find fitting solutions. During this process they have developed the technical use and vocabulary of knots somewhat further while getting those transactions recorded in the best of ways of their time. Knots entered respectable seamanship manuals for a number of reasons. Just like everything else at sea one has to save expensive rope, since it quite obviously is not an immediately replenishable resource. In the aftermath analysis of a dangerous situation apparently trivial things get consideration and nomenclature to describe them is generated. Eventually all of this hard-gained knowledge ends up in the literature where it hopefully will be studied by aspiring Mariner officers. This process implements the essential parts of the propagation of technological knowledge on the Mariner's scene. Conservative maritime traditions comprise the relevant parts of that social group's cultural identity. This results in a relatively fixed set of solutions to the rope problems which Mariners encountered.

Mariners are introduced for a number of reasons. They are representative of an Occidental group having many aspects, in a knotting context, in common with the Inuit. Shared elements cover the sea, their states of relative isolation which result in the necessity to be thrifty with resources, and lastly their frequent confrontation with rope problems and knottable media, though they differ mainly in a materialistic manner. This discrepancy manifests itself in the varying knotting techniques, but the basic thinking required in the solving of the rope problems encountered is maintained. It will become evident that, in an abstract sense, many ideas from both groups are equivalent solutions.

Some Arctic Knot History

The Inuit were among the last of Stone Age peoples. Understanding their use of knots would greatly benefit from comparison with those used by other Stone Age peoples. Hence it would be obvious to gather knot-knowledge about other known prehistoric peoples all over the world and attempt to synthesize an analogy between them and the inhabitants of the Arctic. Alas, for a number of reasons, this is quite impossible. In the global knotting literature not many writings are to be found on aspects of prehistoric knotting. Whatever is available of contemporary so-called primitive peoples' knotting is very fragmentary. For these reasons we shall only review the accessible relevant knot history of the Arctic.

Knowledge of the history of Inuit knotting converges on two intervals of a time scale which spans from 4500 B.P. to the present day. These two intervals are situated at both ends of this scale. We know a little about knots from the Saqqaq Inuit period, which started over 4000 years ago, and slightly more about knots from A.D. 1700 and onwards. Everything in between is extremely difficult to access. No definite answers, for instance, can be given about possible Viking influences, or interactions during the period of whaling by Europeans around Greenland; both Western groups of established rope users.

In the literature we have two major sources which offer indications about

Inuit knotting, ethnographical and archaeological. The methodology we wish to apply will be one derived from their combined approach, that is:

1. A survey of the records of ethnographical observations in the literature. These sources, Sollas, Murdoch and Boas [30], [25], [3] commence rather late, i.e. from ca. 1880 and onwards. This sudden ethnographical interest in the subject was due to the emerging realisation of the importance of technology to the structure and functioning of society, as knots can be seen as a rudimentary form of technology. However, knots seldom stand alone and are usually incorporated into more sophisticated constructions. Porsild discussed the construction and use of knots in a baleen sealing net [28], Ford showed a baleen net knot [10], while Birket-Smith has described the incorporation of knots in implements from the Egedesminde district [2].

2. Accessible hard archaeological evidence in the form of excavated knotted artifacts and either artifacts described in the literature or available at the National Museums of Denmark and Great Britain. The oldest set of known proper knots are those excavated at the Saqqaq settlement of Qeqertasussuk tucked away in the southeastern corner of Disko Bay in central west Greenland [15]. On the other hand, Karen McCullough found knot samples from around A.D. 1000 during her Ruin Island project and on which she reports by means of photographic illustrations [24]. Erik Holtved [22] described excavated Thule Inuit Knots from about 200 years later.

Excepting some further snippets, which are few and far between, the above mentioned cover almost all currently accessible knowledge about Inuit knots. Fortunately there remain the serendipitous sources. H. C. Gulløv, from Denmark's National Museum's Etnografisk Samling in Copenhagen, was so kind as to supply me with descriptions of knots, which he had excavated at a site in the vicinity of Nuuk and dated to be from A.D. 1700. Around 1985 G.Budworth surveyed some knots on a Polar Inuit sledge, which is kept at the Museum of Mankind, the ethnographical department of the British Museum in London [7]. The sledge was brought back by Sir John Ross from his 1818 exploration to find the Northwest Passage. It represents a sample of Polar knotting skills uninfluenced by outsiders. Jørgen Meldgaard at Denmark's National Museum has informed me about material involving knots from A.D. 600 to appear in an In Memoriam paper he is producing on Helge Larsen's excavation of an Ipiutak Inuit site in the western part of the arctic region.

A few words follow on the kind of knots we shall be discussing. The foregoing examples are all of practically applied knotting, but like many other peoples in the world, the Inuit also had creative pastimes involving string. East Greenlandic cat's cradles, for instance, were studied extensively by Victor [31]. Old samples of decorative knotting have been excavated on Ruin Island [24] and a more recent one can be found on the Ross sledge [7]. However, in this paper I will be primarily concerned with simple knotted structures. Hence I will not consider rawhide splices, lashings or any other elaborations.

Concerning the objects under discussion there are a number of possible views which affect the mode of classification:

Structures: One can consider the pure structure, allowing deformation but not cutting or gluing, of the knot we are looking at. In mathematical terminology one speaks of considering structures up to isotopy.

Applications: Roughly, knots' functionality can be split into three categories, namely: Hitches, Bends and Loop Knots. They cover respectively the attaching of a knottable medium to another, usually rigid, object; lengthening a rope; and forming a loop at the end or in the middle of a rope. One and the same structure may have different functionalities. For example, the Sheet Bend structure may function as a bend and also feature in one of the many Bowline forms as a loop knot.

Materials: Here, one groups knotted structures according to the materials in which they are realised. There may be backgrounds in tradition, superstition or other reasons, for using specific knots for certain materials. Although Inuit culture often demands the separation of elements from hunting on the sea and elements from hunting on the land, the Inuit do use baleen for snaring land animals.

Due to the confined scope of this paper we will see knots as application-specific structures. Hence the focus will mainly be on the structures and application. However, we shall commence with a brief introduction to the materials employed. They justify a comprehensive study in their own right.

Description of Materials

Real-life knots cannot be realised without a medium. The choice of knots is affected by the material properties of the medium in which they are realised. In turn this affects the types of knots observers may chance to encounter. For this reason we discuss aspects of the materials in which the Inuit have tied their knots.

Baleen

Baleen is a horn-like plating found in the rear of the whale's mouth cavity, used as a plankton filter. It appears that John Davis was the first to mention

Inuit use of baleen for the making of fishing lines. Excepting that note, from around 1565, there is not much known about prehistoric applications and processing of this material. The Qeqertasussuk find suggests the Inuit used considerable amounts of it for a wide range of applications [16], [17], [35]. A small discussion of a baleen processing experiment can be found in [15].

Baleen is water repellent, which makes it the Inuit's preferred material for fishing lines. On the other hand, its production is cumbersome in the extreme. In the form used, the Greenland whale's baleen must be split into long thin strips. The length of these strips has a natural limit, which, depending on the source, can be 1–2 metres. However, artifacts at the National Museum of Denmark show that lengths down to 0.4 metres were not discarded.

Unless required for very small lashings, where the strings can be almost round with diameters of 1 mm or less, split whale baleen is monofilament and rectangular in cross-section with dimensions 4 mm × 1.5 mm. Knots tied in flat material, like baleen, behave differently from those tied in a cross-sectionally round medium in which there are no edges.

The surface smoothness of baleen is high. In many ways, this material can be compared to long thin flexible strips of hard PVC. Knotting thus requires structures with a high resistance to slippage. Such devices have been found and will be discussed later. However, like most knots, they reduce the breaking strength of the medium in which they are realised.

No information on tensile strength of baleen is available, yet rough indications do exist. It is well-known that some Inuit baleen fishing lines were used to catch pelagic species from the bottoms of fjords over 1000 metres deep. Every knot on a fishing line in use is subjected to the friction of water. This gives rise to a force, which along a length of line may accumulate to a considerable strength. Although a hydrostatic correction would be required, it still remains quite impressive that the limits of baleen's mechanical properties, or even the breaking strengths of the knots tied in it, were rarely exceeded. Apparently, in most of the cases, these limits were not reached.

Topologically, the principle behind the rawhide splice, to be explained in the section on thong, is by far the best solution to make a non-slipping connection between two cross-sectionally flat media. However, one cannot split baleen to make a join reminiscent of such a splice. Hence knots must be used.

Sinew

Sinew has been used for making thread. The fibres were obtained from the sinews of sea mammals, such as seals. The sinews were first dried, then split, crushed and finally twisted into yarns. Inuit women traditionally did the twisting against their cheeks or thighs [27]. The lengths which these threads may reach has no natural limits. No objective data on breaking strengths exists, but sinew thread with a diameter of around 1.5 mm is exceptionally hard to break by hand. Sinew thread is sufficiently rough for relatively simple knots to hold in it.

Inuit sinew threads may have a diameter of 1–2 mm and are often single stranded, but thread made up of two strands has been found at Qeqertasussuk. This remarkable composite consisted of two strands twisted in a left-handed fashion, each of which was made of yarns, which were twisted in a right-handed way. If this non-trivial inversion did not occur, then the thread would be less strong. Much as an untensioned spring can be made to grip firmly onto a rod when stretched, the alternating direction of twisting increases cohesion between fibres.

Although no Inuit laid rope has been found so far, the Qeqertasussuk people unambiguously show that they were aware of the principles of constructing it. Apparently the Inuit refrained from developing this skill. At a later stage, such as shown by the single-strand thread used for stitching in the clothing of the Qilakitsoq mummies from approximately A.D. 1475, it appeared that their descendants did no longer care or may even have lost that knowledge altogether [27].

Thong

Thong is a long strip of rawhide, which is cut in a spiralling fashion from a piece of skin. Hence its length, width and crossection depend on the skill of the cutter. Wet thong displays an extreme surface smoothness. It follows that knotting requires special and very secure structures, unless further precautions are taken [3, fig. 44]. On the other hand this material is excellent to "splice". Rawhide splices can be obtained by making a simple longitudinal slit, close to the working end of one piece of thong, and hitching another length onto it through a similar slit close to the working end of the other piece of thong. Thong is strong and therefore well-suited for this method of splicing. Of all organic materials found at archaeological sites in the Arctic, thong seems to be among the first to perish. It follows that rawhide splices would probably not survive in a form beyond slit thong. Also for this reason not much is known about prehistoric Inuit rawhide splices.

Grass

Virtually nothing is known about this knotting material. I am indebted to H. C. Gulløv for telling me about an observation made by the merchant Anders Olsen in 1764 in southern Greenland, on Inuit use of plaited deep-sea fishing lines made of this material. The Thule Inuit plaited grass to make coiled baskets [22, pl. 34]. As I have not yet encountered any grass knots, they will not be discussed any further in this paper.

The Knots of the Inuit

In this section we shall first discuss the Saqqaq knots retrieved at Qeqertasussuk, and then turn to the later periods, when interaction with Europeans has known to have occurred. A more detailed discussion follows in the next section.

The Qeqertasussuk Knots

The Qeqertasussuk site is dated to be the oldest known settlement of the Saqqaq people in west Greenland [16], [17], [35]. It represents a fascinating find which has spun off a wealth of new knowledge about prehistoric Inuit culture. Among the excavated artifacts were several knotted structures [15].

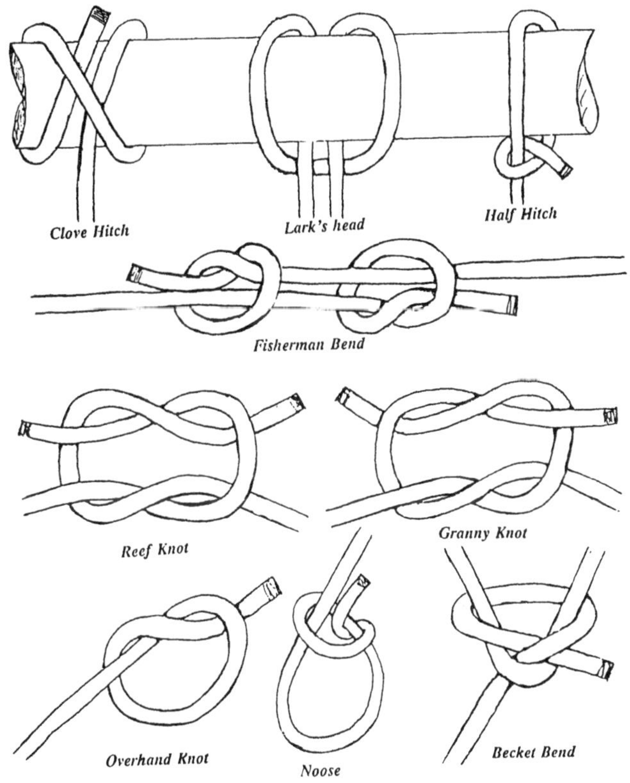

Fig. 2. An overview of the knots found at Qeqertasussuk

The majority were minuscule objects, which had once been tied in whale baleen and thong, while some scraps of clothing had remnants of sinew thread knots. All samples had been preserved in the permafrost for about 4500 years. The whale baleen had hardened so much that the knots it held had become rigid. Hence with reasonable certainty we can say that the knots were Clove Hitch, Lark's Head, Reef and Granny Knot, Overhand Knot, Half Hitch, Becket Bend and a Noose based on a parallel centrally pierced Overhand Knot (see Fig. 2).

In a strip of sea mammal hide a very recognizable Sheet Bend structure was found [35] (see Fig. 3). Its function could not be determined unambiguously. Hence it may have been a bend, part of a loop knot, or something else.

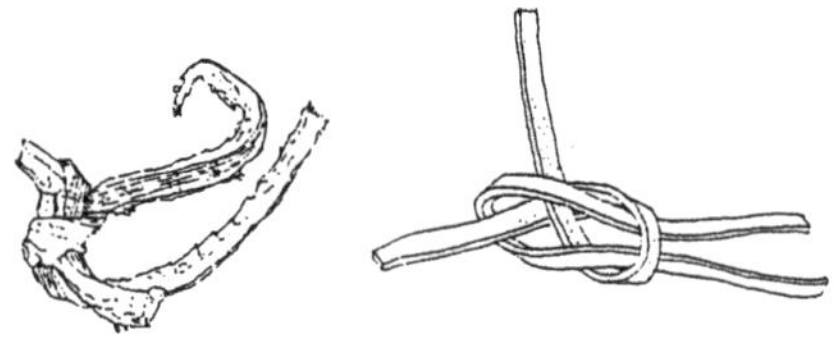

Fig. 3. A Sheet Bend structure excavated at Qeqertasussuk: courtesy B. Grønnow

The knots in the thread of the excavated clothing samples were very hard to identify. The clearest were Overhand knots, which had functioned as stopper knots. There was also a structure which obviously had functioned as a bend, but had become so congealed that it was only possible to see that it consisted of two interlocking Overhand Knots. It presumably was a Fisherman's Bend [1, #1414].

It is truly remarkable to note that the knots found at Qeqertasussuk were in fact so diverse and real knots, i.e. not merely some haphazard conglomerations of Half Hitches and Overhand Knots. This indicates that the Saqqaq people possessed an impressive knot repertoire and knew how to use it.

The Later Periods

It is not until Viking settlers began to colonize Greenland that we could reasonably have expected influences from the east. In contrast with the period from which the Qeqertasussuk Knots stem, we now have come to times of which we know with certainty that cultural interaction on several fronts between the Inuit and Europeans had already taken place. In the next section we shall be concerned with the question about the extent of a possible mutual influence in knots and knotting techniques.

The first accessible archaeological evidence of knots comes from the Ruin Island project [24]. Ruin Island lies not far off the coast of Inglefield Land which forms the southeast shore of the Kane Basin. Ruin Island is situated slightly to the northwest of Inuarfigsuaq. Around the year A.D. 1000 pioneer settlers inhabited the island. Karen McCullough shows several artifacts with remnants of knotwork made by the Ruin Islanders. The photographic evidence is often difficult to interpret. On some pictures a number of Half Hitches can be discerned. The clearest knotted object is a remarkable piece of square plait of which a tying method (or algorithm) is illustrated in Fig. 4 [24, pl.68, figs.x,y].

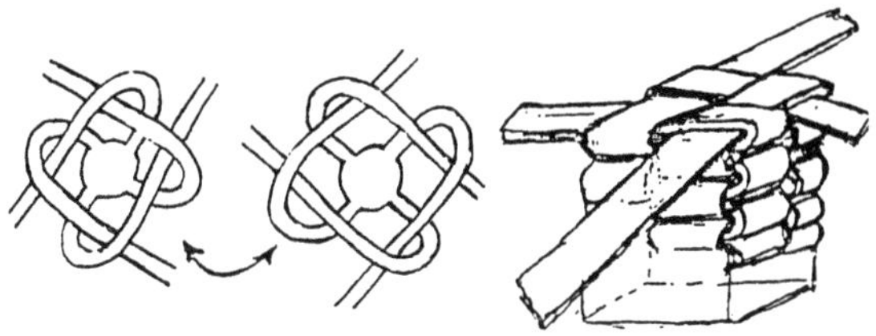

Fig. 4. Square Crown Sennit algorithm

The next possibility to see some old Inuit knots occurs at Qilakitsoq on the north shore of the Nuusuaq Peninsula north of Disko Island on the west coast of Greenland. At the beginning of the 1980s a grave with the mummified remains of six people was found [21]. C-14 examinations revealed that they had lived around A.D. 1425–1525. A number of knot fragments has been investigated. We know that their clothing was sewn up with single-stranded sinew thread. Photographs revealed that they tied complicated plaited Codline-Knot-like structures in the strings which kept their kamik (boots) on their feet [27].

At the beginning of the 18th century extensive trading and a whaling industry were firmly established around Greenland. Places such as Nuuk (Godthaab) were frequently visited because they remained ice-free year round. It would therefore be interesting to see what kind of knots were used in such centres of activity. Fig. 5 shows some truly exciting knots tied in baleen, which are carbon dated back to A.D. 1700. They were excavated by H. C. Gulløv in the vicinity of Nuuk [18]. Among some indistinguishable knots we also notice two newcomers, namely a Bowline variant and a Triple Fisherman's Bend. We shall have more to say about these knots in the next section; for now, we pose the interesting question—were the knots indigenous, or were they brought to Greenland by Mariners?

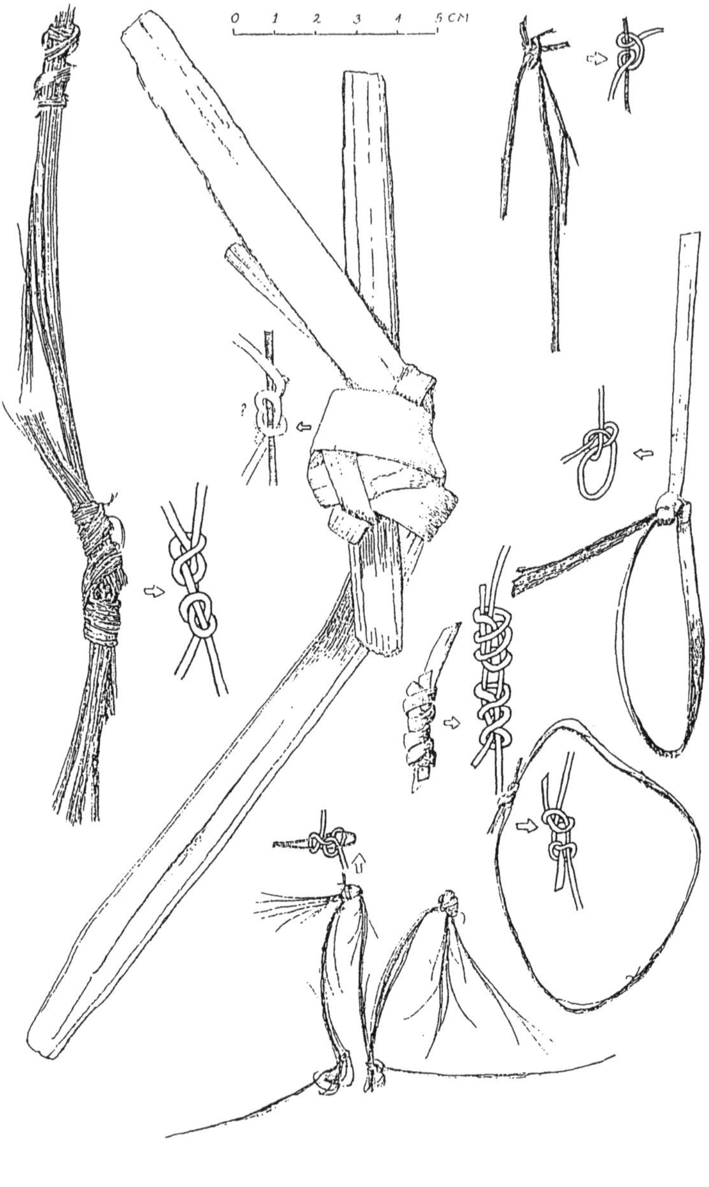

Fig. 5. Inuit knots from 1700 Nuuk: courtesy H. C. Gulløv

Before we turn to the discussion and ethnographical descriptions of knots, there remains the sample of Polar Inuit knotting which Sir John Ross brought back from his 1818 expedition. High up north on the Greenlandic coast he met an Inuit tribe, which had come south for their summer hunting of narhwals. During this encounter they presented Ross with a sledge made of bone and wood and lashed together. On this sledge several knots and rawhide splices have been found [7].

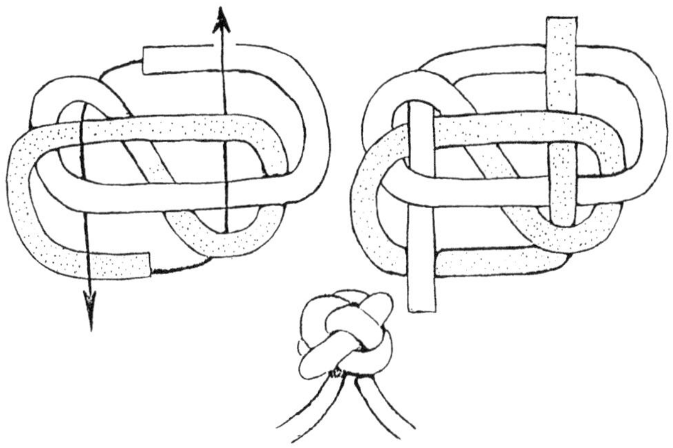

Fig. 6. The Diamond Knot from the Ross Sledge

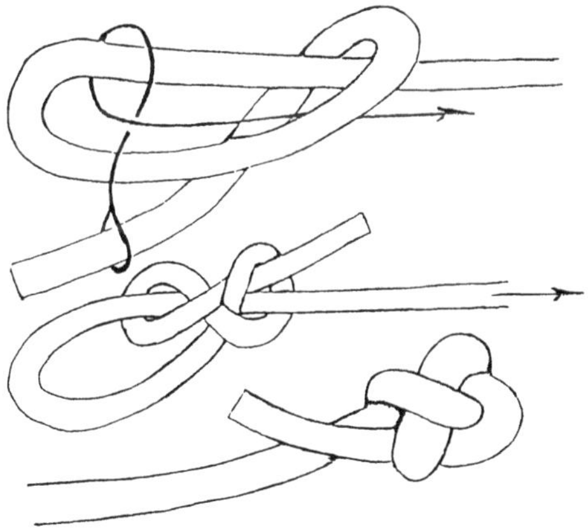

Fig. 7. Qitdlaq

Three of them are of interest to our study. The first, a variant Bowline, we have just encountered among the knots from 1700 Nuuk. The second, a Diamond Knot (Fig. 6), because it is one of the very few samples of decorative Inuit knotting. The last of the interesting Ross Knots is a Groundline Hitch. So far the structure has only turned up on one other occasion in my studies of Inuit knots. The Groundline Hitch structure was shown in a description of a knot used by the people on the west coast of Upernavik in narwhal hunting. Keld Hansen, then curator of the National Museum of Greenland, wrote to tell me that it was called qitdlaq, from the verb qilerpâ, which means ties with a knot [20]. Unfortunately Qitdlaq is also a well-known man's name among Polar Inuits, indicating that a certain Qitdlaq might have been responsible for this knot name. The knot no longer functions as a hitch, but as a stopper knot (Fig. 7), holding the harpoon line inside the float (Fig. 8), and as it is supposed to hold a narwhal it has to be absolutely safe.

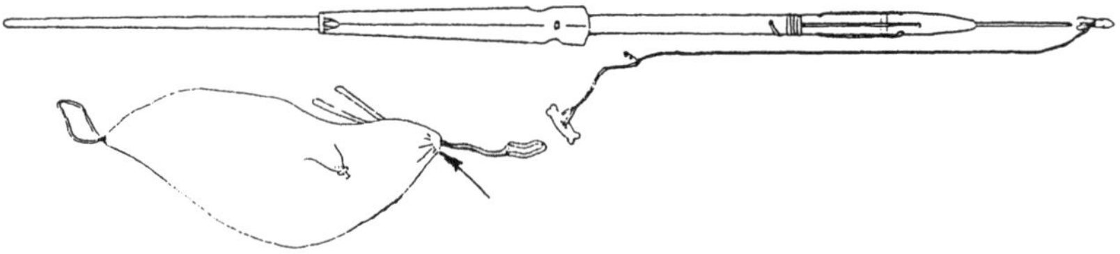

Fig. 8. An Inuit narwhal spear with float

Discussion

From the overview in the previous section it becomes obvious that during the past 4500 years the Inuit have not changed their simple knot repertoire substantially. Of the roughly 15 knots we have met so far, most have been used since palaeo-eskimoic times. By and large they are also the simple knots to be found in use among other cultures. However, two of the more 'recent' knots spring out of that picture, the variant Bowline and the Multiple Fisherman's Bends. This has also been noted by Day [9, p. 84–85]. Furthermore the archaeological evidence dating the apparent first usage of these two structures coincides with intensified contacts between the Inuit and Europeans. Mariners, established rope users, have been known to use similar structures for a longer recorded period. It has been suggested that these knots therefore were elements of western knotting traditions, which had been absorbed by the Inuit culture. For these reasons we shall confine ourselves to a discussion of the variant Bowline and the Multiple Fisherman's Bends and argue that these knots are most likely part of at least some of the aboriginal Inuit cultures.

We shall commence this discussion by first giving a more detailed description of these knots and their possible tying methods.

The Fisherman's Bends

Although the occurrence of the Single Fisherman's Bend (Fig. 9) among the Qeqertasussuk Knots is not totally unambiguous, it is obvious that it must have been a structure the later Inuit were pleased to know. It occurs among the Gulløv Knots (Fig. 5), and the same structure was employed by the Point Barrow Inuits around the 1880s at the far western side of the Arctic [25, p. 279, fig. 265], applied as a loop knot hitched onto a fishing hook.

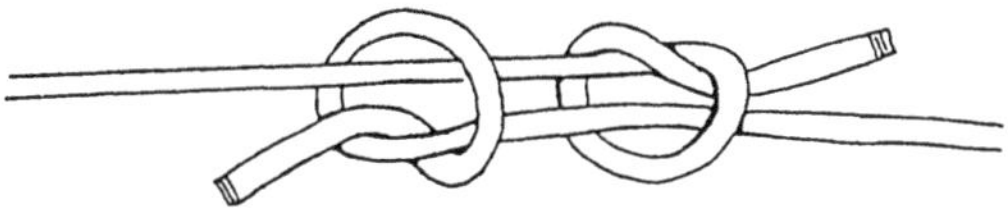

Fig. 9. Single Fisherman's Bend

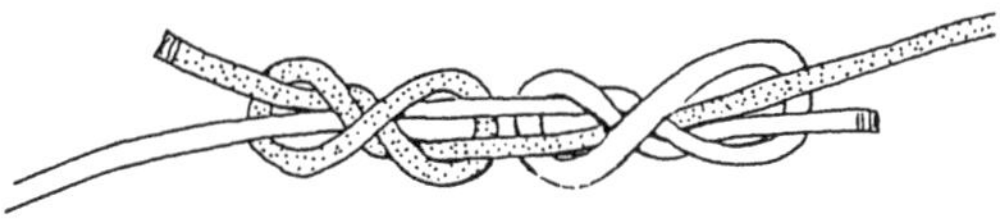

Fig. 10. Double Fisherman's Bend

Many peoples, apparently independently from each other, have come across this loop knot. It has been described by the Greek physician Heraklas almost two millennia ago [8], [9, pp. 101–151]. The two knots are structurally equivalent, though they differ in functionality. However, knowledge of the loop knot version may well imply familiarity with the structure as a bend.

The Fisherman's Bend structure, based on two symmetrically positioned Single Overhand Knots, occurs as a bend on the baleen fishing lines. Baleen is a tough smooth material, which is very resistant to a transverse load. The edges of the flat monofilament may be deflected under a reasonably strong perpendicularly applied load, such as those obtained with a pair of narrow beaked pliers. However, to achieve this via Single Overhand Knots, requires quite a strong pull, which will cause the composing Overhand Knots to pull through under tensioning. Hence to ensure knot stability, care in knotting is plainly insufficient. It follows that one needs a more secure structure with better friction-amplification properties. Without sacrificing the symmetry aspects, one may try using Double Overhand Knots instead of single ones.

On the Ross sledge no form of Fisherman's Bend was reported [7], probably because no baleen had been used. Nearly all of the thong knots shown by Boas [3, fig. 44] have been made secure by means of toggles, seizings, slits and even sewings, but the Double Fisherman's Bend (Fig. 10) was used without any additional precautions. This is not surprising as there is more surface on which friction can become operative in the composing Double Overhand Knots to yield a greater gripping effect. This feature is typically exploited in the knotting of smooth material. In knotting modern monofilament synthetics, structures based on Multiple Overhand Knots are quite common. The underlying principle is one of which the Inuits would have needed to be aware in knotting their slippery materials.

A negative aspect of the Multiple Overhand Knots would be the bulkier result. It can be argued successfully that use of the Double Fisherman's Bend would require too much of a precious material.

The reasons why the Inuits preferred the more elegant rawhide splices for connecting two thongs must probably be sought in these facts. However, as said, it is not possible to make a join reminiscent of a rawhide splice in baleen. If it did, then the foregoing should effectuate an elimination of the Double Fisherman's Bend from the ordinary Inuit's knot repertoire. In fact, the opposite has happened, making the strongest argument in favour of them being indigenous. The bulkiness can also be counteracted by giving the knots a barrel shape. More wrapping turns form a cylinder to stabilize the knot. This already occurs on the Triple Fisherman's Bend from 1700 Nuuk (See Fig. 5). An artefact (Lb.23) from Disko Bay, donated to Denmark's National Museum in 1854, also incorporates Fisherman's Bends based on Triple Overhand Knots. Nowadays the symmetric barrel idea is also to be found in bends used by anglers for the synthetic supersmooth monofilaments. In rupture tests these bends often reach breaking strengths of close to 100%, i.e. they are about as strong as the medium in which they are realised.

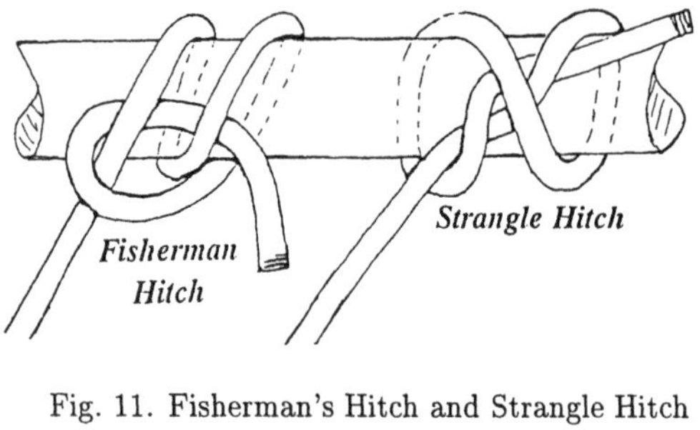

Fig. 11. Fisherman's Hitch and Strangle Hitch

In Europe the Double Overhand Knot structure was certainly known during Carolingian times [33, fig. 47], which spanned the final quarter of the first millennium of the Christian era. Mariners have thus probably known it from their inception and have certainly employed it in numerous applications, for instance in the Fisherman's Hitch or as the Strangle Hitch used by fishermen to attach snoods to hooks. However, to my knowledge the first recording as a bend was not until the 19th century when longline fishermen used the Double Fisherman's Bend to mend snapped horsehair lines [34, p. 38]. Together with the Fisherman's Hitch (Fig. 11), which in effect is nothing but a spilled Strangle Hitch (Fig. 11), which in turn is structurally equivalent to a Double Overhand Knot, it is also used a lot on a bird snare (L17.90) donated to Denmark's National Museum in 1934.

Bowlines and Boas Bowlines

One frequently requires a non-sliding temporary loop knot at the end of a medium. A first approach is to attach the working end onto its own standing part with a hitch. However, hitches that do not spill usually slide. This can be prevented by trying to adapt a simple bend solution instead. For this reason many of the well-known loop knots can be related to well-known bends. Bends are loaded on two standing ends, whilst both working ends remain tension-free. In a loop knot there is only one unloaded working end. In order to convert a bend to a loop knot we must select one of the working ends and connect it to the appropriate standing end.

A sufficiently simple bend to commence from is the Sheet Bend. Just as the Sheet Bend structure gives rise to a number of different bends, the same structure may be used to form 16 different solutions to the loop knot problem. The four parameters which determine these solutions are:

1. Choice of placement of the standing end, i.e. on bight or gooseneck.
2. Choice of orientation of the gooseneck, i.e. the direction in which it is intended to work.
3. Handedness of the gooseneck, i.e. left- or right-handed.
4. Placement of the working end, i.e. inside or outside of the loop.

We shall only discuss the types of loop knots arising from parameters 1 and 2, as 3 and 4 comprise non-relevant mirror symmetries and placements of the working end. Parameter 1 divides the set of resulting loop knots into two distinct major classes. Planar projections of these two classes, in which the crossing information is not considered, are shown in Fig. 12. They are differentiated by whether the change from working end to standing end occurred on the gooseneck or the bight. The latter leads to loop knots with undesired qualities, therefore we shall not be concerned with them. An interested reader is referred to [13].

The useful Bowlines are determined by the orientation of the gooseneck. This yields two distinct types, which have been called the Bowline and the Boas Bowline [6]. We shall first focus on the proper Bowline types.

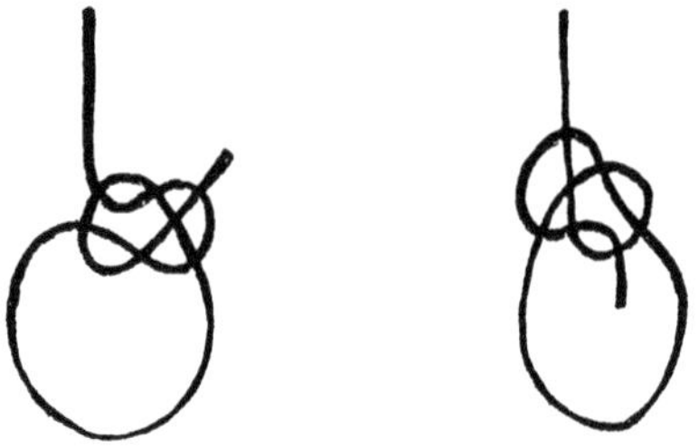

Fig. 12. Shadows of possible Loop Knots based on the Sheet Bend structure

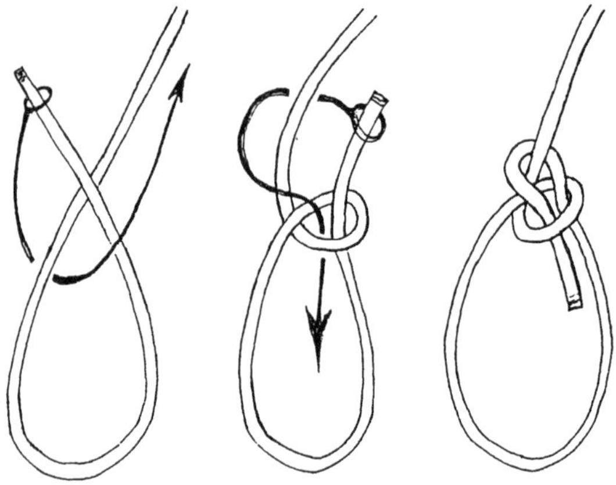

Fig. 13. The Bowline flip-over tying method

There are too many ways to make a Bowline for all of them to be discussed here, but we shall focus on two methods. In the Western knotting literature and knotting world, which are dominated by strong Mariner influences, one of the commonest methods for tying the Bowline is the flip-over method. This method is easy to learn and is excellently suited for relatively thick media, such as rope (see Fig. 13).

Another method of making a Bowline is by spilling a Noose (see Fig. 14). This method is not very common in Mariner spheres, but it is excellent for tying a Bowline in flexible thin diameter media, such as thong.

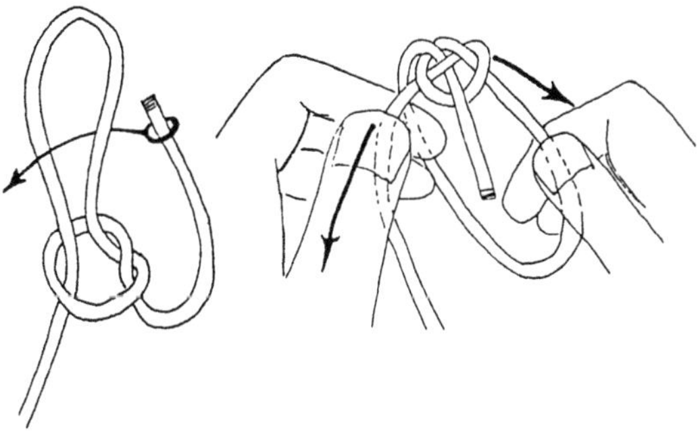

Fig. 14. Bowline tying method based on spilling a Noose

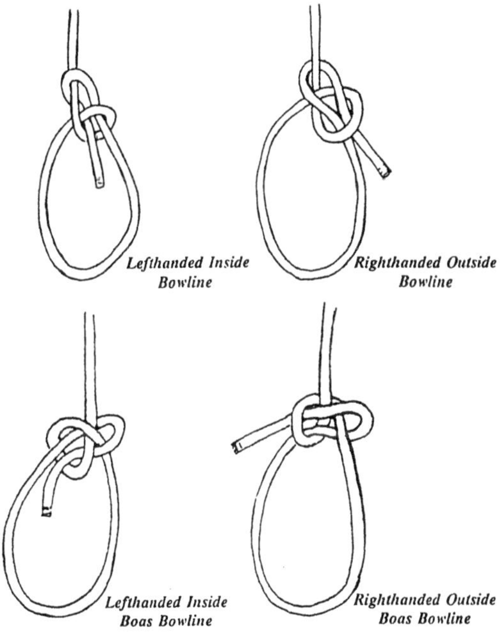

Fig. 15. Various Loop Knot forms based on the Sheet Bend structure

The Boas Bowline is the other type of reliable loop knot, which is based on the Sheet Bend structure. Like the Bowline, there are four types of Boas Bowlines. They are determined by placement of the working end and handedness of the gooseneck. Two types of each form are illustrated in Fig. 15.

This knot owes its name to the fact that F. Boas reported its use by the Inuits of Baffin Island and Hudson Bay at the end of the last century [3, p. 35, figs. 44a,b,h,i,j and p. 86, fig. 124a].

The Bowline proper is usually hailed as being the typical Mariner loop knot. Yet it is not referenced in their literature before 1627. The Boas Bowline has been very rarely recorded anywhere until this century. Accounts independent of Boas seem to be references where the structure is used as a hitch. [23, p. 27, fig. 180a/b]; [12, p. 600, pl. 320]; [34, p. 165, #14]. However, the past decade or so has shown that as a loop knot the structure is actually used elsewhere too, albeit very rarely. In the summer of 1978 it was shown to me by a Dutch boatman, who learnt it on board a barge and used it daily. This, however, is no evidence that Mariners know it well. It has been discussed to a somewhat greater extent in the more recent literature [6], [9], [14] and [29].

Day [9, pp. 84–85] raised the question about this knot's being indigenous in the Inuit knot repertoire. In his book Boas had already indicated its use by two distant groups of users. The Boas Bowlines occurring on the Ross sledge and those tied in baleen from 1700 found at Nuuk (Fig. 5) along with numerous identical knots to be found in hunting equipment from the Thule District dating from the beginning of this century, on display at Denmark's National Museum in Copenhagen, all seem to confirm their being indigenous. Furthermore, a not totally unambiguous drawing [25, p. 228, fig. 219] appears to be yet another Boas Bowline, making it plausible that this structure was known to the Point Barrow Inuits in Alaska too.

If one assumes it was not indigenous, then it would have had to have spread fast after Viking settlers set foot on Greenland, as they were the first (unless they were preceded by the Irish monks, of course) to bring elements of Western culture to large parts of Greenland. The ships on which these settlers came to Greenland were fitted with basically the same kinds of running and standing rope rigging of the early Mariners. We have seen that Mariners have very little or no tradition of using the Boas Bowline. There is thus no reason to believe that Vikings should have had a tradition of using a Boas Bowline, which they could have left in Greenland.

The Boas Bowline may be easily tied by spilling a Slip Knot. I have collected empiric data on the tying of Nooses and Slip Knots, which suggests that irrespective of user handedness, people are more inclined to tie Slip Knots rather than Nooses. This seems to be due to the fact that the forming of a Noose from an Overhand Knot Starting Configuration requires taking the standing part in the preferred hand, instead of the customary working end. In

[3], F. Boas indicates Inuit manual habits to be similar to those of westerners. The preference for tying Slip Knots, would thus extend to Inuits. In other words they have taken the most natural way to produce a loop knot, which suited their needs. They probably shortcut the method somewhat further by not forming a complete Slip Knot, but a version as shown in Fig. 16.

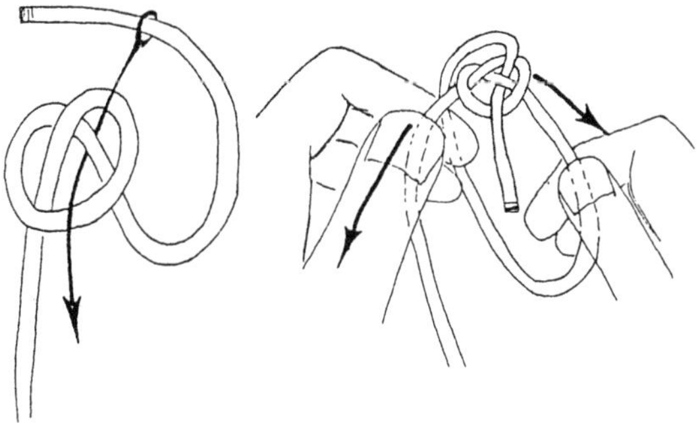

Fig. 16. Boas Bowline tying method based on spilling a Slip Knot

Conclusion

All of the foregoing structures have been used by Inuits, the majority of them since palaeo-eskimoic times. From the material which has been covered in this paper, we can see that the knots familiar to the Inuit are efficient, relatively simple, and in most cases globally well-known and quite reliable.

We can see how Inuits use part solutions, such as (Multiple) Overhand Knots, to produce more involved, often symmetrical composites, such as (Multiple) Fisherman's Bends, when required. The subtle nature of a knot is quite treacherous, i.e. it is never approximately correct [1, ##77–79]. In fact, as a rule, it requires some conscious algorithmic effort to get it completely right, with all crossings correct. The finding, and maintained use, of composite solutions shows active analytical skills.

The Inuit seldom seemed to reach the mechanical limits of their media, but we cannot infer that they pondered on knot strengths. Knot security, however, is certainly reflected by their choice of knots. Most artefacts do not show haphazard conglomerations of Overhand Knots and Half Hitches. The presence of such constructions would indicate that the rope users have not understood (or found) simple solutions to their rope problems.

In order to retain the superior solutions to their rope problems, it appears they must have had a mechanism to eliminate the inferior ones. The Inuit undoubtedly had access to a high level of accumulated understanding of the mechanical and structural properties of the knots they chose to use.

Their knots are well-suited to their nomadic life, in which the ubiquity of rope problems suggests that all elements of Inuit society needed to know how to tie reliable knots. Whether or not there exist cases where the knot reproduction was done by a select group of individuals, would be interesting to know. To my knowledge very few proper knot names have been recorded. The small number of frequently recurring knots suggests they worked their knots anonymously into their artefacts, thus rendering names irrelevant.

Inuits used a number of structures which have not often been recorded in other peoples' knot repertoires. This does not mean the latter cultures have not known them, merely that the frequency of these solutions was not high enough to ensure detection, as in the Inuit case. On the other hand, the comparative lack of decorative knots in the Inuit artefacts and of symbolic knots is very striking.

They possessed knowledge of ropemaking techniques, based on twisting fibres in alternating spiralling sequences, but they did not develop this art. They have produced fishing lines based on a braiding principle, but these have not yet been studied. All the evidence seems to suggest that there has been no interaction between Inuits and Mariners with respect to either knots or rope making. Inuits did not take over any of the latter's knowledge, because they did not need it, not because they lacked skill in manipulating knottable media. The question *Why should they take over the knots if they did not take over the rope?* must be answered by them not needing the rope and having had the knots beforehand.

Acknowledgements

I thank both Dr. Hans Christian Gulløv at Etnografisk Samling of the National Museum of Denmark and Prof. Bjarne Grønnow at the Institute for Arctic Archaeology of Copenhagen University for numerous stimulating discussions, their encouragement to write about Inuit Knots, and above all their shared fascination for the prehistoric Inuit way of life. I also thank dra. Gerre van der Kleij M.A. for many an enjoyable exchange of ideas on the topics touched upon in this paper.

References

1. C.W. Ashley, *The Ashley Book of Knots* (Doubleday, New York, 1944).
2. K. Birkett-Smith, 'Ethnography of the Egedesminde district, composition of implements'. *Meddelelser om Grønland* **66** (1924) 78–79.
3. F. Boas, *The Eskimo of Baffin Land and Hudson Bay.* (From notes collected by Capt. George Comer, Capt. James S Mutch and Rev.

E.J Peck. Part I, plates I-IV and 172 text figures. New York 1901. Reprinted New York 1975).

4. J. G. Bourke, *The Medicine Men of the Apache* (Smithsonian Institute, 1887–1888, Washington, 1892) 550–580.
5. A. von Brandt, *Fischnetzknoten* (Berlin, 1957).
6. G. Budworth, 'The Boas Bowline', *Knotting Matters* **27** (1989) 9–11.
7. G. Budworth, (Unpublished report on sledge knots: Museum of Mankind, London, 1985).
8. U. Bussemaker and C. Daremberg, *Œuvres d'Oribase*, (Paris, 1861) 253–270.
9. C. L. Day, *Quipus and Witches' Knots*, (University of Kansas Press, Lawrence, 1967).
10. J. A. Ford, 'Eskimo prehistory in the vicinity of Point Barrow', Archaeological papers of the American Museum of Natural History, **47** (1), (1959).
11. H. Frith, *The Romance of Navigation* (London, 1893).
12. R. Graumont and J. Hensel, *Encyclopedia of Knots and Fancy Rope Work* (Cornell Maritime Press, Cambridge, USA 1939).
13. P. van de Griend, *Knots and Rope Problems* (Privately published, Århus, Denmark, 1992).
14. P. van de Griend, 'Survival of the Simplest', *Knotting Matters* **39** (1992) 23–27.
15. P. van de Griend, 'The Qeqertasussuk Knots', *Meddelelser om Grønland, Man and Society* (in press).
16. B. Grønnow, 'Prehistory in Permafrost. Investigations at the Saqqaq site, Qeqertasussuk, Disko Bay, West Greenland', *Journal of Danish Archaeology* **7** (1988) 24–39.
17. B. Grønnow and J. Meldgaard, 'De første vestgrønlændere. Resultaterne fra 8 års undersøgelser på Qeqertasussuk-bopladsen i Disko Bugt', *Tidskriftet Grønland* **4–7** (1991) 103–144.
18. H. C. Gulløv, Etnografisk Samling, National Museum, Copenhagen, Denmark. (Private correspondence with the author).
19. L. Hacquebord, 'There she blows—a brief history of whaling', (*North Atlantic Studies* **2**, No. 1–2, (Århus, 1990) 11–20.
20. K. Hansen, Viking Museum, Roskilde, Denmark. (Private communications with the author).
21. J. P. Hart Hansen and H. C. Gulløv (eds.), 'The Mummies of Qilakitsoq—Eskimos in the 15th century', *Meddelelser om Grønland, Man & Society* **12** (1989).
22. E. Holtved: 'Archaeological investigations in the Thule District', *Meddelelser om Grønland* **141**, No. 1, (1944).
23. J. Lehman, *Systematik und Geographischen Verbreitung der Geflecht-*

sarten (Abh. u. Ber. d. k. Zool. u. Antr. Ethn. Mu. zu Dresden. Bd XI, No. 3, 1907).
24. K. McCullough, 'The Ruin Islanders', (Arch. surveys of Canada, Mercury Press, paper 141, Canadian Museum of Civilization, Quebec, 1989).
25. J. Murdoch, 'Ethnological Results of the Point Barrow Expedition', *Ninth Annual Report of the Bureau of Ethnology*, (Washington 1892).
26. T. Møbjerg, Dept. of Arctic Archeology, Moesgaard Museum Århus, Denmark. (Private correspondence with the author).
27. G. Møller, National Museum of Denmark, (Private communications with the author).
28. M. P. Porsild, 'Other hunting implements; a sealing net of whalebone strings'. *Meddelelser om Grønland* **51** (1914) 176–179.
29. J. Smith, 'Invention, Accident or Observation?', *Knotting Matters* **28** (1989) 20–21.
30. W. J. Sollas, 'On some Eskimos' bone implements from the east coast of Greenland.' *Journal of the Anthropological Institute of Great Britain and Ireland* **9** (1880) 329–336.
31. P. Victor, 'Jeux d'enfants et d'adultes chez les eskimos d'Angmagsalik', *Meddelelser om Grønland* **125** (1940) 1–213.
32. J. C. Wilcocks, *The Sea Fisherman* (Guernsey, 1865).
33. U. Zischka, *Zür sakralen und profanen Anwendung der Knotenmotivs als magischen Mittel, Symbol oder Dekor* (München, 1977).
34. *Opfindelsernes Sejrgang* (Copenhagen undated, circa 1925).
35. *Qeqertasussuk, de første mennesker på Grønland* (Copenhagen, 1992).

PART III. WORKING KNOTS

Vines used as ropes, and knots tied in them, may have been the first technical devices used by humans to help them with their daily tasks. Part III presents historical accounts of two fields of knots where their working uses and importance in the service of mankind are clearly defined and acknowledged.

The knots and ropes used by seamen and fishermen play a significant role in the history of knots. The first chapter describes some of this history, and discusses the rope problems encountered by persons who sail boats and bring fish from the seas.

The second chapter deals with a broad class of knots which the author calls *life support knots*. Under this heading he groups the cordage and knots used in activities such as mountaineering, rock climbing, rescue work, abseiling, and caving. He traces the historical development of these activities, and describes many of the knots used in them, paying special attention to the properties required of them in life support tasks.

CHAPTER 8

KNOTS AT SEA

Pieter van de Griend

Introduction

Even though knots were discovered and firmly established in their uses on land at a very early date, the sea plays a significant role in their history. In this chapter we shall view the phenomenon *knot* in the widest possible perspective and see how Man's knotting traditions became affected after his transition from (pure) land-dweller to Mariner. In the context of this chapter, by *Mariners* I shall understand seafaring people who regularly venture out of the sight of land. On a global scale this yields a wide spectrum, which may range from ocean crossings by so-called primitive peoples [13], [14], [25], [26], to contemporary off-shore fishing. With respect to time we shall be extremely liberal: Mariners occur during all ages Man has been at sea.

Focussing on a single phenomenon, that of a knot, in the complex process of Man's metamorphosis to Mariner and tracing it through all of times is virtually impossible. There is an infinity of parameters which comes to affect this development. However, the most important ones seem to be influenced by the technological progresses achieved by Man. They directly affected his ability to produce cordage and have eventually enabled him to venture beyond the horizon via the development of ships and sail.

Physically speaking, knots depend entirely upon the existence of a medium in which they can be realised, such as rope. Archaeological evidence leads to the conclusion that even before palaeolithic times primitive man had acquired the knowledge that fibres could be bundled together and intertwined to extend the aggregate into much greater length, capable of sustaining a pulling force. Certainly by neolithic times most peoples had struck upon a firm understanding of the effect of cohesion between twisted fibres. The level of rope making, by twisting natural or synthetic fibres in order to make a stronger, longer and more flexible material, equates to cultural development. Moreover, rope has always had a geo-political significance of some kind or other. A few highlights of the history of rope-making readily demonstrate that rope is more than a mere commodity of commerce. In the 13th century the Vikings were supplying the European market with *svardreip*, walrus hide rope, which was held in great esteem for its immense natural strength. They shipped this product from Greenland's west coast via their (regular) trade routes to the great fair of Köln in Germany [20]. During the post-war years resource shortages forced synthetic fibres to be developed, showing that Man will make rope with the available materials and accessible technology. Equally well we can state that the very first introduction of stronger knottable materials affected ropeworking techniques. When rope came into being the technicality of knotted structures steeply increased, for splices and seizings emerged. Rope-working techniques got developed virtually simultaneously all over the world. Due to empiric filtering they resulted in a globally near-identical set of structures. Alas, much of the initial stages of this parallel evolutionary process went unrecorded.

According to [29], the oldest known rope, made from palm fibres, is a sample which was found in Kafr' Ammar; that rope dates from 3rd–6th dynasty (approximately 2500 B.C.). However, in this book there are details of earlier grass rope (e.g. *Desmostachya bipinnata*), and there is mention of much older material. At Bandari, grass rope from approximately 4500 B.C. has been found [21], [31].

As rope making was discovered and mainly developed on land, it follows that rope-working techniques were well-established before further technology enabled Man to turn Mariner. It becomes obvious that when primitive man took to the sea his knot repertoire was already synchronised with the knottable media to which he was accustomed. However, life at sea differs from that on land. Irrespective of how fascinating an element the ocean may be, one does not go out there to waste time, as life on relatively small vessels is one of perpetual motion and, moreover, not without danger. In that context the dependency upon rope came to rely on a totally different dimension: that of security. In other words the nature of the encountered rope problems changed [11]. Furthermore when Mankind took to the sea, he not only entrusted his life to the product of his rope-making knowledge, but one other very significant change occurred; the magnitude of his exposure to rope problems exploded. We shall come to see that, under influence of a drive towards differentiation within societies, these two things in fact caused a rope-working discipline to become interwoven with, at least parts of, the various identities of the Mariner.

In our definition of Mariners we chose to view them as a single global phenomenon (over all time). However, knots at sea split into roughly two classes, which depend on the rope-users. Therefore I would like to distinguish between Sailors' and Fishermen's Knots. Professional demands posed by both groups are fundamentally different, though they do have the Mariner's knots in common, which could be regarded as forming a separate, or third class. Due to my severely restricted access to Oriental sources, the Sailors' and Fishermen's Knots to be discussed here will be those of the Western world from the past millenium.

Mariners' Knots

Driving even the most primitive of ships demands relatively advanced rope. If we assume this was (already) the state of affairs when Man took to sea, then Man the Mariner took with him knowledge of the core of *applied* rope-working techniques. Until quite recently not only Mariners' knots, but indeed all knots, were tied mainly in rope made from coarse natural fibres. This caused relatively simple knotted structures to suffice as solutions to the rope problems encountered. Knotted structures split into two classes, according to structural complexity:

I. Temporary 'simple' single-stranded structures, which we shall call *knots*. This class comprises the classical knots such as Reef and Granny Knot, Bowline, Sheet Bend, Two Half Hitches, Clove Hitch and Lark's Head. They are quite material independent, and are still used world-wide. Less well-known, but equally old knots are the Timber Hitch and Anchor Bend. We shall speak a little about the archaeological evidence and literary evidence of the four last-named knots.

   a. It is of great interest that already in antiquity the **Clove Hitch** was appreciated and considered a knot belonging to the sailors. Oreibasius (A.D. 326–403) calls it *Nautikos Brokhos* meaning Seaman's or Nautical Knot [6], [33].

   b. Likewise the **Lark's Head** was known to Oreibasius by the name *Ertos Brokhos* [6].

   c. The **Timber Hitch** is a knot which can be readily derived from the Figure of Eight Knot by dogging the working end. Throughout time many people have noticed this, so it is a very old knot indeed. The Swedish name *timmerstek*, for instance, is mentioned [5] in J.F Dahlman's *Utkast til et Sjøleksikon* from 1765. It is rather telling that the *name* was already so internationalised in the western world at such an early date.

d. The **Anchor Bend** is a knot mentioned in Dahlman's remarkable book too, but there it is named *fiskarestek*, meaning Fisherman's Hitch [5].

II. (Semi-)permanent 'non-simple' multi-stranded structures, which we shall refer to as *splices*. This class of achievements was established when laid rope entered the scene. This occurred very early, as is illustrated by the discussion of two millennia old samples of Egyptian rope-working by Daryl Domning [7]. To my knowledge the illustration of the oldest recognizable Mariner splices is that displaying some of the rope work found on the Oseberg Viking burial ship [35], i.e. an eyesplice and an (earing) cringle [1, #2835, #2842]. Rope-working feats such as the making of seizings, worming and parcelling are relatively late elaborations.

The foregoing selections are modest and far from complete. However, they shed a strange light on the mentioning by John Smith in 1627, which is representative for many of the older Western seamanship manuals, of the use of a mere Bowline, Sheepshank and Wall Knot at sea in his days [27]. We can safely say that Mariners had impressive knot repertoires from a very early date onwards.

Like everything else in this world, knotted structures and their uses became affected by technical and social developments. In very general lines, aboard ships certain tendencies can be observed.

With regard to structure, we can say that there was a drive towards more secure and stronger knotted structures. The perpetual motion of the ocean caused the dynamics to which Mariner Knots were subjected to rise in importance. To illustrate: the Sheet Bend became doubled. See various illustrations in Orazio Curti's book on medieval rigging [4]. With respect to the workings of the Double Sheet Bend the second tuck is plainly redundant. This can be observed by straining the knot beyond its medium's breaking strength. It is always the first tuck which bites itself into the hook, not taking the second tuck along with it. The dynamics of the ocean placed increased security demands on knots. Hence knots of a more secure but also more permanent nature, such as splices, are seen to occur more frequently. This change was also influenced by the rigging of sail driven ships, which became more intricate and incorporated increasingly complex technology, such as blocks and wire, in turn posing more refined demands on the solutions to rope problems. Meanwhile the ships grew larger and the encountered mechanical forces increased. The category of everyday forces to be experienced on ships is not comparable to those encountered in common applications to which primitive man put his knots and rope to use. Simple knots have a shearing property; that is, they mechanically weaken the medium in which they are realised. Hence relatively weak, temporary knots had no longer any place while stronger splices and more secure knots entered the picture.

The perfecting of tying methods is another aspect to be considered. As information about this is irretrievably lost, we must turn to contemporary analogies. When the number of times a knot must be tied increases, a seeking for an optimalisation of its tying method results. This is strongly affected by the medium in which the knot must be realised. The Bowline provides a point in case. The most frequent method which professional rope-users can be observed using is called the fingertip method [30]. It is commonly employed by sailors, fishermen and yachtsmen. It is rarely used on shore, where one mostly observes the rabbit method [9]. A small study of a Bowline diagram will naturally cause this method to be found. One first forms a little loop and takes the working end up through it, around the back of the stend and down into the loop again. It is naive, but it works. Of course, this needn't have been the case before knotting manuals entered the scene, as knots were then handed down orally. Improvement of tying method was also the case for the Sheet Bend, a knot which is structurally related to the Bowline. Ashley provides acknowledgment of discrepancies in tying methods between land-dwellers and Mariners for the latter [1].

As we have seen, differing from life on land, rope problems at sea do not only occur in all-round utility problems. With developments in sail and (running as well as standing) rigging, a quantitative increase and a qualitative change in their nature occurred. This caused the dependency upon knots and rope to alter. Already in the simplest of ships there is a confined environment pervaded with the presence of rope. Under sail, such environments become absolute and cannot be escaped. The concentration of rope problems on ships was, perhaps, just as high as in certain locations on land; but on the whole the marine dependency upon rope was much greater and all the more visible. In life at sea, *anticipation* is a key word. This forced the Mariner to reflect on the use and improvement of his tools. As rope is a non-replenishable and expensive commodity at sea, this caused rope-working techniques to become affected by the introduction of *discipline*. Mariners became more traditionalised in their knot tying, because experience was often gained the hard way, that is by loss of life; which was an expensive way of proceeding and simply had to be avoided. During countless generations the propagation of steadily accumulating knot-experience had been a practical part of an aspiring Mariner's training, gradually shaping parts of the Mariner's cultural identity.

Even though a ship at sea represents a micro-society in itself, Mariners have not always been a separate social group within the societies from which they originated. Initially social differentiation caused Mariners to become distinguishable as a clear-cut social subset. However, professional Mariners were not merely an amorphous social group, but they had their own further specialisations in which *sailor* or *fisherman* was a trade with rope and knots as its tools.

Sailors' Knots

The period and setting we are considering here is characterised by dramatic changes in shipbuilding. Vessels are seen to evolve from Viking longships and knarrs via sluggish carricks to complex sailing machines like Baltic schooners, Yankee clippers and German P-liners, which are generally considered to represent the summit and the end of the era of commercial sail. The role of rope evolved along with the ships that depended on it. As we have already seen, in its days, Viking cordage technology was an export product. By the time P-liners managed to run regular trade routes between Hamburg in Germany and Valparaiso in Chili around stormy Cape Horn, steel wires and chains had taken over the natural fibred sheets and rigging. On bigger ships the introduction and continued use of wire rope eventually killed the natural fibre-knotting tradition. These were large changes, but the most significant change struck the people who handled the ropes and tied the knots before this turnover occurred. The technical evolution aboard sailing ships should not be observed separately from the increase in sophistication of societies which took place simultaneously. Shipping was not only a pillar of many Western economies, but the sailors it involved had become social entities within those societies. They differentiated themselves by means of their social and cultural identitities. Since knots are technical artifacts we shall focus on the latter.

Fig. 1. Wale Knot

A demand for terminology had caused sailors to concoct quite a number of names, sometimes exotic ones (e.g. see Fig. 1), for their knots. Many of them have been collected and supplemented with further names, by Clifford Ashley or Raoul Graumont and John Hensel in their landmark knot monographs [1],

[10]. Knots also featured in the art of sailors, expressing a reinforcement of their cultural identity. This outburst of creativity resulted in artistic elaborations of splices and fancy knots. These developments are perhaps best documented by the aforementioned monumental works. As Sailors' Knots laid the basis for future knotting manuals it is worthwhile to pause and consider this development's origin.

We may assume that the genesis of nomenclature must have been due to the intensified exposure to rope problems, and its obvious function to improve communication. As personal observations show that even for intensive (professional) rope users the tying of a knot is often an anonymous act, we cannot assume that Mariners developed an oral nomenclature discerning all their specialised rope problems and their even more sophisticated solutions. As a rule they have no names for the specific knots they use. It is difficult to determine the exact state of affairs as far as the common sailor of bygone days was concerned. Romantically inclined traditionalists say they were people taking pride in knowing much about knots, but official sources reject this assertion [1, p. 425].

> *'Splicing' is fastening two Ends of a Rope together, with uncommon Slight, to execute which requires no ordinary Skill: as I can venture to say not one Seaman in twenty can perform it.*
>
> A Naval Repository, 1762.

The earliest mention of knots and rope work as tools of the sailor occurred in the first half of the 17th century [27], [19]. Initially these naval sources merely gave some names, but were not illustrated at all. Manuals with illustrations occurred a century later [2], but still showed no knotted structures. The latter began to turn up in the flurry of seamanship manuals, headed by William Falconer, a short time later [8], [16], [22]. By the beginning of the 1800s seamanship literature was abundant, well-illustrated and knotted structures were documented by relatively consistent nomenclature all over North Western Europe. Showing that the Occidental seafaring community was sufficiently tightknit to accept the introduction of international standard words.

The presentation of knots in the maritime literature represents a significant change. It is noteworthy that nomenclature appeared in print around A.D. 1625. Why the lack of incentive before, it may be asked? Until then rope-working techniques had been passed on as 'natural' skills as part of a trade, not as formal theoretic knowledge. Obviously knots entered books for the benefit of people who had access to the literature, i.e. those who could (or had to) read them. The main objective did not seem to be an attempt at displaying *dynamic* techniques by means of *static* images, which would be predestined to fail. Hence the people that read these books were essentially not interested in the dynamics of knots, but only needed to know that such things as knots existed and occasionally had names. It is obvious that this knowledge entered respectable seamanship manuals because it *existed*, as words spoken among mariners. For a long time Mariners were virtually the only ones to display functional knots. In an overall perspective this manner of propagating technological knowledge added to an awareness of knotting discipline, i.e. to a reinforcement of the cultural identity of the Mariner, but it also confirms that otherwise apparently trivial objects or operations have an interest and extent that are by no means commensurate with the estimation in which they are ordinarily held. Not much changes it seems. However, the present day solution to these problems is more refined. Nowadays, rope problems on merchant and naval ships are reduced to the putting-in of an infrequent splice in wire or multibraid rope and on smaller vessels occasionally making fast around a bollard. Whatever is left of nautical rope problems has been solved on shore by specialists before it is delivered on board. At sea, the intensive use of rope by fishermen, seems to remain the sole exception.

Fishermen's Knots

As there are many ways to catch a fish, the topic of fishing knots is a large one [3], [34]. However, I shall restrict myself to a very modest part of the North Atlantic fishery scene's interesting history, and say something about knots in the industrialisation of the exploitation of its fish resources. A process which commenced as early as the 1430s when British cod fishermen were lured over to the rich banks off Iceland and possibly even visited Greenland. Nowadays huge trawler fleets of the nations bounding the North Atlantic basin are pressing its ecosystem to depletion. Between these two extremes a wide range of other fish-catching methods is being practised. Here I will only consider longlining and aspects of gillnetting as fish-catching methods with ancient roots, and conclude with a few words on trawling as a contemporary development. I shall focus on the functional role of knots in both.

In contradistinction to the recording of Sailors' Knots in the seamanship literature this field is characterised by a remarkable lack of recording. Unlike early naval literary sources discussing knots, there are very few fisherman seamanship manuals which do so [23], [32]. Fishermen's Knots are, however, specialised tools of an ancient trade. Even though prehistoric finds are biased due to chemical preservation, the oldest of knotted archaeological artifacts are related to fishing [15]. Another aspect in which fishermen differ from sailors is their ship-life work patterns. To fishermen *boats* are just things which take them out to the fishing ground. The *real* work is the fishing. Hence the emphasis on the most used knots, is shifted from a knotting discipline directly related to security of driving their ships safely across the sea to that of work and money. At sea their time is sparse, which is witnessed by the lack of knotted arts produced by fishermen.

Concerning their knots, general observations to be made are three-fold:

(i) Knots have to fulfil criteria in their *production*, *man-handling* and *maintenance*. In fishing it is generally the sheer numerical magnitude of the times a knot has to be realised, which plays a role. Longlining involves thousands of hooks attached to a groundrope, and trawling demands at least equally many netknots to create meshes. Hence the knots fishermen tend to employ are usually extremely effective solutions to encountered rope problems. During their careers they would have to tie them innumerable times, thus creating a harsh 'simplicity criterion', which is an optimalisation of structural and algorithmical simplicity combined with functional effectiveness.

(ii) The dynamics of a fish catching system causes the extensive use of slipped (temporary) but reliable knots. The Slip Knot commonly appears as an elaboration of other knots and is meant to ensure easy opening. It is as such usually not more than an extra tuck with a bight of the working end, often followed by a 'locking security'. In general these operations do not affect the working of the knot negatively, but result in some other useful actions. First of all the slipped segment toggles the knot, preventing it from drawing up tight into a nipped compact mass. Secondly it provides a part on which one can exert tension when opening the knot after it has been put to use. Thirdly it offers 'locking facilities'. The Codline Knot is an advanced solution to all of these demands and also it suitably illustrates which evolutionary forces knots used by fishermen are susceptible to.

(iii) The impact of synthetics on knotting has been great. The nature of artificial fibres requires knots with a high resistance to slip. This is an area of fishing where some rope problems have undergone quite spectacular changes to their solutions, but we shall see that in industrial fishing, frictional problems have been solved in a quite different manner.

Fishermen's knots constitute an enormously broad subject and the context of this treatise is too confined to give a complete picture. We shall restrict ourselves to a few typical older knots such as used in longlining and gillnetting, which the author collected from various museum fishing gear exhibits and learnt during talks with senior fishermen.

In the days before grapnels, the longlines and gillnets were anchored to the seabottom by means of ropes which were Timber Hitched onto boulders. It is of interest to note that this hitch has been used for a similar purpose by the fisherman society of the Faroe Islands in preventing their hay from being blown away during the haymaking season. At the Faroese National Museum in Tórshavn another application of stone weights is to be found. In Bátasavnid there is a traditional hand fishing gear with *vadsteinur* on display. The *vadstein*, Fig. 2 which is a sinker, is connected to the gear by two interlocking Timber Hitches. In his discussion of Faroese line fishery Jens Christian Svabo has a description of these stones, but tacitly assumed it to be known *how* the knots ought to be tied. [28, 345–359].

There are numerous bends used by fishermen to interconnect (long)lines. For example the Carrick Bend plays a role in gillnetting. Another traditional knot (Fig. 3) is the (English) Fisherman's Knot [33]. This is an old structure which was discussed by Oreibasius, who called it *Haplous Karkhesios* [6]. In its doubled form (Fig. 4) it is first recorded in the fishermen's literature at the end of last century [32].

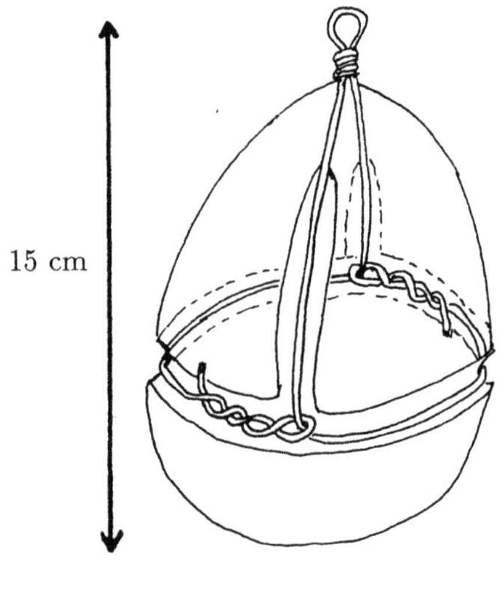

Fig. 2. The slinging of a Vadstein

However, here I wish to discuss a typical locking elaboration of all sorts of slipped knots performed by fishermen. Longlines (and gillnets) which are set in the ocean are several miles long. On deck they are kept in sections. The attachment of these sections to each other requires temporary knots, which have to withstand the dynamics of man-handling, sea currents and fishload. Hence they are not only slipped, but also locked constructions. One of their many variations is illustrated below (Fig. 5). Further examples of slipped longliner knots can be found in [12].

If we follow the lines out towards to the hooks then the next problem is the attachment of snoods to the groundrope. Traditionally this has been done with the Groundline Hitch (see Fig. 6), [1, #277], [36]: an ingenious solution, which exploits the lay of the rope to anchor part of its structure, and which is resistant to tugging under any angle. In actual practice this is what a hooked fish would do. The Groundline Hitch has also been used to attach the floats to the head rope of a gillnet.

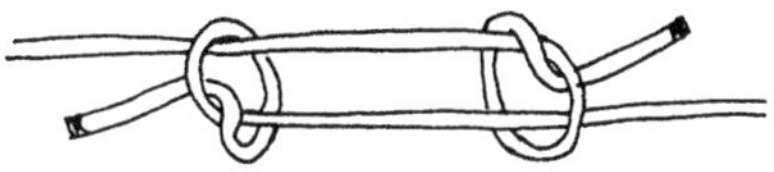

Fig. 3. Fisherman's Knot

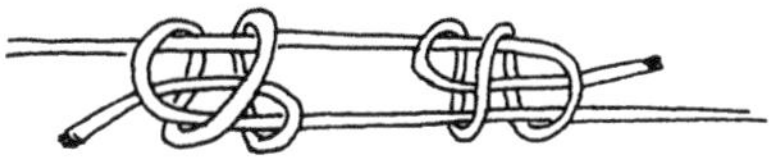

Fig. 4. Double Fisherman's Knot

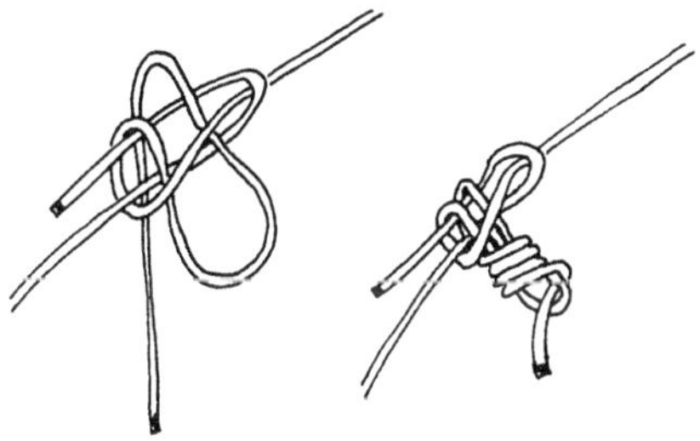

Fig. 5.

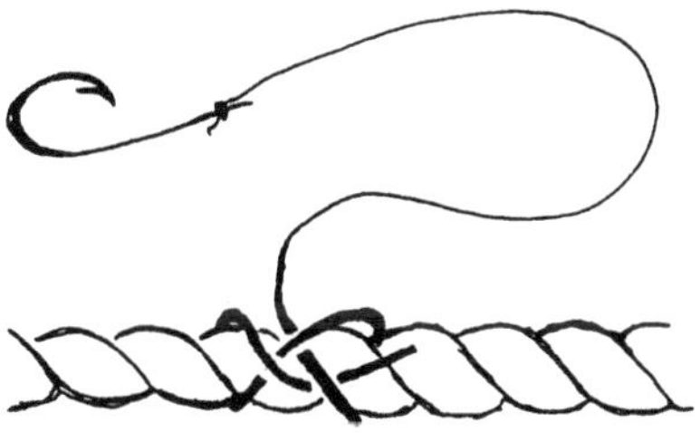

Fig. 6. Groundline Hitch

Attaching hooks to a snood has been done by means of a Double Overhand Knot. The drawings in Fig. 7 illustrate the process. Further examples of longline knots can be found in [24].

Nowadays longlining and gillnetting still comprise a significant sector of industrial fishing, but trawling is equally large. Trawlermen encounter knottable media which range from 2 mm twine via all sorts of rope up to 42 mm diameter steel wire. Hence they have to know a great variety of knots. Many traditional knots are in fact still being used. All fish catching methods have experienced the transition to synthetics, but knot-wise speaking the impact on trawling has not been so great as might be expected. The monofilamentous fibres which are extremely smooth are braided into twine, which has a sufficiently rough surface to enable the continued use of simple knots. Simplicity is not only demanded for the sake of the netmaking machines, which churn out the actual netknot by the billions, but also for the inevitable repair, which still relies heavily on the manual skills (see Fig. 8) of the fishermen [17].

The softness of the artificial fibres has the further ideal property that under tension it readily deforms into compact stable knots. After all, why should simplicity, the result of generations of experience, be sacrificed to technological progress?

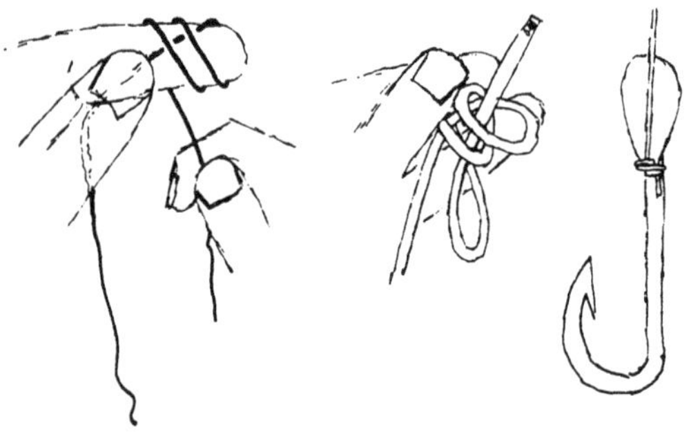

Fig. 7. Attaching a hook to a snood

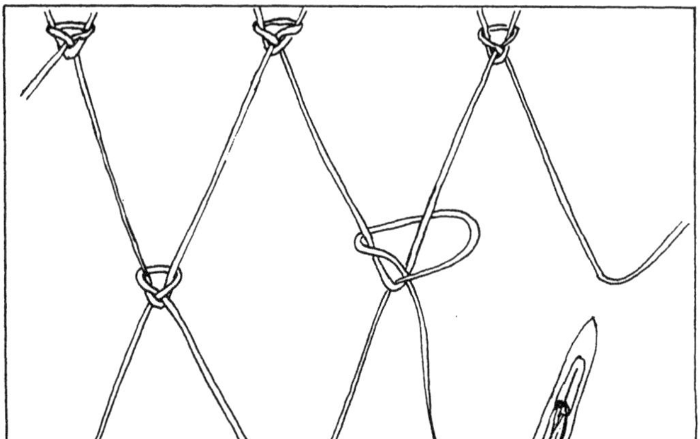

Fig. 8. Making a Net

References

1. C. W. Ashley, *The Ashley Book of Knots* (Doubleday, New York, 1944).
2. T. R. Blanckley, *A Naval Expositor* (London, 1750).
3. A. von Brandt, *Fish-catching Methods of the World* (Fishing News Books, Farnham 1972).
4. O. Curti, *Masten, Blokken en Tuigage* (De Boer Maritiem, Bussum, 1980).
5. J. F. Dahlman, *Utkast til et Sjøleksikon* (Örebrö, 1765).
6. C. L. Day, *Quipus and Witches' Knots* (University of Kansas Press, Lawrence, 1967).
7. D. Domning, 'Some Examples of Ancient Egyptian Ropework', *Chronique d'Egypte* **52**, (Brussels, Belgium, 1977) 49–61.
8. W. Falconer, *An Universal Dictionary of the Marine* (London, 1769).
9. S. Grainger, *Creative Ropecraft* (Norton, New York, 1975).
10. R. Graumont and J. Hensel, *The Encyclopedia of Knots and Fancy Rope Work.* (Cornell Maritime Press, Cambridge, USA, 1939).
11. P. van de Griend, *Knots and Rope Problems* (privately published, Netherlands, 1992).
12. P. van de Griend, 'On a Few Longliner Knots', *Knotting Matters* No. 36, (1990) 20–23.
13. T. Heyerdahl, *Kon-Tiki Ekspedisjonen* (Oslo, 1948).
14. T. Heyerdahl, *Early Man and the Ocean* (London, 1978).
15. G. van der Kleij, 'On Knots and Swamps—Knots in European Prehistory', *History and Science of Knots* (World Scientific Publishers, New York, 1996).
16. D. Lever, *The Young Sea officer's Sheet Anchor* (London, 1808).
17. L. Libert and A. Maucorps, *Mending of Fishing Nets* (Fishing News Books, Farnham, 1978).
18. P. Lipke, *The Royal Ship of Cheops.* (Archaeological Series, No. 9, BAR, International Series 225, National Maritime Museum at Greenwich, 1984).
19. H. Manwayring, *Sea-mans Dictionary* (London, 1644).
20. G. J. Marcus, *The North Atlantic Conquest* (Woodbridge, 1980).
21. T. Midgley and Sir W. Flinders Petrie, *Qua and Badari* **I**, (1915).
22. G. S. Nares, *Seamanship* (London, 1862).
23. O. T. Olsen, *The Fisherman's Seamanship* (Grimsby, 1885).
24. Y. Paulsen and T. Robertsen, *Linefiske og Lineegning* (Universitetsforlaget, Bergen, 1982).
25. T. Severin, *The Brendan Voyage* (New York, 1978).
26. T. Severin, *The Sinbad Voyage* (New York, 1983).
27. J. Smith, *A Seagrammar* (London, 1627).

28. J. C. Svabo, *Indberetninger fra en Reise i Færø 1781–1782* (Published by N. Djurhuus. Selskabet til udgivelse af Færøske kildeskrifter og Studier. Copenhagen, 1959).
29. V. Täckholm and M. Drar, *Flora of Egypt*, **II** (Cairo Fouad I, University Press, 1950).
30. B. Toss, *The Rigger's Apprentice* (Camden, Maine, USA, 1984).
31. W. Wendrich, private correspondence with the author.
32. J. Wilcocks, *The Seafisherman* (Guernsey, 1865).
33. H. A. Öhrvall, *Om Knutar* (Stockholm, 1916).
34. *FAO catalogue of small-scale fishing gear* (Edited by C Nédélec, Fishing News Books, Farnham, 1975).
35. *Norsk Sjøfartshistori,* **1** (Oslo, 1935).
36. *Saturday Magazine* (**4**, No. 126, Guernsey, 1834). 236–237.

CHAPTER 9

A HISTORY OF LIFE SUPPORT KNOTS

Charles Warner

Probably since the invention of rope, people have been suspending themselves on ropes when climbing trees, cliffs and so on. It is likely that the Australian Aborigines who entered Koonalda Cave on the Nullarbor Plain to obtain flints and create their rock art about 20 000 BP used rope in some form to get to the bottom of the 24 m steep or overhanging entry sinkhole. With or without rope, people have been climbing mountains and entering caves since time immemorial, but doubtless only for utilitarian or ceremonial purposes, not recreation. There are no readily available records of any special knots or tricks of ropework associated with these activities. Over the last couple of centuries the adventure sports of mountaineering, caving and rock climbing have grown and flourished, generating many books and magazines. They have been joined more recently by the sport of canyoning; and also the activities of specialist rope rescuers and abseil engineers have developed considerably. These have demanded specialist ropes which are increasingly being called life support ropes. In the absence of any other common name for the group of knots used in these and similar activities, I am calling them *life support knots.*

The European Alps were climbed in searches for transport routes for anything from armies to smugglers; chamois hunters were sometimes lured even higher. Cliffs were climbed to obtain birds' eggs or the birds themselves. Early records of climbs just for fun are rare, but it is known that Mt Etna was climbed during the Roman Empire for the sunrise view; and the first recorded rock-climb, up Mt Aiguille near Grenoble in 1492 seems to have had little more serious purpose. Early visits to caves were to obtain flint or gypsum, later guano or saltpetre or, in SE Asia, edible birds' nests for sale to China. Scientific interest in how caves and cave formations developed, and in tracing underground water, started in classical times; scientific interest in mountains and glaciers seems to have arisen a bit later. The early development of the sports is described, with bibliographies, in the General References section of my Bibliography below; no other citations are given in this paper for the history of the sports, as distinct from the history of the knots.

The first records of the use of rope in caving date from 1535, and in mountaineering from 1574; in both cases it seems that the ropes were used as simple handlines. Rope ladders were first recorded as being used in caving in 1780; the rungs were also rope, making for difficult climbing, but from 1821 wooden rungs provided much easier use. Some form of winch may well have been used earlier. All this refers to written records; it is very likely that the unrecorded use of rope was much earlier and happened in many countries.

For a long time, interest in either mountaineering or caving was only sporadic. Few people climbed more than one high mountain or explored more than one cave system, and often long periods elapsed before a second party climbed that mountain or explored that cave. No one made climbing mountains or exploring caves anything like a major part of life.

The Beginnings of the Sports

In the late 18th and early 19th centuries, the Romantic Movement changed attitudes to natural scenery, and people began to explore and enjoy themselves in wild natural areas. Recreational mountaineering began in the Alps of Europe. Horace-Bénédict de Saussure of Geneva was probably the first to proclaim that his sole aim in climbing a snow-covered mountain was to reach the top; after unsuccessful attempts to climb Mont Blanc, the highest mountain in Europe, he offered a prize in 1760 for the first ascent; it was not claimed until 1786; the first ascent by a woman was in 1808. Thereafter there was a growing interest, at first amongst the Swiss, French and Germans, but later especially amongst the English. Guided climbing eventually spread all over the world. Nearly all the early climbers, or 'travellers' as they were known, were 'gentlemen'—a few were 'ladies'; but they were by no means all wealthy. Initially, the guides were local chamois hunters or smugglers, but soon they became more or less full-time professionals, helping their 'travellers' in many ways, including carrying loads and performing the routine duties of the journey. The first guides' organisation, the Syndicat des Guides de Chamonix, was formed in 1823, at first to conduct tourists and scientists over the glaciers, where the guides were often familiar with the routes. Later they often led parties up the high mountains, where no-one had been before; but at least the guides were skilled and strong. The formation of the Alpine Club in England in 1857 marked the recognition of mountaineering as a sport. Other European mountaineering clubs followed soon after; the Ladies' Alpine Club was not formed till 1907. Early leaders of the sport were Edward Whymper, who first climbed the Matterhorn, and Albert Mummery, perhaps the first to regard mountaineering as 'unmixed play'.

Organised tourism in caves began about the middle of the 19th century; Mammoth Cave in the USA was one of the most popular. But cave exploration never became as popular as the quest for 'first ascents' of mountains. Édouard Martel, a Paris lawyer, was an enthusiastic cave explorer who did much to spread interest in caving, both as a sport and as a science. He explored caves all over Europe, Turkey, Russia and America, a few weeks to a few months every year between 1888 and 1913. He did much work on the development of cave formations and on underground hydrology. He usually involved local people in his explorations and many of them continued after he had gone. Several women had been part of cave exploring trips run by men from at least the 1870s, but two American women, Ruth Hoppin and Luella Owen did several independent explorations in the Ozarks in the late 1880s and early 1890s; Owen's book of 1898 did much to make American caves known in Europe. The early caving societies, for Vienna (1879), Trieste (1883), Yorkshire (1892) and Paris (1895) had a predominant interest in the various speleological sciences, and in exploration and mapping as aids to those sciences. The sport of negotiating caves for its own sake has never attained the popularity of mountaineering and rock climbing; most caving clubs retain some scientific interests.

Rescues using rope must have occurred ever since the invention of rope of adequate size, but the methods have been restricted locally: rescues on inland cliffs were made by miners or quarrymen, on coastal cliffs by coastguards or mariners, rescues from trees by loggers, rescues from bogs by farmers. When cities developed, firefighters and police became involved in rescues. When the sports of mountaineering, rock climbing and caving developed, rescues were made by fellow sportspeople, using the equipment and techniques of their sports. Local guides and other villagers played a large part in many rescues in the Alps, and there was usually some regular way of calling out teams of rescuers; but it is claimed that the first organised mountain rescue service was formed in 1896 in Vienna.

The First Knots Used

I should state here that the allocation of names to knots by climbers and cavers is as chaotic as that by other users of knots. Many knots have had several names given to them; and often the same name has been given to several different knots. Many knots are named after a person, but there is confusion about these names also: the name applied is often different in different regions or languages. I have not attempted to sort out the contribution of each person. I use only one name per knot, the one that seems to me the most satisfactory.

It is not now possible to identify several of the knots used by early climbers, since they did not illustrate them and the names they used were given by others to several different knots.

At the time of the establishment of the sport of mountaineering, the only special equipment used was a rope and an alpenstock, the forerunner of the modern ice-axe. In the earliest days, long metal rods were held in the hands to give assistance; they were soon replaced by ropes, and then the practice of tying the rope round the waist or onto a belt was adopted. The climbers were linked by the rope, with the chief guide in front and sometimes another in the rear. The rope kept the party together and gave confidence. It was sometimes used to haul the patron physically up difficult pitches, and it served as a safety measure in case of a fall from a cliff or into a crevasse; however it was not unknown for some guides to unrope the party at a particularly dangerous place so that in the event of a slip only one would be lost, not the whole party. Often the only knots used were the Overhand Loop (Fig. 1) tied round the waist or chest, in the end or the middle of the rope, and the related Overhand Bend (Fig. 2) to join two ropes. Usually the whole party moved together, separated by a few metres; but sometimes only one person climbed the more difficult pitches at a time, while the rest payed out the rope or drew it in as required.

Fig. 1. Overhand Loop

Fig. 2. Overhand Bend

The Alpine Club took an interest in the ropes and knots used. A special manila rope was manufactured to the Club's specifications, and in the first volume of the Club's *Alpine Journal* was a report of tests on the ability of ropes to withstand the jerk imparted by a falling weight (see Chapter 10): it was shown that any knot weakened the rope. The knots recommended [1] for use were the Fishermans Knot (Fig. 3) for joining two ends, the Fishermans Loop (Fig. 4) for making a loop at one end, and an Overhand Noose (Fig. 5) for making a loop in the middle of the rope. It is clear that both the end loop and the mid loop were meant to be tied round the waists of climbers, but sometimes they were tied to a spring hook or passed round a belt [8]. The Overhand Noose seems a surprising knot to tie round the waist as being particularly likely to draw very tight under load; indeed I have found no other reference to this knot in this application in the climbing literature.

Readers of the *Journal* were informed that the Overhand Loop (Fig. 1) is 'one of those [knots] which most weakens the rope' and it was mentioned only 'in order that no-one may imitate it'. I do not know what evidence convinced the members of the Special Committee that the Overhand Loop was so weak, since they also say 'how great a loss of strength results from a knot we cannot undertake to estimate'. This is the first of many complaints by British climbers over the next hundred years or so that the Overhand Loop, so favoured by Alpine guides, was unsatisfactory, because of either weakness or a tendency to jam under load. Swiss guides in particular continue to use the knot to this day: indeed, it is known as the *'nœud de guide'*. They claim that in actual practice the rope never seems to break at that point. Quite a number of climbing manuals, though few from Britain, still recommend this knot, both as end loop and as mid loop because of its simplicity and ease of use.

Fig. 3. Fishermans Knot

Fig. 4. Fishermans Loop

Fig. 5. Overhand Noose

It is noteworthy that the illustrations of both the Fishermans Knot and the Fishermans Loop have discordant Overhand Knots: i.e. they are of opposite handedness. A brief note in the following number of the *Journal* [8] recommends the Fishermans Loop as a mid loop, and describes how to tie it in the bight without using the ends and finish with the Overhands concordant. All subsequent illustrations by British climbers of the Fishermans Knot and Loop have concordant Overhands (as shown in my drawings, Figs 3,4), though only a few comment on it. However, Wright and Magowan [34] showed by tests that the concordant knots are stronger than the discordant, as is now generally held by most knot tiers.

Mountaineers, especially in Britain, started to gain practice for alpine expeditions by climbing local rock outcrops, without guides. This became the start of rock climbing as a separate sport. Some people who had no intention, or no opportunity, to visit the Alps took up rock climbing, and the sport became more democratic in its membership, attracting many working-class people. The use of a rope on British rock was regarded initially as almost unsporting and did not become common until the 1890s. Even then, the rope was not tied on to anchorages, just held in the hands of the person supposedly safeguarding the climber, though it was sometimes passed round a rock knob. Once into the twentieth century, however, the use of a rope became universal and more and more use was made of anchorages, initially with just a turn or two round natural rock knobs and chockstones. When knots were used, they were just standard hitches which I will not discuss further. Knots and rope techniques developed in pure rock climbing were, when suitable, adopted in general mountaineering and vice versa.

Some caves are more or less horizontal and can be explored by scrambling and crawling only, without need of ropes. Other caves have vertical sections, at first tackled with ladders and winches and needing only very basic knotting. Yet other caves include steep and difficult terrain; rock climbing techniques were adopted, often with minor modifications for the different conditions. However, it was a long time before cavers developed any special knots not already in use by climbers.

Fig. 6. Bowline and Half Hitch

Fig. 7. Figure Eight Tie

The second report of the Special Committee of the Alpine Club [2] gives the first account of the Bowline used for climbing (Fig. 6). Note that the running end is secured to the nearby leg of the loop with a Half Hitch. All early climbing manuals that mention the Bowline stress the need for this, and all manuals published after the introduction of nylon rope stress the necessity for securing the end in some way, as discussed later, but some manuals written between the Wars omit this point. However Wright and Magowan [34], in one of the best accounts of climbing knots ever written, discuss the tendency of the Bowline to 'spring loose', of the end to 'spring out', and recommend having a long end, which is twisted round the adjacent leg of the loop several times and fastened with one or two Half Hitches. They recommended that the twists and Half Hitches should have a handedness opposite to that of the turn of the Bowline. The Committee [2] also recommended the Fishermans Loop (Fig. 4) for a mid loop and the Fishermans Knot (Fig. 3) for temporarily bending two ropes of the same size together, but a Figure Eight Tie (Fig. 7) for joining two ropes for prolonged security. This Figure Eight Tie is made by the follow-through or reweaving technique, making a simple Figure Eight Knot in the end of one rope and then following all of its turns with the other end. This method of tying is used for quite a number of climbing knots (see later), though it is uncommon in general knotting. They also tested several other knots, said to be 'likely to be of interest and importance to mountaineers', but it is not stated that the knots were in actual use by mountaineers in any application. They did not recommend the knots specifically, and they will not be further discussed here.

This report [2] seems to be the first mention of what became something of an obsession amongst British mountaineers—the claimed advantage of tying all knots 'with the lay of the rope'. Unfortunately, there seems to be considerable disagreement about what constitutes 'with the lay', though everyone agreed that knots should be tied that way. Wright and Magowan [34] discussed this point at length; to them, the phrase seems to mean that any twists and turns of the rope within the knot should have the opposite handedness to the lay of the rope.

Climbing down a cliff or other difficult slope is often more difficult, more dangerous and much slower than climbing up. If any substantial height needs to be down-climbed, it is a considerable advantage to use the climbing rope to descend, putting all or part of the weight on the rope and controlling the rate of descent by friction against the rope. There are reports that this method was in wide use in the Chamonix area by the 1870s. The first record of using friction to slow a descent on a mountain seems to have been on the way down from a climb of Mont Blanc by de Saussure in 1787. The climber glissaded down the slope, sitting on the rope anchored at the top and held tight at the bottom. The rope passed between the legs and over an alpenstock held across the body at thigh level. Friction could be increased and the climber slowed by raising the alpenstock or increasing the tension on the rope [6]. Sailors, firemen and the like have long used ropes for a quick descent, but climbers have need for much longer descents, often under more difficult conditions, so that they have developed novel techniques, calling the activity 'abseiling' from the German or 'rapelling' from the French. The arrangements of the rope may be called 'flowing hitches' (see below). All techniques before World War II involved braking with friction round parts of the body, such as the thigh, arms or trunk [5],[18],[27]. Braking on the body alone is now rare (see later) but is still used for quick short descents and for emergencies where no extraneous equipment is available. The commonest method still in use (Fig. 8) is usually called the Classic Abseil [10, p. 142],[14, p. 159]. It is a variant of a technique practised in Chamonix in the 1920s [5].

Fig. 8. Classic Abseil

Developments before the Second World War

Climbers have always been strong individualists, so that even in 1907 it was claimed that 'an almost endless variety [of knots] are used by climbers', though relatively few were recommended in any one manual. This position continues, and in the preparation of this paper well over a hundred knots were found, and this was in mainly British and American books; more knots are known to be used elsewhere. However, most people recommend that climbing parties should limit the number of different knots used in the interests of safety through uniformity; indeed, just about all climbing purposes can be met with fewer than ten knots. The only problem is getting agreement on just which those ten should be! I propose to discuss from here on only those knots or usages that might be unexpected by knotters brought up in the transatlantic nautical tradition. Many ordinary knotting books contain few if any of them, and even such a comprehensive book as Ashley [3] fails to mention many.

Development of climbing techniques, including knots and ropework, continued more or less steadily until World War II. Throughout this period, a climber falling on a rope from any substantial height was liable to either break the rope or suffer considerable body damage from the rope alone. The dangers were reduced a bit from the 1920s by replacing the 'direct' belay, where the belayer passed the rope round some rock feature or the like and kept it tight between the climber and the anchorage, with an 'indirect' or 'resilient' belay, where the rope was wrapped round an arm, shoulder, thigh or the whole body of the anchored belayer. A falling climber caused these ropes to tighten and moved the body of the belayer over some distance, reducing the shock on both rope and climbers. Techniques for a 'dynamic' belay had been developed to some extent during the 1930s but they by no means eliminated the problem; the rope to the falling climber was deliberately allowed to slip past the belayer under controlled friction, using at that time a rope arrangement as in the resilient belay, to give a gradual arrest. The leader had to pay attention to the rule that 'a leader does not fall'. On the other hand, a 'second', a climber following the leader, supported from above by the rope from the leader, could fall only a very short distance till the rope took up the tension, and this was normally safe.

Increasing use was made of artificial anchorages, particularly pitons or steel pegs driven into cracks. To avoid threading the rope through the eye of the piton, a karabiner, or metal snap link, was used, first adopted around 1910 by Munich climbers from their local fire brigades. 'Artificial' or 'aid' climbing began in the 1920s; instead of putting all the weight on the rock, artificial aids such as pitons which sometimes held slings were used. All these practices were for a long time regarded by many, particularly in Britain, as unsporting, and were not generally adopted until after World War II. The knots used with these devices were mainly the loops already discussed, together with standard hitches such as the Clove Hitch (#53,[3]) or Girth Hitch (#56,[3]).

Fig. 9. Bowline on a Bight

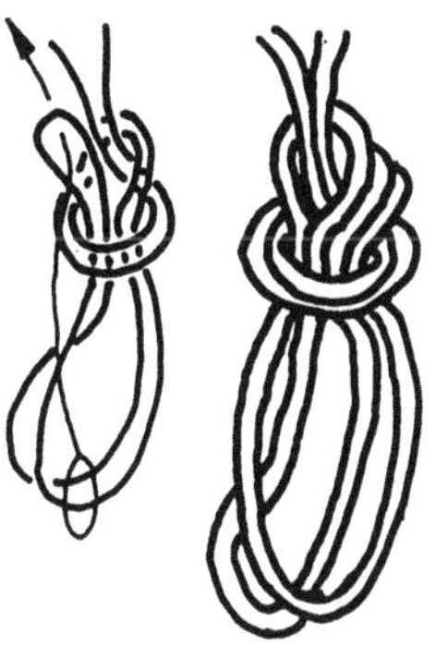

Fig. 10. Overhand Loop on a Bight

Fig. 11. Figure Eight Loop on a Bight

The first mention of the Bowline on a Bight (Fig. 9) that I have found in the mountaineering literature was in a book by Dent in 1900 [7], who recommended its use as a chair knot when rescuing an injured climber from a crevasse. Its use as an end loop, providing two turns round the waist, was mentioned by Wright and Magowan [34], but dismissed because they found the knot weaker than many. However, several manuals right up to the present date continue to recommend its use as a mid or end loop [12, p. 75][32], or as a sit sling or improvised shoulder harness [14, p. 15]. The common method for tying this knot (Fig. 9) has been adopted for two other knots, both uncommon in general knotting. The Overhand Loop on a Bight (Fig. 10) was recorded in 1920 [28] and is still in use [28, p. 30]. The corresponding Figure Eight Loop on a Bight (Fig. 11) appears to have been developed more recently [14, p. 16],[15], [29, p. 37].

Those three knots have never attained widespread use in climbing, but two papers were written in this period that have had a major influence on the knots used by climbers and cavers. Wright and Magowan [34] developed the Alpine Butterfly Loop and the Bowline and Coil, amongst several other novel knots that never became popular, and Prusik [25] developed his Hitch; according to Prohaska [24], this paper influenced climbers' rope techniques probably more than any other single event before or since.

Wright and Magowan

In 1928 Wright and Magowan [34] published a two-part paper in the *Alpine Journal*, discussing in critical detail the kinds of knots needed by climbers and criticising many of the knots in common use at that time. They then developed and named a series of 'new knots' (saying 'no earlier record has been traced'), described how to tie them in the practical situations of actual climbing, and estimated the time needed to tie them and their effects on the breaking strength of the rope. I know of no other description of the knots used for climbing that gives so much detail of so many aspects of the knots.

Their paper arose from a dissatisfaction with many of the knots recommended at that time for climbers; many of them could not withstand the 'pulling about by alternate straining and slacking' so commonly found in climbing practice. They were particularly dissatisfied with the Fisherman Loop (Fig. 4) as a mid loop, though it was the knot most frequently recommended in manuals. They pointed out that in one direction it was a slip knot, though optimists ('a term necessarily including all climbers') claimed that no troubles were ever found in practice. The knot also suffered from having the ends emerging together from the knot in the same direction, rather than in opposite directions, in line. That meant that under load one end must be subjected to a very sharp angle, an unsafe procedure. This disadvantage is also shared by the Overhand Loop (Fig. 1), which was also criticised because it jams, wears badly and has no spring in it. The authors claim that they developed all the following knots (Figs. 12,13,14,15,16,17).

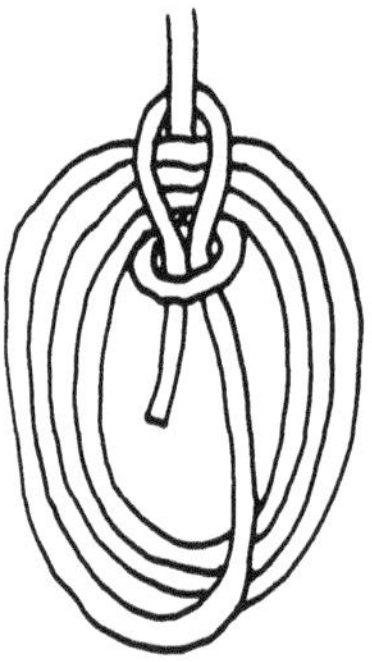

Fig. 12. Bowline on a Coil

Fig. 13. Rover Loop

The Bowline on a Coil (Fig. 12) was recommended as 'the most useful and adaptable' end loop, despite being a little weaker than the Double Bowline (Fig. 22). The several coils made for a more comfortable waist loop, spreading the load, particularly in a fall. This knot became very popular in North America, where it is still recommended [10, p. 102][14, p. 14][33] for emergency use when no harness is available. The present almost universal use of special nylon webbing harnesses by all climbers has made this kind of knot unnecessary.

Two bends were developed, the Sennit Bend (Fig. 14) and the Reever Bend (Fig. 15), which were said to be very strong, stable to fluctuating loads and springy under a shock load. Neither became popular, indeed I have seen no other reference to them. The knots look complicated to tie. Knotting is usually of minor interest to climbers; they do not wish to spend much time learning new knots unless the advantages are substantial and manifest.

Fig. 14. Sennit Bend

Fig. 15. Reever Bend

The remaining knots were developed as mid loops for attaching the middleman to the rope. The Rover Loop (Fig. 13) was stronger in one direction than another; indeed at its best it was the strongest knot tested. While it might be possible for the middleman to make a reasoned guess about the most likely direction of pull, the extra fuss seemed a bit of a nuisance. Moreover, while the knot made an excellent end loop if applied in the right direction, again care had to be taken in tying the knot. The authors did not therefore recommend it very strongly and it has never been much used in climbing; the name was given to the knot because it could be used anywhere in the rope. Ashley [3] lists the knot as #1043, saying that it is very strong and secure as either end or mid loop, and labels it 'probably original', having apparently missed Wright and Magowan's paper; of course, even earlier reports may yet be found, since it is very difficult to sustain a claim to have originated a simple knot.

Fig. 16. Half-Hitch Loop

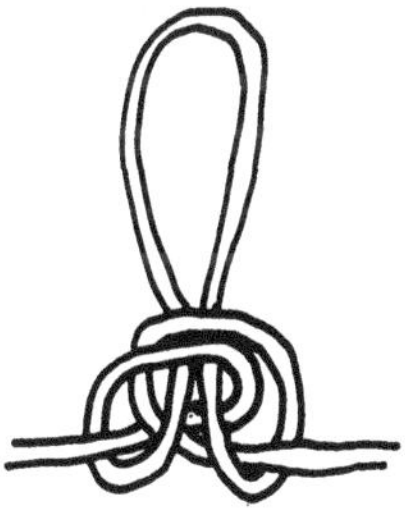

Fig. 17. Alpine Butterfly Loop

The Half-Hitch Loop (Fig. 16) and the Alpine Butterfly Loop (Fig. 17) are very similar in structure; the difference is that in the latter the two Half Hitches are interlaced. The authors found the Alpine Butterfly, named for a fancied resemblance in some configurations to the insect, to be close to ideal, much the stronger of the two, and able to withstand a pull on the two ends. There is much confusion in the climbing and caving literature about these two knots, many attributing the name and the excellent performance of the Alpine Butterfly to the Half-Hitch Loop. Despite Wright and Magowan's claim to originality, several general knotting books list one or the other of these knots, usually under another name, stating that they were used by American electrical linesmen; I have not seen any clear indication that in fact both knots were in use, rather than that the linesmen shared the climbers' confusion about them.

As noted above, the authors were impressed with the advantages of taking the pull of the climbing rope on several turns round the waist to lessen the discomfort of holding a fall. They recommended the Bowline and Coil for an end loop and devised a Three- or Five-fold Butterfly Loop for a mid loop, having three or five turns, respectively, round the waist. These knots also looked complicated, and I have never seen them mentioned anywhere else. However, the authors investigated ways of attaching the bight of the climbing rope to a separate length of rope wrapped round the waist. This also does not seem to have come into general use, perhaps because it only allowed for two passes of rope round the waist. I have not illustrated either of these knots. It is possible, however, that this paper, though it did not itself provide a popular way of taking the load of the middleman through several turns of rope, did call attention to the idea. Certainly, the wearing of a 'waist length' of several turns of light rope round the waist became popular around this time. It was easy to attach the end climbers to this waist length, but attachment of the middleman only became practicable when a karabiner was used as a link between the waist length and a mid loop in the rope, such as the Alpine Butterfly. This practice continued until climbing harnesses came into common use in the 1980s.

Prusik

Ashley [3] in his great book of mainly mariners' knots recognised a class of knots for tying on to the middle of a rigid rope that 'may be slid up and down with the hand but remain firm under a pull on the standing part'. Climbers and, more recently, cavers have found some novel applications for some of these movable or sliding hitches that can withstand many and frequent repetitions of the load-unload-move cycle, and have come to call their knots either 'prusik hitches', after the first such knot to achieve widespread use, or, more generally, 'friction hitches'.

E. Gérard [9] seems to have been the first climber to recognise this need, though he said only that it would allow the climber to climb up a smooth rope of 10 mm diameter. He used three 'rope rings' around 70-80 cm long, made of plaited cord of about half the diameter of the rope. These rings were tied one above the other onto the climbing rope with his special knot (Fig. 18). One foot was placed in each of the lower rings, and the head and arms put through the upper. To climb up the rope, loosen, push up and retighten each ring in turn (Fig. 19). Gérard's note was quoted practically in full in a long article about the use of the 'spare rope' by Blanchet [5], under a heading suggesting the method could be used to 'climb a vertical or overhanging and holdless step', though without discussing how to get the main rope into position. Neither Gérard nor Blanchet seem to have given much thought to just how the knot could be used in mountaineering practice. These unenthusiastic accounts did not lead to any widespread application; indeed, I have seen no other mention of Gérard's knot until Thrun [32, 10] mentioned it nearly 50 years later. Apparently he had discovered it independently, but he condemned it, without testing, as inferior to the Prusik Hitch. I have also seen Gérard's knot in some French general knotting books, but called a Prusik Hitch. Note, however, that Gérard's system of manipulating slings is still in occasional use [14, p. 127], though simplifications using two slings (one for one or both feet, one to waist or chest harness) are much more common, and are described in many climbing, caving and rescue books world-wide.

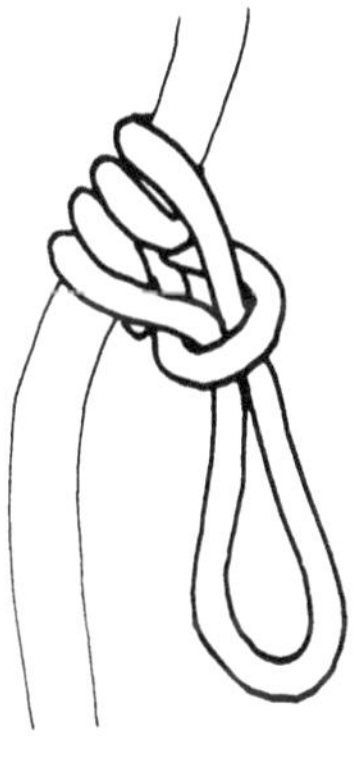

Fig. 18. Gérard Hitch

Fig. 19.

Fig. 20. Prusik Hitch

Fig. 21.

A couple of years after Gérard, Dr Karl Prusik [26] described a similar knot (Fig. 20) and, in contrast to Gérard, discussed several uses for the knot that any mountaineer could recognise as valuable. Prusik was a professor of music in Austria, and a keen climber and writer of mountaineering books [24]. He is said to have developed his knot from experience in mending broken guitar strings during military service in World War I. He described a method of self-rescue from a crevasse using two slings, each attached to a foot, passing through a chest harness, and tied with his knot to the main rope at about head level; each sling was moved up in turn (Fig. 21). This precise technique is also still in occasional use [11, p. 74], [27, p. 133]. Prusik also discussed other uses of the sling and Prusik Hitch: to safeguard a rope under load; to aid hauling on a rope; and to attach components of several tackle systems to raise a helpless person. All these applications are still in use. It may be of interest to note that experts in a hurry can climb up a rope, using friction hitches and slings only, for 30 m in a little over 1 min, and for 120 m in a little over 9 min.

These clearly described applications caught the imagination of mountaineers, particularly in Europe. The method of ascending a rope has become known as 'prusiking', and the slings held ready for emergency use are known as 'prusiks'. Additional applications, mainly developed since the War, include the attachment of a middleman to a climbing rope by way of a small sling attached to a waist length; or the attachment of a solo climber to a fixed rope, as for a practice climb; the ascent of a fixed rope traversing difficult country in expedition mountaineering; more elaborate tackle systems, some with automatic locking; and in caving. A few European cavers experimented with an abseil-down, prusik-up technique in vertical caves in the 1930s [21, p. 310], avoiding the slow and cumbersome use of ladders.

However, the idea did not catch on at that time. Meanwhile, mountain rescue teams were becoming better organised, and began to investigate techniques using rope and knots not needed for ordinary climbing or caving. Crevasse rescue inspired the adoption of the Bowline on a Bight as a chair knot and development of the Prusik Hitch for self-extrication, as has already been discussed. Several ways of constructing emergency stretchers were developed that used nothing but a single climbing rope; no special knots were used in the construction.

Post-War Period

This might be even better called the Post-Nylon Period, because the effect of the War most significant to climbers and cavers was the invention of nylon ropes in 1941. The US Army has long had a mountain section. The entry of Japan into the War abruptly terminated the supply of manila fibre, up to that time the fibre of choice for climbing rope. Sisal proved inferior, but nylon proved both stronger than manila and to have much greater elongation before breakage. The shock loading on the climber at the end of a fall was much less than when using manila. In addition, nylon was less affected by moisture, frost or mildew; but it was more slippery and it melted at relatively low temperatures, so that friction between two ropes could cause failure. Proper specifications were developed for nylon climbing rope, which at that time was all of three-strand, laid construction. After the end of the War these ropes became the choice of climbers and cavers. Other synthetic fibres have been tested for climbing and caving ropes, but have not removed nylon from its overwhelmingly dominant position.

These new properties influenced rope use and the knots tied. For example, an early report states that 'a foot of nylon will often slip through a knot before it tightens enough'. Traditional knots had to be either abandoned and substitutes found, or at least modified, such as by securing the running ends to standing parts with Overhand Knots or the like.

At this time, increased mobility and affluence brought about great increases in the numbers of people taking part in adventure sports, including rock climbing and caving. All this caused a great burst of experimentation with knots. This was increased when, beginning in the mid 1950s, kernmantle rope (sheath and core) started to replace laid nylon rope for climbing.

Kernmantle ropes can be made with different properties for different climbing purposes. These specialist life-support ropes are quite different from the sheath and core ropes that have been developed for yachting and the like. Though many have been tried, only two main designs of climbing rope are now found: dynamic ropes for ordinary rock climbing and mountaineering, which are normally unloaded but have good elongation under shock loads; and static ropes used for ascending and descending on the rope, as in caving and some rescues, with minimal stretch under a person's weight but still ample stretch for emergency shock loads. Kernmantle ropes are even more slippery than laid ropes, and some knots or rope usages cause separation of sheath and core.

Environmentally friendly artificial anchorages began to be developed, that could be inserted into cracks and crevices and removed after use with much less damage to the rock than pitons. At first, in the early 1960s, engineering nuts with the threads reamed out were tried; then optimally shaped irregular hexagons and other shapes were used; and more recently highly sophisticated mechanical devices were introduced. Dynamic belays were made more controllable, having all the slippage of the rope to a falling climber go past a metal device, rather than the belayer's body; consistent friction was applied, allowing a suitable compromise between too large a shock loading on a falling climber or the belay, and too long a run out of the rope, increasing the danger of hitting an obstacle or running out of rope. Nylon tape and webbing allowed the construction of special climbing harnesses, at first improvised and knotted, but increasingly permanent and sewn, allowing safer and more comfortable connection to the rope. Karabiners and other devices are now much stronger and lighter than they were.

With these new stronger and more elastic ropes, and with new techniques and mechanical devices, it now became possible to hold falling leaders with minimal damage to rope and climber, so that now many climbers on routes of the greatest difficulty almost expect to fall, perhaps several times, confident of being held safely, to repeat the move or try elsewhere.

The greatly increased numbers of people taking part in all these sports have led to a great increase in the numbers of books and magazines catering for them. This causes a problem when investigating the history of techniques, including knots. It is rare for climbers or cavers describing the knots they use to say whether they developed the knots themselves or copied from someone else. It is impracticable to try to trace back all publications to find the first mention of a knot because few general or club libraries stock any substantial fraction of all the publications covering these activities. In the following pages I give a specific reference when I know the origin of the knot, otherwise I give usually a couple of references to technical manuals. It is rarely possible to say whether any new knot was developed by a climber, or copied from some source in some other occupation. I first discuss minor modifications of knots already well known in climbing or caving circles before the introduction of nylon, and go on to discuss more novel knots.

Securing the Ends of Knots

The frequently fluctuating loads imposed on most knots used in climbing, caving and the like, together with frequent contact with an often jagged rock face, often cause the running ends of the knots to work back through the knot, perhaps right out, if not prevented. This tendency was noted even in the days of manila rope, but was much more marked with nylon. Several accidents have been attributed to this kind of failure of a knot.

Preventing the running end from emerging by using a stopper knot such as an Overhand Knot has been used only very rarely; in very slippery conditions it may not prevent the end working out of the knot. It is much more satisfactory, and common, to secure the running end to its standing part. Securing the running end of a Bowline with one or more Half Hitches has already been noted (Fig. 6); this was common practice in the early days, but sometimes neglected between the Wars. Occasionally an Overhand Knot was used instead of a Half Hitch; an early manual [28] has each end of a Fishermans Knot secured with Overhands. With the introduction of nylon, some manuals recommend the securing of the ends of all knots tied in life-support ropes, no matter how secure the knot itself or how stable its application. The most popular security knot is probably the Overhand Knot shown in Fig. 22; but the Grapevine Knot (Double Overhand) is also used a lot. An interesting application is the Secured Reef Knot [12, p. 76][29, p. 22] (Fig. 23). This bend is said to be very easy to untie even after prolonged heavy loads, and of course it would not slip or capsize in the way of an unsecured Reef Knot used as a bend.

Instead of securing the running end to its standing part as above, some people tuck it back through the knot, sometimes following this with the usual security Overhand. Perhaps the commonest such knot is the Tucked Bowline (Fig. 24) or Tucked Double Bowline [21, p. 38], believed to have been developed by American climbers in Yosemite Valley. The Tucked Figure Eight Loop [29,

p. 33] (Fig. 25) is less common; I have seen the Tucked Girth Hitch (Fig. 26) only in a French caving book [14, p. 65]; it was used instead of an end loop to attach a rope to a karabiner fastened to a climbing harness.

Fig. 22. Double Bowline and Overhand

Fig. 23. Secured Reef

Fig. 24. Tucked Bowline

Fig. 25. Tucked Figure Eight Loop

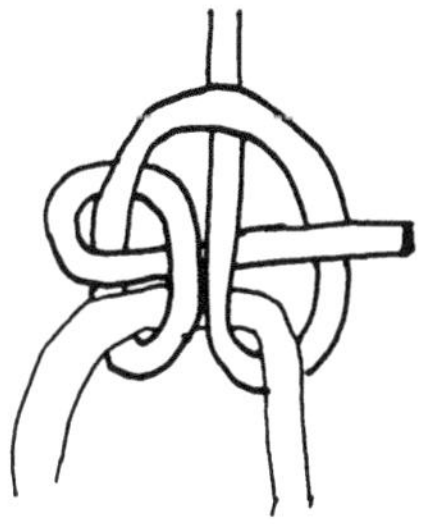

Fig. 26. Tucked Girth Hitch

Fishermans Knots and Loops

Many of the early mountaineers were 'gentlemen' who would have been familiar with the sport of fishing or angling and its traditional knots. Is this the reason for the popular adaptation of the Fishermans Knot (Fig. 3) and Fishermans Loop Knot (Fig. 4) to climbing ropes instead of fishing lines right from the start of the sport? Or were they in contact with some other group who were already using these knots in substantial rope? Nearly all general knotting books list these knots for line only, and it is puzzling why they were used in climbing ropes right from the beginning; attempts to introduce bends other than the Fishermans Knot were uniformly unsuccessful.

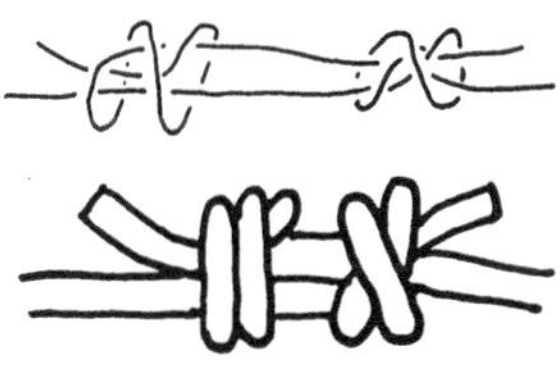

Fig. 27. Double Fishermans Knot

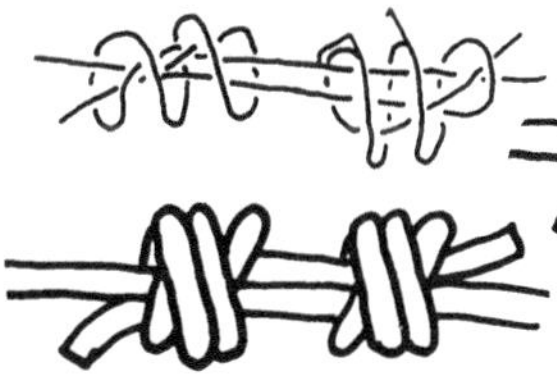

Fig. 28. Triple Fishermans Knot

Fig. 29. Swami Loop

When nylon rope was introduced, the Single Fishermans Knot was largely dropped, and the Double Fishermans Knot (Fig. 27) became the bend of choice for life-support ropes. The Triple Fishermans Knot (Fig. 28) has also been recommended [21, p. 48]. Some people secure the ends with Overhand Knots.

The Single Fishermans Loop was used in manila rope for a long time as both end and mid loop, but alternative knots were always available and often more popular. When nylon rope seemed to require something more secure, it was found inconvenient to tie a Double Fishermans Loop; tying the Grapevine Knot a long way from the end is a somewhat fiddly operation of a kind unpopular with many cavers and climbers. However, a Swami Loop [12, p. 76][29, p. 37], sometimes called a One and a Half Fishermans Loop or a Three Quarters Fishermans Loop, has been developed, apparently initially in New Zealand. This seems to have most of the advantages of the Double Fishermans Loop and yet is easily tied.

Reweaving or Follow-through Techniques

Several simple end loops can be made by tying a simple stopper knot in doubled rope; these include Single and Double Overhand, and Figure Eight and Figure Nine Loops. All of these have found considerable application in climbing and caving; the last two are specially suitable for nylon rope. But it is quite often desired to tie these knots round an object, such as the loops of a climbing harness, in situations where the ready-tied knot is inapplicable. The solution is to tie the simple stopper knot in the standing part of the rope in the position where the knot will be wanted, lead the end round or through the object and then take it back to the original knot and carefully follow all the turns of the knot with the end, finishing with it beside the standing part, pointing away from the loop. This reweaving or follow-through technique for tying knots can also be used in other situations, and is now frequently applied to life-support ropes. It is uncommon in general knotting.

Fig. 30. Double Overhand Loop

Fig. 31. Figure Eight Loop

Fig. 32. Figure Nine Loop

The Overhand Loop was the first end loop used in mountaineering (Fig. 1), and can also be made by reweaving [20, p. 29][27, p. 75][29, p. 25]; it is still in use, specially in continental Europe. The Double Overhand or Grapevine Loop (Fig. 30) is a bulky but secure loop, prone to jamming under load, used more in caving than climbing [12, p. 76][13, p. 65]; it can be tied by reweaving, but is more commonly tied direct with the bight. The Figure Eight Loop (Fig. 31) is now used as an end loop [10, p. 96][14, p. 28][27, p. 75] much more than the Bowline, however secured; most applications require reweaving. The Figure Nine Loop (Fig. 32) is mostly used by cavers [12, p. 75][13, p. 86]; it is stronger than the Figure Eight Loop as an end loop, but weaker as a mid loop in static kernmantle rope; when subjected to a shock load, the knot compacts a bit, absorbing a part of the shock.

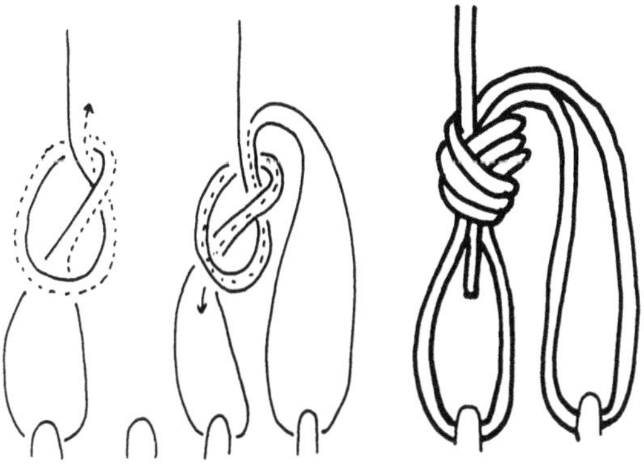

Fig. 33. Overhand Double Loop

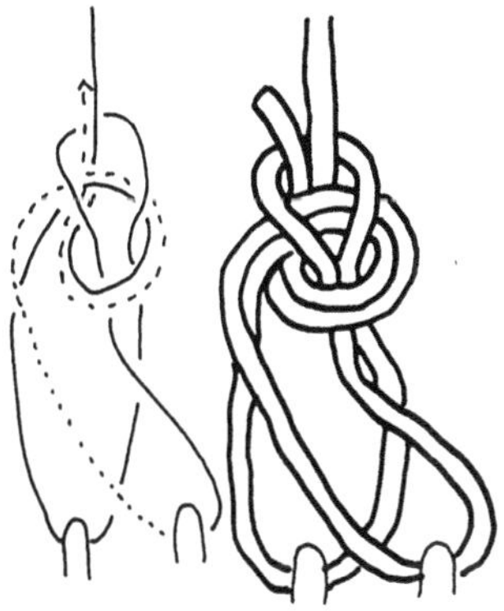

Fig. 34. Rewoven Bowline on a Bight

Twin loops can also be tied round two separate objects by this reweaving method. An Overhand Double Loop (Fig. 33) may be used [20, p. 40][29, p. 77] or alternatively a rewoven Bowline on a Bight (Fig. 34); Wright and Magowan [34] first used the latter, tied this way, to make a double loop around the waist without having to make the loops first and then step into them, but it is now also used to attach to two points of a harness [10, p. 97]. Some combined chest and waist harnesses have their attachment points well separated, and it is better to have separate attachments to avoid too much movement in a single large loop.

Fig. 35. Tape Knot

A reweaving technique is also used in some bends. The rewoven Figure Eight Tie (Fig. 7) has been in use since 1892, and has already been described. A similar bend based on the Overhand Knot is sometimes tied in rope [27, p. 68][29, p. 26], [33, p. 16] but is widely recognised as the only bend suitable for the nylon tapes now used in so many climbers' and cavers' slings and other accessories; the knot is thus best known as the Tape Knot (Fig. 35) [10, p. 95][13, p. 113][20, p. 30][29, p. 26].

Novel Knots

The post-war knots considered so far have been only minor variations of knots either already known in the climbing or caving worlds or at least fairly well known in other applications. But there were some quite novel knots developed.

The first was the Tarbuck Hitch (Fig. 36), developed by Ken Tarbuck [30] soon after the introduction of nylon climbing rope to Britain. This was a shock-absorbing end loop, claimed to be as strong as a splice, able to hold several times the climber's weight without slipping but, when subjected to the shock load of a falling climber, it would slip until tight and absorb some of the shock. The knot could be used tied round the waist if steps were taken to limit motion when shock-loaded, but was more commonly tied into a karabiner. Tarbuck recommended that the loop should have a diameter of around 30 cm, so that at most about 60 cm of rope would pass through the knot to absorb some of the shock; I have not seen any estimates of the forces involved in this process. The knot became quite popular for a time. As kernmantle gradually replaced laid rope in the late 1950s through to the 1970s, it was found that when the Tarbuck Hitch held falls in the way it was designed to, the sheath of the rope became separated from the core over some distance, making the rope unsuitable for further use. Once climbers realised that this effect was occurring, the Tarbuck Hitch came into disfavour and is now no longer seen in climbing activities. A few general knotting books, however, continue to list the knot and it seems to be finding new applications.

Fig. 36. Tarbuck Hitch

Fig. 37. Mariner Hitch

The Mariner Hitch (Fig. 37), named after Wastl Mariner who developed it [15, p. 99], allows the use of a sling to take the load off a rope for a period, perhaps to allow knots in the rope to pass some device, using a knot that can be readily untied and allows the load to be returned to the main rope in a controlled smooth way, without jerks. The original version had several turns round the karabiner and the bight tucked directly between the legs of the sling; it is now generally agreed [14, p. 67][29, p. 73] that it is best to share the load between turns round the karabiner and twists round the sling. Usually there is a full round turn on the karabiner and perhaps a couple of turns round the sling, as shown.

Friction Hitches

The first friction hitches, specially the Prusik Hitch (Fig. 20) have already been described and it was mentioned that a few European cavers experimented with an abseil-down, prusik-up method of traversing vertical caves. This technique has developed a lot since the War. The various ways of manipulating the slings to ascend a rope constitute one of the major contributions by cavers to the use of life-support ropes.

Starting in 1952, Bill Cuddington developed this system for use in American caves. It has since been adopted and adapted world-wide, becoming known as the Single Rope Technique or SRT. From the 1950s on, many different knots have been suggested as alternatives to the Prusik Hitch; Thrun [32] in 1971 described more than 25, most developed by American cavers. Novel knots continue to be developed all over the world, by both climbers and cavers, despite the introduction of mechanical devices which grip the rope and can be used in similar ways to the Prusik Hitch. The first mechanical ascender was invented in 1958 by two Swiss mountaineers, Jusy, a game warden, and Marti, an engineer, and given the name Jumar [21, p. 311], so that climbing with ascenders is sometimes known as 'jumaring', though 'prusiking' is also used to describe all ways of ascending a rope using slings. Improvements and other devices followed later. These mechanical ascenders have practically replaced friction hitches in SRT in caving, in the use of fixed ropes in expedition mountaineering, and in many of the activities of full-scale rescue teams. But in other applications, the knots are still used extensively, partly because of the cost, weight and bulk of the ascenders, partly because of the versatility of the knots and the ready improvisation of the equipment. Only the most popular of the friction hitches will be mentioned below; note that many of them are named after a person.

It must have been discovered quite early in the experience of the Prusik Hitch that in some conditions it did not hold. Mountaineers with iced-up rope or cavers with muddy rope could expect trouble. Cuddington is said to have developed the Double Prusik Hitch (Fig. 38) to extract himself from a solo trip to a muddy cave in the 1950s [21, p. 311], and the knot is now commonly used when the ordinary Prusik Hitch slips. All the friction hitches need to have the number of turns adjusted to suit the conditions; experience and trial and error are needed. My drawings show the number of turns most frequently called for.

There is a trend towards simplicity in the structure of the more successful friction hitches; they seem easier to tie and less likely to jam. The Klemheist Hitch (literally: clamp hoist) (Fig. 39) should have the upper loop as short as possible. The knot is also satisfactory when tied in climbers' tape or webbing [11, p. 36][20, p. 34][29, p. 46]. If the lower loop is half-hitched round the upper in a sort of Sheetbend configuration, the knot is less likely to jam and easier to loosen [32]. The French Prusik Hitch (Fig. 40) is the only knot in this series that is readily released under load [29, p. 481]. Indeed, Shepherd [29] in particular uses this knot in many applications, including as a safety rope for abseiling, providing emergency braking or ready passage past a knot, or as a safety backup when hauling on a rope.

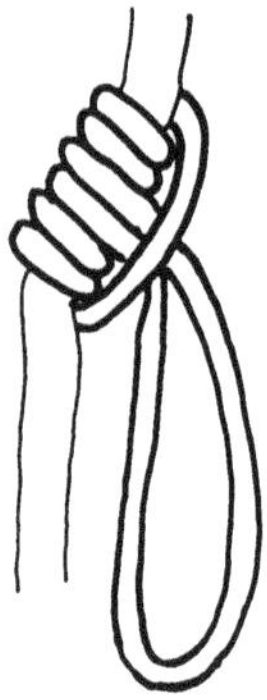

Fig. 38. Double Prusik Hitch

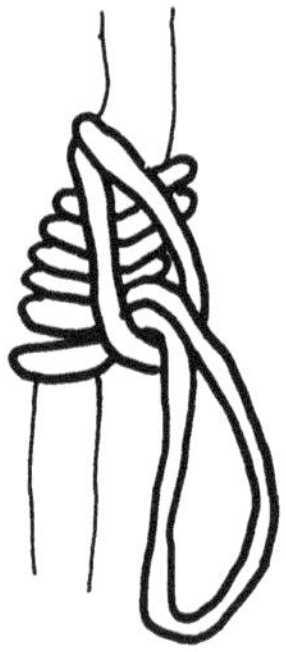

Fig. 39. Klemheist Coil Hitch

Fig. 40. French Prusik Hitch

Fig. 41. Wend Prusik Hitch

Fig. 42. Tucked Coil Hitch

Sometimes people wanted to tie a friction hitch with a single rope rather than a sling. Perhaps they had no sling and either too short a rope or insufficient time to make one; or perhaps they wished to attach the other end to somewhere special; or perhaps there was a need to adjust the friction very delicately. If an ordinary Prusik Hitch was simply tied in a single rope, making the Wend Prusik Hitch (Fig. 41) taking the load on one of the ends, it was found that with repeated load–unload–move cycles the wend (running end) worked back through the knot. Even making an Overhand in the end will not always prevent this, though making the Overhand round the standing part (Fig. 41) will usually work. With this knot, there is the choice of having more or fewer coils above or below the ends, and of taking the load on the upper or the lower rope. Unfortunately, authorities disagree on how these choices should be made. Someone needs to do some careful tests if the knot is to be used with confidence and efficiency. Possibly the most adaptable of the single-rope friction hitches is the Tucked Coil Hitch (Fig. 42), developed by Prohaska [22]. This shows minimal tendency to work the end through the knot, and a simple Overhand in the end will prevent it. Prohaska says that if the knot slips because the cord is too stiff, then more coils should be added to the lower part, but if it slips because the load is too big, the extra coils should go to the upper part.

Friction hitches can be difficult to manipulate if the fingers are stiff and cold; this can be made easier by incorporating a karabiner inside the knot, against the main rope. Both the Prusik and the Klemheist Hitches have been used in this way. Note that the karabiner is not used as a handle; the knot enclosing the karabiner is simply moved in the usual way.

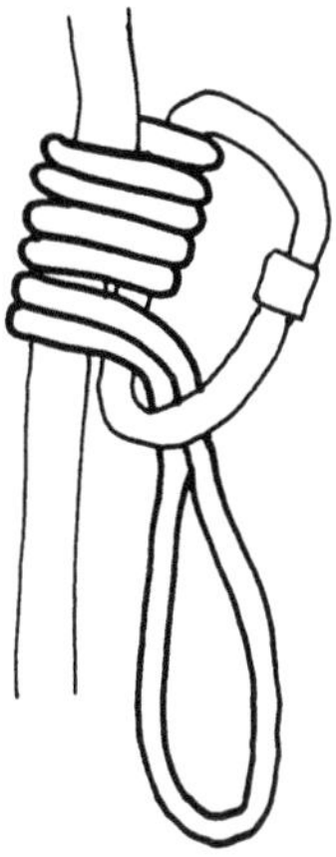

Fig. 43. Bachmann Hitch

Fig. 44. Karabiner Hitch

The best-known of these semi-mechanical hitches, and the oldest, is the Bachmann Hitch (Fig. 43), specially recommended [17] for use with wet nylon rope. It is usually claimed that it also works well with icy ropes, though Shepherd [29, p. 48] disagrees. The knot is often used with pulley systems and for rescue work, with many applications similar to the French Prusik [14, p. 134][21, p. 100].

The Karabiner Hitch (Fig. 44) was originally developed by Meier [10], who said that it could work very well with dry laid rope with only one turn, if the karabiner was held in the horizontal position. It does not seem to have had much use until Prohaska [23] independently developed it for use with tape slings, claiming that it needed fewer turns and was more reliable than the Klemheist Hitch; he found it best to have the karabiner in its more natural, sloping position as shown in Fig. 44.

Recent experience using this knot with tape seems to have made it more popular using rope [19, p. 16].

Flowing Hitches

A hitch is defined as a knot attaching a rope to an object. With an ordinary hitch, the knot, rope and object are held in place, stationary relative to one another. With a friction hitch, the rope and the knot are stationary relative to each other, but the two together move past the object (the standing rope). I give the name flowing hitch to a group of knots in which the knot and the object are stationary relative to each other but the rope moves, flowing through the knot. Obviously, the knot must be very simple to allow the rope to flex its way through it. Indeed, the definition of a knot must be somewhat strained to include some of these structures as knots. The best known example is the round turn taken round a bollard or the like when paying out a rope under load.

Climbers and cavers use other simple flowing hitches to provide friction when abseiling down a rope. Even the Classic Abseil technique (Fig. 8) can be considered a flowing hitch; but today descenders are commonly used, mechanical devices intended to provide the friction of the rope during abseiling. Pre-war systems of abseiling, such as the Classic, involved wrapping the rope round parts of the body to provide friction during a descent. This was not too uncomfortable when wearing clothing suitable to alpine conditions, but decidedly warming when wearing clothes for a hotter climate. A descender allows the rope to be removed from contact with the body; the braking friction derives from the movement of the descender along the rope. Early descenders consisted of karabiners in various configurations, of which the simplest is the Twisted Hitch (Fig. 45). A karabiner is attached to the body harness usually by another karabiner. The rope, often doubled, is anchored at the top and then taken in one or, for greater friction, two round turns (up to three turns when using a single rope) round the back of the karabiner [4][10, p. 141], and is grasped in the gloved hand below the descender before dropping to the bottom of the pitch. The other hand helps to keep the body in balance by lightly holding the rope above the karabiners. Friction can be varied by the tightness of the grasp and the descent can be stopped by bringing the lower rope up beside the upper, and then either holding the two ropes together or fastening with an appropriate knot.

Several other arrangements of karabiners have been devised, some with accessories, but most modern abseiling is done with purpose-built devices, forcing the rope through a more complicated curved path. Some provide means for varying the friction through a wide range, some have automatic safety devices which stop the descent if the grasping hand is released. Probably the most popular device is the Figure Eight Descender (Fig. 46) made in several configurations [4][10, p. 140][27, p. 123]. This has two apertures of different sizes, allowing use to be adjusted to the circumstances [30, p. 127]. Care must be taken to avoid capsizing the knot in front of the descender, forming a Girth Hitch (#56,[3]); some designs of the descender are made to avoid this [14, p. 163].

Many descenders can be also used to allow the controlled lowering of a load [15], [29, p. 71][33]; the descender is anchored at the top and the rope is threaded through it, with the load hanging below; the friction substantially lessens the force needed to hold the load.

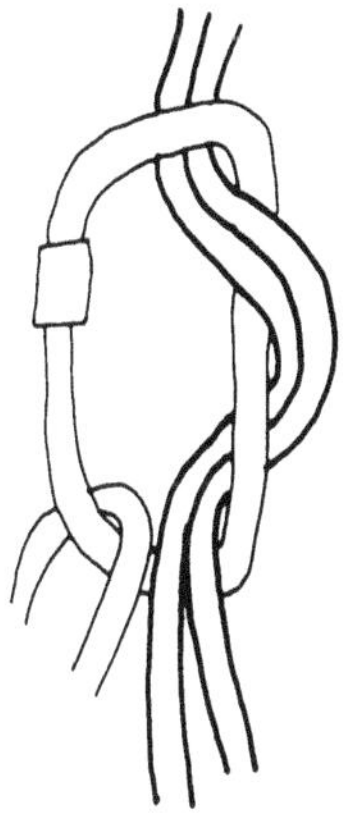

Fig. 45. Twisted Hitch

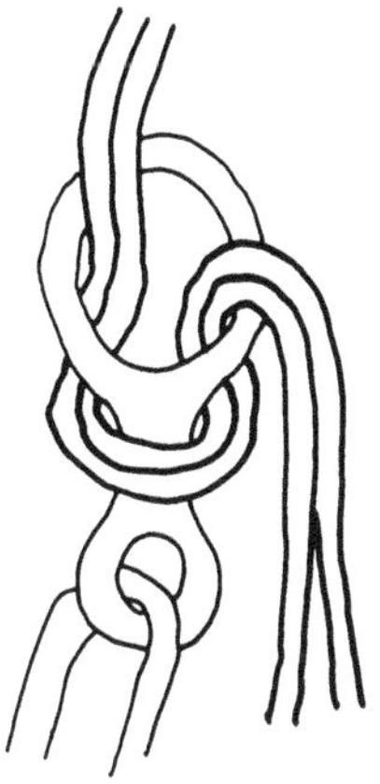

Fig. 46. Figure Eight Descender

Such an arrangement of rope and descender is not usually suitable for use belaying a climber. A belay system must freely pay out the rope to the climber but be prepared to hold instantly if the climber falls. A dynamic belay reduces the impact force on both anchorage and falling climber by allowing the rope to pay out under controlled friction to bring it to a halt gradually. Before World War II the friction was usually provided by wrapping the rope round the arms and across the back of the belayer. This system brought problems after the falling person was held, perhaps swinging freely at the end of the rope: the belayer needed to disentangle himself from the system in order to take appropriate action but without further endangering the victim. Using friction round a device held in front of the belayer, instead of round his body, minimised this problem. The simplest device is a Belay Plate (Fig. 47) or even the smaller eye of a Figure Eight Descender. This constrains the rope as it passes round a belaying karabiner. In normal use the rope passes freely, guided by the gloved hands, but in the event of a fall the free rope, the end not leading to the climber is pulled back hard. This tends to jam the rope against the device and provides friction [10, p. 122][14, p. 58][27, p. 84][29, p. 54]. Once the victim has been held, it is simple to make the rope fast with an appropriate hitch and disengage the belayer. Several similar devices have been developed, leading the rope round more or less complex curves.

Perhaps the belay system most interesting to knot tiers is the use of a karabiner with an Italian Hitch (Fig. 48), developed by Münter in the 1970s. The karabiner is anchored at the belay point and the rope attached to it as shown. The rope is paid out in the direction shown by the arrow; the belayer may need to feed the slack into the knot to minimise the load on the climber.

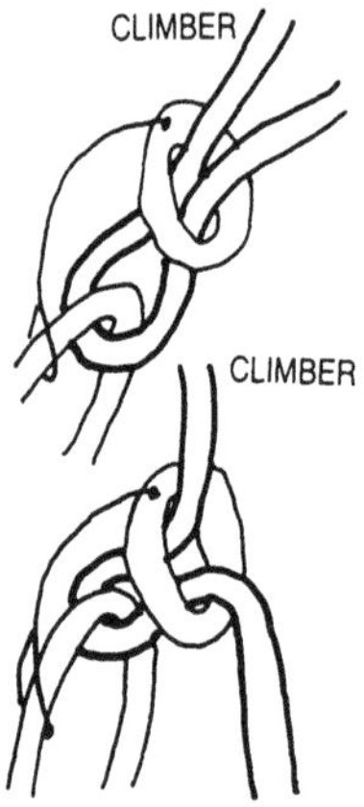

Fig. 47. Belay Plate

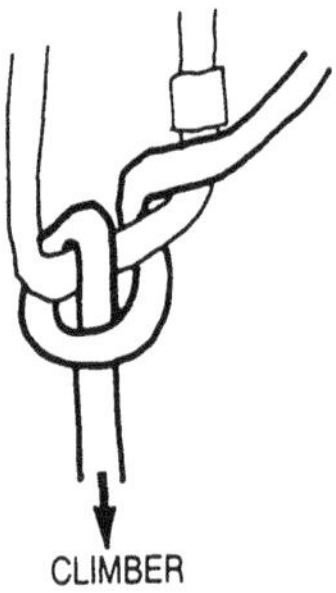

Fig. 48. Italian Hitch

If the belayer wants to take in the rope, as would happen if the climber reverses his moves, the free rope is pulled, the knot capsizes into the reverse formation, allowing free movement, though again it may be necessary to feed in the slack. In the event of a fall, the free rope is grasped by the belayer (gloves are essential!) and, for greater friction, swung forward until nearly parallel with the live rope leading to the climber. Under ordinary climbing conditions, the force provided by this knot is about 2.5 kN, providing a valuable lessening of the shock force. Once the fallen climber is held, the rope may be tied off with an appropriate hitch. Under some conditions the friction in the knot may generate enough heat to lightly glaze the nylon sheath of the rope, but this is in no way dangerous; twists or kinks may also develop in the rope [10, p. 123][20, p. 35][29, p. 129]. Greater friction can be obtained by making an extra turn [25], forming the Double Italian Hitch (Fig. 49).

The Italian Hitch or Double Italian Hitch on a karabiner may also be used for abseiling or lowering a load [13, p. 236][25][27, p. 134], in a similar way to the descenders; but the twists formed, specially in a long abseil or lower, are a disadvantage, indeed a danger in some circumstances. It seems best to use these applications of the knot only when no other method is available. Some other belay systems can also be used as descenders for abseiling and the like.

When making a long hoist by hand, it is often convenient to have a device which will automatically hold the load to allow changing the grip or the attachment point of tackle. Some of the friction hitches have been used for this purpose, but they seem to need constant watching.

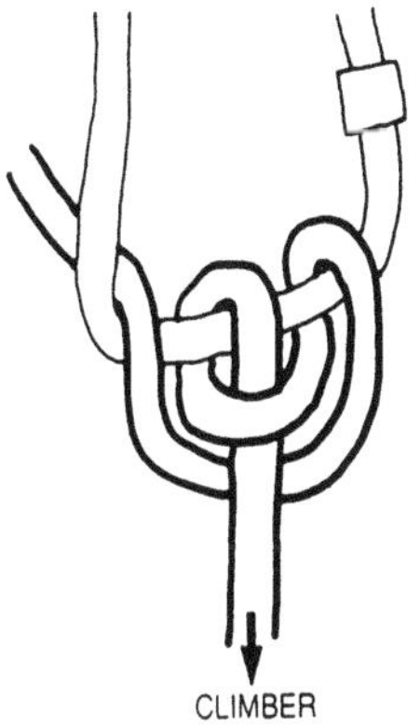

Fig. 49. Double Italian Hitch

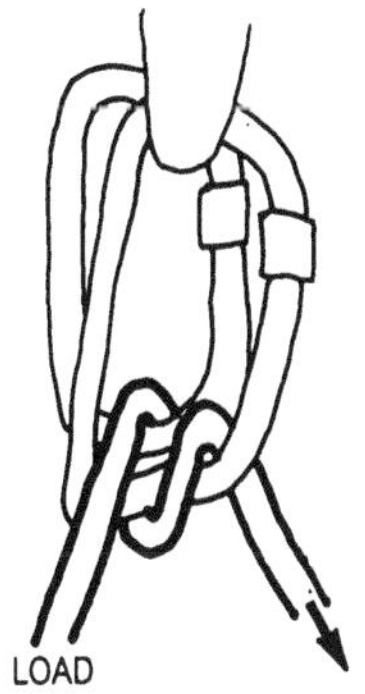

Fig. 50. Alpine Clutch

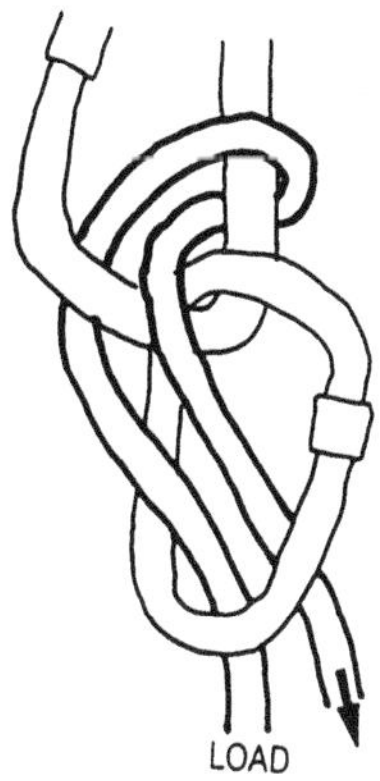

Fig. 51. Stuflesser Hitch

Two hitches that seem to need less attention have been devised by climbers: the Alpine Clutch (Fig. 50) [13, p. 244][14, p. 137][29, p. 52] and the Stuflesser Hitch (Fig. 51) [13, p. 243][14, p. 138]. Both use a pair of karabiners which need to be selected to have the appropriate size and shape. If the end is pulled in the direction of the arrow, the rope will move fairly freely, though the friction developed will reduce the efficiency of any tackle used. If the end is released, the load pulls down on the rope, jamming it; the jam is readily released when hauling is resumed.

The Alpine Clutch, in particular, can be used in the place of a Prusik Hitch and sling in a number of applications [29, p. 143], but it must be remembered that it works in one way only and the movement of the rope through the knot cannot be reversed without a very great degree of difficulty.

Other Applications of Life Support Ropes and Knots

Abseiling has developed as a sport in its own right in some places, though few can maintain interest in it for long if it is not combined with other activities. However, abseiling is the major ropework component of canyoning, a growing adventure sport in the French Alps and Pyrenees, the Spanish Pyrenees, the Australian Blue Mountains and perhaps elsewhere where appropriate terrain exists. This sport involves following down steep and narrow watercourses, abseiling down beside or in waterfalls, jumping into and swimming along cliff-girt pools, traversing boulder blockups and other activities, all in narrow gorges from which there are often few escape routes, so that the canyoner is definitely committed to the route. Most canyons can be traversed in a few hours. To date, ropework and knots have been adopted from climbing and caving, but perhaps traversing waterfalls may spur invention of something new.

Rescue work for the adventure sports has become more and more technological since World War II, making use of lightweight stretchers adapted to use in mountainous areas or in caves, helicopters, electronic navigation and communication apparatus and the increasingly technical equipment and skills of rock climbing, mountaineering and caving. The various technical experts in these activities are being increasingly integrated into special emergency services, available for rescue work not only in the adventure sports but also in some kinds of accidents on roads, industrial and construction sites, and in natural disasters and the like. Often people from the police, fire and ambulance brigades and the armed services are closely involved, and given training in the use and application of life-support ropes. Most use is made of static kernmantle rope, because of its minimal stretch under moderate load. There seem to be few knots used exclusively in rescue work. However, there have been developments in the preparation of lightweight tackles, operated by hand alone and involving sometimes long hauls when only short tackles can be used, and all the manipulations are confined to small and exposed areas. There has also been a need to develop anchor systems where no single anchor point is strong enough for the load, so that several weaker anchor points have to be used. Anchorages have to be linked so as to spread the load and cope with any failure of a single point. These tackles and anchorages are now also used in ordinary climbing, caving and canyoning.

Finally, I will describe a new industrial technique, sometimes called abseil engineering, which is now being developed. This involves aspects of the construction, maintenance and repair of industrial sites such as oil rigs, bridges and high buildings where access by ordinary means is particularly difficult, time consuming, unsafe or even impossible, but where access using the life-support equipment and techniques developed by climbers and cavers is relatively safe, easy and quick. Practitioners are sometimes called spidermen; most are recruited from the pool of experienced climbers and cavers, though some are industrial riggers acquiring new skills. Perhaps the night climbers of Cambridge, and other university students who surreptitiously scale the outsides of their colleges (and also their more recent non-academic imitators), can now be regarded as training themselves for respectable occupations, not just being irresponsible pranksters.

All the usages of life-support ropes are clearly closely related. The present considerable interchange of information between rescue workers, cavers, mountaineers, abseil engineers, rock climbers, canyoners, mountain army corps and others must lead to the development of simpler and safer techniques. So far, there have been few contributions to techniques or equipment from outside Europe, North America and Australasia; but others, Asians in particular, are becoming more interested in and more skilled at the adventure sports, and there is little doubt that they will make their mark in techniques also.

Note that in this paper I have described knots only. I have not spelled out details of how they are to be applied, of the precautions needed for safe and efficient operation, or of the risks involved when using them. If any readers wish to try any of these knots in practice, they should read further in the appropriate literature and consult people experienced in their use.

BIBLIOGRAPHY

General References

J. Bernstein, *Ascent* (Simon and Schuster, New York, 1989).
T. G. Brown, 'Early Mountaineering' in *Mountaineering*, ed. S Spencer (Seeley Service, London, ?1935) 17–39.
T. R. Shaw, *History of Cave Science* (Sydney Speleological Society, Sydney, 1992).

Specific References

1. Alpine Club Special Committee, 'Report on ropes, axes and alpenstocks', *Alpine Journal* **1** (1864) 321.
2. Alpine Club Special Committee, 'Report on equipment for mountaineers', *Alpine Journal* **16** (1892) Special Report 1.
3. C. W. Ashley, *The Ashley Book of Knots* (Doubleday Doran, New York, 1944).
4. A. Blackshaw, *Mountaineering* (Penguin, Harmondsworth, 1965).
5. E. R. Blanchet, 'The spare rope in theory and practice', *Alpine Journal* **41** (1929) 63.
6. D. Busk, 'The first rappel: a study in vanity', *Alpine Journal* **66** (1961) 365.
7. C. T. Dent, *Mountaineering* (Longmans, London, 1900).
8. H. B. George, 'Knots for roping travellers', *Alpine Journal* **2** (1865) 95.
9. E. Gérard, *La Montagne* (1922) 322; seen in [5].
10. D. Graydon (ed.), *Mountaineering, The Freedom of the Hills* (The Mountaineers, Seattle, 1992).

11. K. Hoi and E. Jenny, *Behelfsmäßige Bergrettungstechnik* (Bergverlag Rudolf Rother, München, 1988).
12. D. Judson (ed.), *Caving Practice and Equipment* (David and Charles, London, 1984).
13. G. Marbach and J-L. Rocourt, *Techniques de la Spéléologie Alpines* (Techniques Sportives Appliquées, Choranche, 1986).
14. B. March, *Modern Rope Techniques in Mountaineering* (Cicerone Press, Milnthorpe, 1985).
15. W. Mariner, *Mountain Rescue Techniques* (Österreichischer Alpenverein, 1963).
16. D. Meier, 'The carabiner prusik', *Tech Troglodyte* **3**(1) (1964) 3; seen in [34].
17. *Mitteilungen des Österreichischen Alpenvereins* Sept/Okt 1958.
18. P. Montandon, " 'Abseilen' ", *Alpine Journal* **33** (1920) 209.
19. W. Muller, *Rettung in de Seilschaft* (Verlag Edition Alpin, Mastrils, 1989).
20. W. Muller, *Alpinisme d'été: Escalade Sportive* (Editions du Club Alpin Suisse, Chur, 1990).
21. A. Padgett and B. Smith, *On Rope* (National Speleological Society, Huntsville, 1987).
22. H. Prohaska, 'Gesteckte Klemmknoten', *Der Bergsteiger* (Nov. 1981).
23. H. Prohaska, 'Ein Klemmknoten für Bandschlingen', *Die Alpen* (Okt, 1982).
24. H. Prohaska, 'The Prusik Knot: a 60 year history', *Nylon Highway* **34** (1992) 1.
25. H. Prohaska, 'Mehrfache Halbmastwürfe', *Die Alpen* (Mar, 1992) 102.
26. K. Prusik, 'Ein neuer Knoten und seine Anwendung', *Österreichische Alpenzeitung* (Dec 1931) 343.
27. P. Schubert, *Die Anwendung des Seiles in Fels und Eis* (Bergverlag Rudolf Rother, München, 1988).
28. Schweizer Alpenclub, *Ratgeber für Bergsteiger* (Art. Institut Orel Füssli, Zurich, 1920).
29. N. Shepherd, *Manual of Modern Rope Techniques* (Constable, London, 1990).
30. Swiss Alpine Club, *Mountaineering Handbook* (Paternoster, London, 1950).
31. K. Tarbuck, 'British safety methods of rope management', in [30], p159.
32. R. Thrun, *Prusiking* (National Speleological Society, Huntsville, 1971).
33. W. Wheelock, *Ropes, Knots and Slings for Climbers* (La Siesta Press, Glendale, 1992).
34. C. E. I. Wright and J. E. Magowan, 'Knots for climbers', *Alpine Journal* **40** (1928) 120, 340.

PART IV. TOWARDS A SCIENCE OF KNOTS?

This Part contains five chapters which deal with aspects of knot study in which respectively behaviour testing, mathematical modelling (two kinds), experimentation, and computer aided design of knots is involved.

The first chapter describes tests and results from many experiments, dating from the mid-nineteenth century, designed to find out how knots behave under load.

The second chapter traces the history of mathematical modelling of knots, and studies of their topological properties, from the time of C. F. Gauss in the early 1800s up to the present day.

In the third chapter the question is asked whether the study of knots can reasonably be called a Science. It is argued that there is much work to be done on the modelling of knots outside the field and the narrow constraints within which topologists choose to work. Knot classifications in various encyclopedia of knots, produced since the 1930s, are briefly described. And the ideas and works of A. G. Schaake, who since the 1980s has developed an extensive new theory of braiding processes, are reviewed.

The fourth chapter describes studies made by D. Mandeville (1910–1992), of a process which he called *trambling*. 'To tramble' is to produce a sequence of real knots by altering them one tuck at a time; the changes are called *tuck transformations*. Mandeville made many discoveries about trambles, and devised shorthand symbols for describing them. He believed that one day his tramble theories would make a contribution to an emerging knot science.

The final chapter is on Crochet work. It has been included in this Part because not only does it describe the techniques and history of this popular form of textile creation, but also it describes steps which its author and her colleagues have taken towards establishing a system of Computer Aided Doily Design. It is possible that their methods for generating and studying classes of doily designs will be paralleled by others for large scale computer studies of braiding problems in future knot science.

CHAPTER 10

STUDIES ON THE BEHAVIOUR OF KNOTS

Charles Warner

Rope and cordage are amongst the oldest human artefacts, and knots are an essential part of their use. Yet even now there are many aspects of their function that are not at all well understood. The main emphasis in this essay is on the behaviour under load of single-stranded knots tied in fibre rope. I have nothing on splices or other multi-strand knots, nothing on decorative knots, nothing on wire rope. The properties of unknotted rope are described only briefly, to aid understanding of the properties of knots. Monofilament lines of various kinds are used principally by surgeons and by anglers. Surgeons' knots are tied in very specialised materials, and are chosen mainly for ease of tying in their specialised conditions; many studies have been made on the practical behaviour of these knots, particularly on their security, but I do not discuss them here because they do not seem readily applicable to the kind of knot that is my main concern. To a large extent, the same applies to anglers' knots, but the study of their behaviour made by Barnes [7] raises some topics interesting to the general knot tier that I will briefly discuss.

Rope

Ropes function primarily to transmit tensile forces from one point to another. The tension may be almost wholly static, as in a binding or lashing, or very dynamic, surging, bending or weaving about as in oil-well drilling, roping steers, or traversing cliff faces. Specialist ropes may be made for specialist purposes, but most have to be able to withstand a wide variety of stresses, including those in knots. Ropes in use suffer a number of basic forces. There are tensile pulls on the whole rope and all its constituents, strands, yarns, fibres etc. Except for the straight longitudinal fibres in the core of some kinds of core and sheath ropes and for monofilament lines, tensile pull is accompanied by some degree of torsion. There is internal friction between constituents, fibre against fibre, yarn against yarn and so on. There is external friction, of the outside of the rope against capstans, spars, rocks or the like. Often these forces fluctuate as the rope flexes round curves, accelerating, jerking, chafing against surfaces, perhaps reversing direction [20, p. 213]. Knot failures are often due to the action of these forces. A fundamental property of rope is that it stretches when loaded. This is best measured by subjecting a known length of the rope to a steadily increasing load, and plotting the load against the resulting elongation. Fig. 1 shows typical curves. Quite a small load causes substantial initial elongation, but later a large load would be needed to produce a similar elongation. Laid nylon rope shows much more elongation compared with natural fibre rope at the same load, and it is also significantly stronger.

The initial flat part of the load/elongation curve (Fig. 1) is mainly due to internal movements of the rope components, the later steep part to the actual stretching of the fibres [20, p. 223] A sheath and core rope with a fairly loose sheath and a core of untwisted fibres will behave more like the pure fibres, with a nearly straight load/elongation curve; a hard-laid rope will have a more nearly straight curve than a soft-laid one. The rate of elongation affects this behaviour.

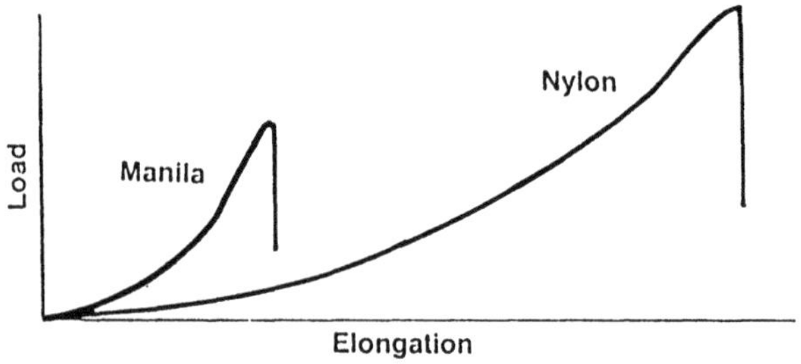

Fig. 1. Typical load/elongation curves for manila and nylon rope, taken up to the breaking point

Slow rates allow more time for the components of the rope to overcome their frictional binding so that the load/elongation curve is more concave, with a prolonged early nearly flat phase. Fast rates of loading, specially the jerk of holding a falling object, may give a slightly convex curve with some complex kernmantle climbing ropes [4]. Consequently for any work comparing different ropes, or comparing rope with and without knots, the testing methods must be carefully controlled.

Rope Breaking Strength

There are several kinds of machine used for testing the strength of rope and knots [15], all originally devised for testing engineering materials fairly early in the industrial revolution. Older machines for the routine testing of rope used either a constant rate of extension (CRE) or a constant rate of loading (CRL), where one end of the rope was clamped and the other was moved either at a constant speed (CRE) or so as to increase the tensile force at a constant rate (CRL). Recent testing machines tend to use a constant rate of traverse (CRT), where one end of the rope is moved at a constant speed and the other responds by moving (more slowly) at a rate depending on the load [15]. The different machines yield somewhat different results. The end fastenings must be stronger than the rope; a wet eye splice usually suffices for natural-fibre ropes, but synthetics have the ends clamped round a bollard. Rates of extension are usually chosen to produce breakage within somewhere around a minute or so. Most studies of the effects of varying the conditions of the test have used the simple criterion of the breaking strength, the maximum load applied at the time of failure.

Small changes of the rate of elongation around the standard rate make little difference to the measured breaking strength, but at slower speeds, the strength rapidly decreases [16], and increased speeds in the usual machines can cause a 20% increase [20, p. 217]. Very low and very high speeds of elongation will be considered later.

While the breaking load varies considerably with the rate of elongation, the amount of elongation at break is nearly constant [20, p. 226][25]. Similarly, old used rope is weaker than new, but the elongation at break is little affected [25]. However, wet natural-fibre ropes may show slightly greater elongation than dry [20, p. 226].

There is a tendency for the breaking strength of rope in proportion to its size (best measured as mass per unit length) to be greater for small sizes than large [25], but inspection of some rope manufacturers' catalogues shows that this effect is neither very big nor very regular.

No rope is truly elastic, the elongation is never strictly proportional to the load; a constant heavy load produces an immediate elongation followed by a slow continuing one (see below under Creep) and when the load is removed, recovery is slow and incomplete, specially with new rope (Table 1). The smaller the rope (Table 1), and the smaller the load as a fraction of the breaking strength, the faster and more complete is the recovery. In another test [20, p. 238], 12 mm diam. manila rope loaded at 50% of its breaking strength recovered 42% of the elongation in 24 hr, but loaded at 10%, it recovered 83%. On the other hand, a nylon rope recovered 78% and 85% respectively. Hard-laid rope shows slower recovery than soft-laid, four-strand than three, wet than dry. Ropes that have been repeatedly loaded are, when unloaded, longer and thinner than at the start; hence, when reloaded, and provided all loads are less than already imposed, they have a reduced elongation for a given load but eventually show complete recovery from this elongation [20, p. 231].

Table 1. ELONGATION AND RELAXATION OF ROPE
Load 20% of breaking strength [20, p238]

	12 mm diam sisal rope	40 mm diam manila/sisal rope
Elongation		
On reaching load	6.0%	14.2%
After 15 min holding load	6.4%	16.2%
Relaxation		
Recovery after release of load		
15 min	62.4%	47.5%
24 hr	70.3%	56.0%
1000 hr	76.6%	58.5%

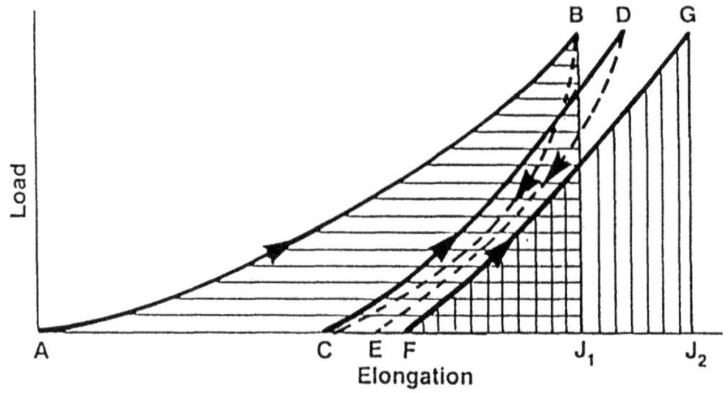

Fig. 2. Load/elongation curves for repeated loadings of a rope. The maximum load imposed was 50% of the breaking strength; shaded areas correspond to the energy absorption of the rope on the first and last cycle

When a rope is tensioned and stretches, the energy imparted to or absorbed by the rope is equal to the area under the load/elongation curve; it may be expressed as metre-Newtons or Joules. As seen in Fig. 1, little energy is absorbed by the rope when the load is light, that is, very little work is needed for the early stretch, but as the load is increased, the energy absorbed or the work needed increases more and more rapidly. Determination of the energy absorption capacity of a rope requires accurate measurements at all parts of the load/elongation curve (see Fig. 2).

A new rope subjected to a large load (e.g., 50% of the breaking strength) shows a curve like AB in Fig 2. The elongation is AJ1 and the energy absorbed by the rope is the area ABJ1. If the load is then released, the elongation at zero load after a short time corresponds to point C. The rope is then subjected to a similar load, and the steeper curve CD is obtained, with less elongation and less energy absorption. If the load is again released, all the elongation is not recovered, and the rope returns to point E. This process can be repeated, with less and less elongation and energy absorption until eventually (usually more than 10 cycles) a stable curve is found, such as FG, and the rope may have about 50% of its original energy absorption capacity FGJ2 [30].

Knot Strength

The relative knot strength, also called knot strength efficiency or knot efficiency, is the breaking strength of a knotted rope expressed as a percentage of the breaking strength of the same rope without a knot. Some of the many measurements that have been made are summarised in Tables 2 to 7. I have entered in my tables only those data for which I am reasonably certain of the identity of the knot (I have given the Ashley [5] number, when available), and either at least two studies have been made on a knot, so that different estimates may be compared (Tables 2, 7) or one person has made measurements on different kinds or sizes of rope, again allowing comparison (Tables 3 to 6). Several published knot names are ambiguous in the absence of an illustration; I have ignored these data.

The first impression of the tables is the wide range of values for some of the knots. In Table 2, for example, the efficiency of the Fisherman's Knot varies from 39 to 81%, of the Bowline from 52 to 78%. In trying to explain this variability we need to look at two things: how reliable the actual values are and whether the different investigators were really looking at the same things.

Table 2. KNOT STRENGTH EFFICIENCIES

Knot and Ashley number [5]	1892 [2] Manila Climbing Rope	1928 [36] Flax cord	1957 [20] Natural Fibre Laid	1967 [34] Nylon Climbing Laid	1970 [16] Manila Laid	1977 [27] Nylon Kernmantle Dynamic	1983 [11] Natural Fibre Laid	1986 [23] Nylon Kernmantle Static	Mean Table 4 Polyprop Laid	Mean Table 5 Kernmantle Dynamic
Knots										
Overhand Knot #519	55%	—	50%	—	45%	—	—	—	—	—
Figure Eight #520	61%	—	—	—	45%	—	—	—	—	—
Bends										
Reef Knot #1402	53%	63%	55%	54%	50%	—	45%	slipped	42%	—
Ring Bend #1412	—	—	—	—	—	60–70%	—	44%	—	65%
Sheetbend #1418	47%	—	—	65%	50%	—	50%	45%	48%	—
Fishermans Knot #1414	62%	81%	—	59%	65%	60–65%	65%	39%	—	61%
Double Fishermans Knot #1415	59%	—	—	—	—	65–70%	—	56%	—	—
Loops										
Bowline #1010	72%	78%	62%	65%	60%	70–75%	60%	52%	57%	70%
Double Bowline #1013	—	73%	—	69%	—	—	—	—	—	—
Fishermans Loop #1038	65%	74%	—	—	—	—	—	—	—	—
Overhand Loop #1009	64%	80%	—	49%	—	60–65%	—	50%	—	67%
Figure Eight Loop #1047	77%	—	—	—	—	75–80%	—	55%	—	—
Hitches										
Clove Hitch #1670	—	—	78%	—	—	60–65%	75%	slipped	54%	65%
Timber Hitch #1665	—	—	80%	—	70%	—	70%	—	67%	—

Reliability of Data

Few investigators give any indication of the reliability of their data. The first people to measure knot strength, the British Alpine Club [2], simply state that their values were the mean of, usually, two tests; Wright and Magowan [36] stated that they made 7–13 tests of each knot, and give values of 1.2–13.2% for the 'percentage variation'; Microys [27] gives a range of values, but does not say how these were calculated (they have obviously been rounded off); and Wildsports [35] state that their values have a 'precision' of ±3%. Neither 'percentage variation' nor 'precision' are standard statistical terms, and their meanings in this context are uncertain. In some of his tests, Barnes [7, p. 136] used six replicates each time; he claimed 'the extreme differences in averages was less than 2%', but also said that he had rejected any 'erratic figures', on the not necessarily correct assumption that the knot had been wrongly tied. Others of his tests had many more replicates [7, p. 123]. But none of these data give any certain guide on how big a difference between values is needed to reach any set confidence level. All other investigators report a single value only, and do not state whether these are results of single tests or means of several. A few anglers' knots have been stated to have 100% efficiency, e.g., [11, p. 86], but in the absence of confidence limits these figures are difficult to interpret. Few investigators give any technical details of their tests; however, I would expect the errors due to machine operation to be much less than those due to the variability of the knot and rope samples and the factors mentioned in the next section. Rope is expensive and the tests time-consuming, so that there are obvious economic limits on the number of replicates of each test that can be made. But there seems little worth in publishing values with no indication of their possible variation or of the differences between values for different knots needed for confidence in the results. Otherwise, all we are saying is that knots weaken rope. In what follows, I take the published figures at face value.

Cogency of Results

It is not certain that all investigators were testing exactly the same things, since the details of the tests were not published. The variables most likely to be important are the way the knot is tied and the load is applied, and the nature of the rope used; sometimes the environmental conditions might have been significantly different.

Knots: One investigator said: [26] 'it was found impossible to tie a Bowline in exactly the same manner, no matter how proficient we were'. Barnes [7, p. 124] said that hours of practice were needed before he could tie his anglers' knots with a uniform breaking strength. No others have commented on any such difficulties. It is important that all knots should be properly dressed, arranged to give all parts smooth curves without twists or kinks, that is, with minimal torque within the cord, and the most direct transfer of internal forces; it has been said [29] that neglecting this precaution may as much as halve the knot strength. All the knot parts should also be optimally packed, tightened to avoid any deformation when the load is applied; this is usually best achieved by working round the knot tightening each emergent end against the knot and repeating until no further motion occurs (complex knots may need to have internal parts worked through to the end). However, care should be taken with very light lines not to overtighten; one recommendation [7, p. 297] was that the tightening tension on knots tied in anglers' line should never exceed 60% of the line's breaking strength, or the line could be greatly weakened. Many knots have asymmetric elements such as twists and turns, which may be 'right-handed' (or 'Z-twisted') or 'left-handed' (or 'S-twisted'); 'right-handed' follows conventional screw or bolt threads. This may be expected to have some effects on the behaviour of the knot, specially when using rope that itself has a handedness, such as the lay. Most investigators ignore this point, but there are exceptions. Day [16] found that a Fisherman's Knot is 20% stronger when the Overhands are tied against the lay of the rope; on the contrary, a Fisherman's Knot with both Overhands of the same handedness was 15% stronger than when they differed [36]. Then Day [16] found that the Sheetbend was equally strong and secure whether the turn was left- or right-handed (this does not refer to what is sometimes called a Right- or Left-handed Sheetbend, depending whether the ends came out on the same or different sides; these may be better called Direct or Oblique knots). Chisnall [15] made a number of comparative tests using improvised methods of loading a light line containing two knots. Full details of these tests are not available, but some results are interesting. Using right-handed line, a Reef Knot tied with the first Half Knot left-handed ('left over right') was stronger than if it was right-handed; a right- handed Ring Bend was stronger than a left-handed; and Single and Double Fisherman's Knots were stronger if the Overhands were left-handed, corroborating Day [16] above.

Fifty or a hundred years ago, mountaineers, specially from Britain, were very emphatic about the value of tying their knots 'with the lay', but appear to have been in some doubt about what that term meant [36]. The most comprehensive study of this topic [36] defined 'right-hand lay rope' in the usual way, twisting as in an ordinary screw, but defined making turns, hitches or twists 'with the lay' when they were made in the direction usually followed by most right-handed people, which in the authors' case generally meant counter-screwwise, called 'left-handed' or 'against the lay' by most knot tiers in the nautical tradition. Most of the discussion is of the Bowline, arguing that it should be tied 'with the lay', defined as with the turn in the direction called 'left-handed' in the nautical tradition, the opposite direction to that found after many tying methods. The authors justify their opinions purely by the way the rope lies, how it packs when tightened, and the supposed likelihood of working itself loose, but neither experimental data nor field experience are cited. This point could do with further study. The Alpine Club [2] reported on two knots of whose exact identity I am unsure, tied 'with' and 'against' the lay; I am unsure of the sense in which they meant these terms. The 'Swiss Loop' had a strength efficiency of 57% against the lay, 71% with it, whereas the 'Openhanded Knot' had efficiencies of 53% and 55% respectively. The way in which the parts of a knot are loaded may also be important. Are the loads on the two legs of a loop equal? Is the load on hitches always at the same angle to the object hitched? Is the Clove Hitch loaded on one end or two? And so on.

Rope: Inspection of Table 2 failed to find any consistent differences in knot efficiency that could be attributed to the kind of rope, except that the thin flax of column 2 had higher efficiency values that any others. On the other hand, Table 3 shows a marked effect, attributed to the differing inherent rigidity of the different fibres and the internal frictional forces [20, p. 243].

Table 3. INFLUENCE OF ROPE TYPE ON KNOT STRENGTH EFFICIENCY
'Overhead Knot' tied in ¾ inch diameter rope [20, p243]

Type of Rope	Laid		Braided	
	Dry	Wet	Dry	Wet
Ramie	58%	94%	53%	84%
Cotton	50%	69%	77%	96%
Nylon	54%	40%	58%	75%
Manila	62%	60%	—	—

Changing the size of the rope changes the knot efficiency somewhat erratically, with a tendency for knots tied in smaller ropes to have lower efficiencies (Tables 4 to 6). The ranking order of knots may also be a bit different in different size ropes. Thus in Table 4, the Overhand Bend has a higher efficiency than the Reef Knot in $\frac{3}{4}$ inch and 1 inch diameter ropes, lower in $\frac{5}{8}$ but equal in $\frac{3}{8}$ and $\frac{1}{2}$ inch rope; in Table 5, a Clove Hitch had a higher efficiency than an Overhand Loop in 4 mm and 5 mm rope, but lower in 7, 9 and 12 mm, while the Overhand Loop was of substantially lower efficiency than the Bowline in 5 mm rope, slightly lower in 4 and 7 mm, and equal in 9 and 12 mm; in Table 6, the Figure Nine End Loop has a higher efficiency than the Double Fisherman's Knot in 9 mm rope, lower in 11 mm. In other work, Prohaska [32] found that the ranking order of the strengths of some loop knots varied with the diameter of the rope used; he did not express his results as efficiencies, so I have not tabulated them. Different sizes of rope are not, of course, simple scale models of each other. For any one kind of rope, the basic fibres are always the same size, whatever the size of the rope. Laid rope and simple braids tend to have yarns of the same size, so that the main differences in different size ropes is in the number of yarns per strand. Sheath and core ropes are more complex; the yarns forming the sheath change in size, but the core may change in several ways, depending on the details of the design. When a rope passes round a curve, as in a knot, the curvature of the rope as a whole, expressed relative to the diameter of the rope, is more or less constant whatever that diameter, but the curvature of the individual fibres, expressed relative to the diameter of the fibre, is much less for large sizes of rope (containing many fibres) than for small (containing few). The effects of all these differences in structure are unknown. Barnes [7, p. 86] found that knot efficiencies were lower in lighter nylon monofilament lines.

Table 4. INFLUENCE OF ROPE SIZE AND TYPE ON KNOT STRENGTH EFFICIENCY
Laid and plaited Polypropylene and double-braided nylon ropes [19]

	Polypropylene							Nylon
	Three-strand laid						Plaited	Double-braided
Knot and Ashley number [5]	3/8 inch diam	1/2 inch diam	5/8 inch diam	3/4 inch diam	1 inch diam	Mean (to Table 2)	3/8 inch diam	1/2 inch diam
Bends								
Reef Knot #1402	36%	44%	44%	43%	41%	42%	37%	slipped
Sheetbend #1418	40%	52%	48%	51%	48%	48%	40%	48%
Double Sheetbend #1434	41%	49%	51%	49%	48%	48%	49%	61%
Overhand Bend #1410	36%	44%	38%	46%	47%	42%	43%	slipped
Loops								
Bowline #1010	52%	61%	62%	55%	56%	57%	53%	88%
Hitches								
Clove Hitch #1671	43%	67%	60%	slipped	45%	54%	61%	73%
Timber Hitch #1665	64%	67%	66%	70%	67%	67%	70%	86%

Table 5. INFLUENCE OF ROPE SIZE ON KNOT STRENGTH EFFICIENCY
Nylon kernmantle dynamic climbing rope and cord [8, p32]

Knot and Ashley number [5]	4 mm diam	5 mm diam	7 mm diam	9 mm diam	12 mm diam	Mean (to Table 2).
Bends						
Ring Bend #1412	53%	62%	68%	59%	63%	61%
Fishermans Knot #1414	56%	66%	71%	62%	68%	65%
Loops						
Bowline #1010	64%	72%	75%	67%	71%	70%
Overhand Loop #1009	61%	60%	72%	67%	71%	66%
Hitches						
Clove Hitch #1670	62%	72%	66%	60%	63%	65%

Table 6. INFLUENCE OF ROPE SIZE ON KNOT STRENGTH EFFICIENCY
Nylon static kernmantle rope [35]

Knot and Ashley number [5]	9mm diam		11 mm diam	
Bends				
Double Fishermans Knot # 1415	78%		88%	
Loops	Mid loop	End loop	Mid loop	End loop
Alpine Butterfly #1053	66%	78%	76%	79%
Figure Eight Loop #1047	58%	72%	63%	73%
Figure Nine Loop (based on # 521)	58%	90%	51%	84%

Table 7. KNOT STRENGTH EFFICIENCY IN FISHING LINE
Monofilament nylon line, diameter not stated

Knot and Ashley number [5]	1951 [7]	1981 [22]	1983 [11]
Bends			
Blood Knot (based on #295)	79%	77%	80%
Water Knot (based on #290)	79%	—	95%
Loops			
Blood Bight (based on #521/1047)	80%	92%	80%
Perfection Loop (#286)	60%	60%	—
Hitches			
Half Blood (based on #300)	>80%	96%	80%
Two-circle Turle (based on #305)	80%	62%	—

The ropes tested in Table 2 seem all to have been fairly close to 12 mm diameter, except for the flax cord used by Wright and Magowan [36], which was only 1/37 inch in diameter. They were using this cord as a model for the manila climbing rope used previously [2]. Knot strength efficiencies were higher in this cord than in any other in these tests. Wright and Magowan said the differences 'may be explained in part by the differences in pliability and roughness, which would greatly affect the nip of the knot, and in part by the great difference in thickness, which may well tell in favour of cord'. They go on to say 'in any case, the discrepancy must be due to some cause which affects all knots in common; in all probability, the relative position of the knots is unchanged'. In fact, by both the tests the Reef Knot had the lowest efficiency, but the other knots appear in different orders. No information is given on the size of the fishing lines in Table 7. The environmental and mechanical conditions (temperature, humidity, rate of elongation) used in the tests in Tables 2 to 7 are not, in general, described, so that their effects cannot be guessed. Table 3 shows the effect of wetting the rope on knot efficiency (I am uncertain of the identity of the knot); Day [16] stated that a Bowline was 50% stronger after soaking the $\frac{3}{8}$ and $\frac{5}{8}$ inch diameter manila rope.

Conclusions

Day [16] said that bends on average seem to be weaker than loops and nooses, which in turn are weaker than hitches; Chisnall [15] said that in general hitches were stronger than loops because in a hitch the standing part passes straight through the knot, with no bight in the load-bearing side of the knot; this is confirmed by some early tests [3] showing that a Round Turn and Half Hitch [5, #1834] and a Fisherman's Bend [5, #1840] were both stronger than a Bowline [5, #1846] when used to fasten a manila rope to a standard pin. Tables 2, 4 to 7 show that these are good generalisations, but with exceptions, even within the few knots tested. The Bowline shares its basic structure with the Sheetbend, but consistently has a higher efficiency (Tables 2, 4); the only difference in practice would seem to be that in the Bowline the load is shared between the legs of the loop, but in the Sheetbend that load is on one cord only. I have seen no extensive accounts of just where the breaks occur in these knots, and these results remain difficult to explain. My overall conclusion from this work is that, if the data may be held to be reasonably reliable, then knot efficiency is not simply a function of the knot, but rather a function of the knot tied in that particular kind and size of rope. Therefore, whenever it is important to choose a knot of high efficiency for any purpose, published figures can only be used as a guide if the kinds and sizes of the rope to be used are similar to those in the tests.

Mechanism of Knot Action: In 1864 the Alpine Club Special Committee [1] claimed that the rope broke at the knot for two reasons: as they cross each other, the parts of the knot are strained suddenly at the point of crossing, and one of them is cut through; and the rope is so sharply bent that the outer side of each curve is much more stretched than the inner, so that nearly all the strain comes on only one side of the rope. One version or another of these reasons are still stated from time to time. Day [16] commented unfavorably on assertions that the breaking strength of a knot depends on the radius of the sharpest curve within the knot and that the outside fibres on a curve are the first to break. Budworth [11, p. 30] claimed that the sharper the curve and the tighter the nip, the point of frictional pressure within the knot where a sharp turn causes the parts to grip each other, then the more chance there is that the rope will break. Blandford [9] stated that the strongest knots have their parts taken in easy curves and use a minimum of them. Chisnall [15] said that knots with greater bight radii tend to be stronger than knots with sharp bends and angles. Ashley [5, #142, 143] quoted one of the 'laws' of knot strength as 'the strength of a knot depends on the ease of its curves' and compared the common method of breaking string by hand with the Bowline Bend [5, #1455], believed to be one of the best of the hawser bends. Both have two cords turning sharply 180° round each other, but one is thought to be weak, the other strong; Ashley concluded that the 'so- called law' did not fit this particular case. An early test [3] of fastenings to a standard pin showed that ordinary hitches, with a sliding knot on the standing part, broke where the standing part met the pin; a Bowline broke at the inward part of the turn of the knot; the strongest fastening was an eye splice. But none of those authors listed knots in order of the magnitude of the curvature or nip, let alone measured them, comparing them with measured efficiencies. And I have found no description of a break actually occurring at the point of greatest curvature or nip within a knot tied in rope. The two Alpine Club Committees [1, 2] said that the rope always broke 'at the knot'; others [5, p. 17][11, p. 30][16][20, p. 243][23, p. 64] that the break occurred where the rope enters the knot; I have not noticed any account of which rope breaks in an unsymmetrical knot like a Sheetbend. Ashley [5, p. 17] said that a rope is weakest just outside the entrance to a knot, seemingly due to the rigidity of the knot; a knot, he said, will be stronger if the nip is well within the structure. Himmelfarb [20, p. 243] stated that as tension is applied to it, the rope is compacted, the compression prevents internal movement of the fibres and the rope begins to act like a rigid bar, and failure occurs where the rope enters the knot. Day [16] speculated that the complex stresses and strains that operate in the rope where it enters the knot are amplified by the rigidity with which the rope is held in place at that point by the knot itself. In his tests on knots tied in gut or nylon anglers' lines, Barnes found that the knots tied in gut broke just outside the knot proper [7, p. 73], as with the fibre ropes discussed above. But if the knots were tied in nylon monofilament, the break tended to be within the knot, at the nip [7, p. 69], the place where a particular crossing first makes the knot secure. Apparently the slippery surface of the nylon and the reduction of its diameter under tension allowed a small fraction of the standing part to be drawn out of the knot, the break occurring at the main nip within the knot [7, p. 130]. Each knot tested was found to have its own specific point of breaking [7, p. 78]. Barnes thought that the loss of strength of the line due to a knot depended on how sharply the weakest coil was bent [7, p. 90]. There seems to be no consistent explanation of where the break occurs at a knot.

Creep

We have all had the experience of tying a rope tightly between two points, only to find a while later that the rope has loosened and requires retightening. The rope has continued to stretch; this effect is measurable after only a few minutes (Table 1). This slow stretch is known as 'creep'.

If a constant load is applied to a rope, such as by suspending a fixed weight, the rope continues to stretch slowly and, if the load is a substantial fraction of the breaking strength of the rope determined as in the previous section, the rope will eventually break. A manila rope given a load of 80-90% of its breaking strength, breaks within a few minutes; with a load of 50% of the breaking strength, the rope breaks within a few days [20, 24, 25]. Some extrapolated figures suggest that, even with a load of 20% of the breaking strength (listed as the safe working load for many industrial applications) the rope would break through creep within a few years [20, p. 242]; in practice, of course, weathering would have weakened the rope sufficiently to cause a break before that. Laid nylon rope is less susceptible to creep; when exposed to a load of 75% of the breaking strength, it took about 10 days to break [25]. It is suspected that the creep effect is influenced more by the ultimate elongation of the fibres rather than the rope structure; manila stretches much less than nylon for a comparable degree of loading and so attains its limiting stretch well before the nylon, and therefore breaks sooner [20, p. 240]. I know of no experiments on the effect of creep on the strength of knots. Could it be that a knot tied tightly in a rope some time before testing would show a lower efficiency? It would be interesting to examine the behaviour of a knot exposed for a long time to a high constant load, as suggested by Chisnall [15], including testing strength and security from time to time and examining what might turn out to be a slow-motion breaking of the rope.

Shock Forces

It remains to consider the effects of a sudden jerk, such as applied by a falling weight. This topic is of particular interest to people who might find themselves falling from a height, with only a rope to arrest their fall; that is, to climbers and other users of life support ropes (see Chapter 9) The earliest tests of this property of a rope that I have found were described by the Alpine Club Special Committee in 1864 [1]. They wanted to be able to recommend appropriate available ropes to their members. They first eliminated all ropes that would not hold a 12 stone (76 kg) weight falling 5 feet (1.5 m) and examined those remaining. They finally recommended a rope which they said would hold a weight of 12 stone (76 kg) falling 10 feet (3 m), or 14 stone (89 kg) falling 8 feet (2.4 m); any rope that would hold 14 stone falling 10 feet was considered to be too thick and heavy for convenient use. No details of the experimental conditions are given, and the results are now difficult to interpret.

It was not until ropes made of synthetic fibres, much stronger and more extensible than natural fibres, were available that climbers felt at all confident that their climbing rope might hold them in a substantial fall. The UIAA, the international association of alpine clubs, sponsored a study by Prof. Dodero of the physics involved in a fall, and eventually devised a standard for climbing ropes described below.

Physics of a Falling Weight held by a Rope

Consider a rigid weight mg where m is the mass and g the force of gravity, attached to an extensible rope of unextended length l fixed at the other end to a rigid anchorage. The weight is allowed to fall freely a distance h before the load is first taken by the rope, at which time the energy in the body can be considered to equal the initial potential energy, mgh.

If the rope is not to break, the energy of the falling weight must be fully absorbed by the rope stretching throughout its length. When the weight stops falling, all its initial potential energy is contained within the rope as a product of the maximum tensile (arresting) force, F_a, which could also be called the shock force or, in popular parlance, the impact force, and the maximum stretch in the rope, S. Thus we can say $mgh = F_a \times S$.

The fall factor f is the ratio of the vertical height fallen to the length of the rope from the weight to the anchorage, that is, $f = h/l$. If the weight is dropped from the anchorage point, the fall factor is 1. If the rope is extended to its full length above the anchorage before releasing, the fall factor is 2. Intermediate values are obtained by starting at some intermediate point, perhaps with some slack in the rope or passing the rope through a directional pulley or the like at some point above the anchorage, limiting the distance fallen for the amount of rope paid out. Fall factors greater than 2 can only be obtained by such exceptional circumstances as snatching the rope close to the falling object after it has already fallen some distance.

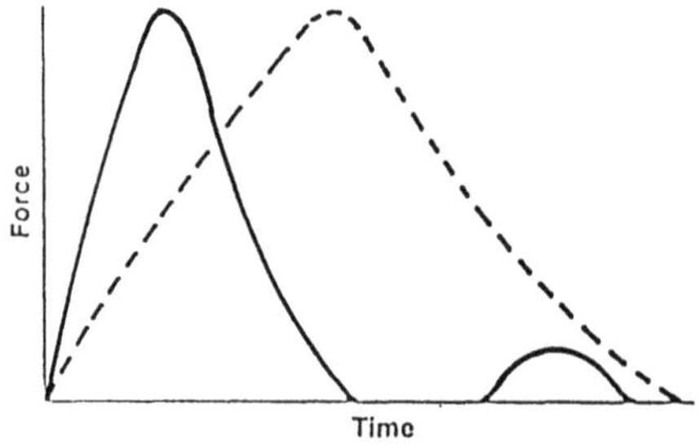

Fig. 3. Duration of arresting forces [8, 18]. A mass of 80 kg was dropped with a fall factor of 1.78. Solid line: rope length 2.8 m, distance fallen 5.0 m; the small curve to the right shows the first oscillation. Broken line: rope length 11.2 m, distance fallen 20 m

It can be shown [33] that, ignoring minor complications,

$$F_a = mg + mg\sqrt{1 + 2(fM/mg)}\ ,$$

where M is the modulus of the rope relating the stretch to the load. That is, the shock force applied to the rope by arresting the falling weight is independent of the absolute distance fallen, h, but is a function of the fall factor, f only. This does not mean that all falls of the same fall factor are equivalent to the falling object. At the same fall factor, the greater the length of the rope l (and therefore the greater distance fallen h), the longer will it take to reach the maximum impact force F_a (Fig. 3) and the longer the time the rope, the anchorage and the attachment to the load will be affected by high forces. If the load is deformable, as is a person, the longer the time a high force is applied, the greater the likelihood of severe damage. Other consequences of longer falls at the same fall factor are a greater absolute elongation of the rope, with a greater chance of hitting the bottom, and a higher speed, with greater damage if any intermediate object is hit.

UIAA Fall Test

The UIAA has devised a standard fall test (Fig. 4). To qualify, a full-sized climbing rope has to withstand at least 5 falls of 80 kg with a fall factor of 1.78, with the maximum shock force of the first fall less than 12 kN. Only the shock force of the first fall must be less than 12 kN; subsequent falls generate higher forces as the energy-absorbing capacity of the rope diminishes, reaching substantially higher values up to the point of failure [18, p. 18][23, p. 77]. A force of 12 kN on a person is only probably survivable, depending on how it is distributed through a harness or the like, but damage could be expected; it is also greater than the breaking strength of some of the smaller anchoring devices. Consequently, climbers take precautions to minimise fall factors, and hence shock force, by placing intermediate running belays at suitable intervals; they are also interested in various kinds of dynamic belay techniques and flowing hitches (see Chapter 9). These are usually intended to limit the shock force to between 2 and 6 kN, considered the best compromise between the force imposed on the falling object and the amount of rope run past the belay [10]. Static kernmantle ropes used in caving, rescue and the like are unlikely ever to need to be called on to meet a high fall factor; they are made to have much less stretch, specially at low loads, than the dynamic climbing ropes. They are therefore tested at lower fall factors, often 1.0 or less, often in a simpler apparatus without the bypass edge.

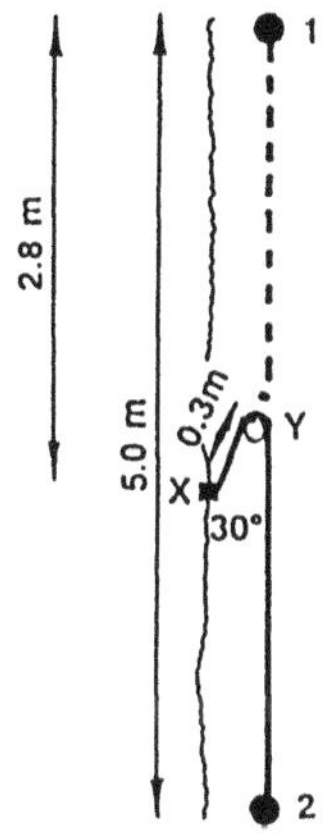

Fig. 4. The UIAA Fall Test. A test weight (1) of 80 kg (for a full-size climbing rope) falls 5.0 m freely to (2), held by a rope 2.8 m long, passing over a bypass edge of 5 mm radius at (Y), 0.3 m from the rigid anchorage (X) and forming an angle of 30°

Knots

Only a few tests of the strength of knots under shock forces have been made; kernmantle ropes, either static or dynamic have been used. Some fall tests have been made on the holding power of friction hitches and flowing hitches and belay devices (see Chapter 9) and will be discussed below in the

Security section. At least in some conditions, knots in a rope may decrease the shock force and increase the number of standard falls sustained, acting as if the strength of the rope was increased rather than decreased. Some climbers use a shock sling to limit impact force [18, p. 52]. This consists of some nylon tapes sewn together with many sequential bar tacks that could each be ripped apart by a fixed force, usually about 3 kN. If a force exceeding that level was imposed, the bar tacks would rip out until the force was reduced. Cannon [12] tested such a device, at a constant fall factor of 1.0, and found that an increase in the distance fallen increased the number of bar tacks ripped: there was a longer time during which the imposed force exceeded the rip point. Fewer bar tacks were ripped if this sling was attached to the rope with a loose knot than if the knot were tight. That is, the loose knot (identity not stated) acted as a shock absorber, reducing the time of exposure to high forces while it tightened. A similar shock-absorbing capacity for knots was noted by Marbach and Rocourt [23], who tested two loop knots tied in the middle of static kernmantle rope subjected to factor 1.0 falls with an 80 kg load. A new 9 mm rope without knots sustained an impact force of 8.7 kN at the first fall and broke at the second. If an Overhand Loop were tied in the middle of the rope, the rope sustained three falls before breaking, with maximum impact forces of 3.7, 5.2 and 6.4 kN. Some used 10 mm rope was similarly tested; impact forces were reduced and there was some increase in the number of falls held. More tests of this kind seem desirable.

Other Tests of Shock Loads

Ashley [5, #68] described a method of testing the strength of knots by subjecting them to a series of jerks 'of increasing force', caused by a weight falling from increasing heights. The tests are not described in sufficient detail to know how the fall factor varied, but this seems unimportant since no results are given. Barnes [7, p. 54] described a 'dynamic shock tester' for monofilament nylon line, in which a weight slid down a 'practically frictionless' wire on to a hinged rod suspended by the test line. The fall height of the weight was progressively increased until the line broke (this is roughly equivalent to a progressively increasing fall factor). The final height fallen is taken as a measure of the dynamic breaking strength of the line. He found [7, pp. 55, 39] that knot efficiencies were more variable and lower in these dynamic tests than in his more usual static tests using a hand pull against a spring balance [7, p. 53].

The Security of Knots

When subjected to sufficiently high loads, knots fail before the main part of the rope breaks. Up till now, we have only discussed failure through breaking, but some knots may fail by capsizing or by slipping; such knots are said to be less secure than others. Close observation of the knots and the load/elongation curves during stretch and fall tests can give information on the slipping of knots. Of course, there are some knots that are intended to slip in controlled circumstances; these include the friction and flowing hitches used with life support ropes and discussed in Chapter 9, and for these one needs to measure the conditions providing slip.

It is to be expected that the nature of the cordage and the dressing and packing of the knot would play an even larger part in determining the security of a knot than in determining strength, but little attention has been paid to these factors in the work reported here. Most observations on the security of knots have been qualitative, reporting that slippage etc has occurred, rather than quantitative, reporting on the magnitude and direction of the pull needed to make that slippage etc. occur.

Capsizing a knot is changing its form, rearranging its parts, usually by pulling on specific ends in specific ways. Few knots capsize spontaneously in normal use, but special circumstances can cause many knots to capsize. For example, one way to untie a Reef Knot is to capsize it to a Bucket Hitch [11, p. 48], but this can also happen when a load is applied in appropriate (or inappropriate!) ways. Bowlines and Sheetbends can be tied by inserting a running end in an Overhand Noose [5, #1788, 2562]; once tied, the process can be reversed, and the knots can be manipulated back to an end through a noose, which slips; this is postulated as explaining at least one fatal fall in climbing [28]. A Carrick Bend [5, #1439] is often tied in the flat form which is then stabilised either by seizing the ends (sometimes done for anchor cables) or by capsizing into a bulky but quite strong form, perhaps the most common way of applying the knot for other than decorative purposes. I do not know of any systematic studies of capsizing as a mechanism of failure of knots in use, but many studies of failure do not mention the mechanism.

Perhaps the first study of slipping in knots was made in 1907 [21]. Unfortunately, details of the experiments are not given, the cordage used was admitted to be non- uniform, and it is not always possible to identify all the knots tested with certainty. The major conclusion that can now be drawn is that the working ends of the rope outside the knot should not be too short, in order to allow for a little slippage.

Ashley [5, #63] tested a number of knots tied in a special mohair yarn when looking for a satisfactory knot for use in weaving some upholstery fabrics using particularly slippery yarns. He devised an apparatus to apply a fluctuating load and counted the number of cycles withstood before failure (no knots broke in his conditions). Some of his results [5, p. 273] were that, on average, a Granny Knot [5, #1442] slipped undone after 3 cycles, a Reef Knot (#1441) after 19, a 'Left- handed' (Oblique) Sheetbend (#1432) after

15, a 'Right-handed' (Direct) Sheetbend (#1431) after 22, a Double Sheetbend (#1434) after 36, a Carrick Bend (ends on the same side, #1428) after 20, a Carrick Bend (ends opposite, #1439) after 70, an Overhand Bend (right-handed in left-handed yarn, #1558) after 23, or (left-handed in left-handed yarn, #1557) after 33, a Fisherman's Knot (#1414) after 43; a Ring Bend (#1412) withstood at least 100 cycles. Ashley called attention to the effects of handedness of knot and lay in the Overhand Bend and how the ends lead out in the Carrick Bend. He claimed that the security of a knot seemed to depend solely on the nip of the knot, the spot within a knot where the end is gripped and thereby made secure, and used as a major argument the behaviour of the Direct and Oblique Sheetbends. He said that the Direct Sheetbend has a good nip and does not slip easily, while the Oblique has a poor nip and is unreliable. Yet the one slipped at 22 cycles and the other at 15, which I should have thought not all that very different, and not all that good; and I can recognise no real difference in the nips of the two knots.

Both Ashley [5, p. 16] and Day [16] denied what they called a common misconception, that to be secure, adjacent parts of a knot should tend to move under load in opposite directions. This does apply to a Reef Knot (not very secure) but not to many other, more secure, knots.

Day [16] in the course of standard strength tests found the Oblique Sheetbend insecure when tied in manila rope, but Chisnall [14], using a variety of more or less improvised tests, found that the Oblique form was more secure than the Direct when tied in Kernmantle and some braided ropes. He also [15] found that, using right-hand-laid rope, a Reef Knot tied with the first Half Knot left-handed ('left over right') was more secure than when the first Half Knot was right-handed.

An interesting example of insecurity is found in the Figure Eight Bend (as [5, #1410], but using a Figure Eight Knot, [5, #524]). When subjected to a heavy load, the knot flypes, that is, turns 'inside-out', perhaps several times if the working ends are long enough, before reaching the breaking point [6]. Because the knot is so little liable to jam in cracks or under ledges when the knotted rope is drawn over rock, some [6] think the knot has valuable applications; others [12] disagree, finding inadequate security.

In many fields, natural fibres have been largely replaced by synthetics for rope. But many knots are less secure when tied with the synthetics; this has often necessitated the introduction of modified or new knots (see, for example, the climbers' knots discussed in Chapter 9). Monofilament nylon fishing lines are more slippery than gut; anglers have often met the problem by increasing the number of twists or turns in the knot. Thus the Blood Knot had $1\frac{1}{2}$ turns in gut [5, #295] but $4\frac{1}{2}$ turns in nylon [11, p. 88]; the Water Knot and the Fisherman's Knot used single Overhands in gut [5, #296, 293] but multiple knots in nylon [11, p. 90].

Mountaineers, cavers, rescuers and other users of life support ropes (see Chapter 9) use friction hitches to provide a movable grip on a rope, and flowing hitches to allow controlled movement of a rope past a point. Prohaska [31] devised a simple test for friction hitches, to see if they would support at least 200 kg, in order to find out how many turns were needed in these knots under various conditions.

Tests using shock loads have been used to study the braking effects of various flowing knots and belay devices (some described in Chapter 9). Experimental details are lacking in some tests [18, p. 53] but fuller descriptions are given by Chisnall [13, p. 378], who used dynamic kernmantle rope and measured the impact force on the belaying devices while they held (with controlled slippage) 81 kg loads, fall factor 1.5, bypass angle 90°, and by Dill [17] who used static kernmantle and simulated conditions used in rescue work: a load of 200 kg, fall factors 0–0.93 (mostly 0.33), maximum impact force allowable 15 k. These results are of considerable interest to many users of life support ropes, but they are specialised and dependent on the actual conditions of use so that I do not discuss them further here.

Jamming of Knots

Many knotting books report that some knots are more prone to jamming than others, that is, that after use they become particularly difficult to untie. This tendency to jam is important in selecting knots that are expected to endure heavy loads yet require frequent tying and untying. It seems that no one has solved the problem of how to test that property reproducibly and objectively; anyhow, no one seems to have reported such tests. This is a pity.

Conclusions

Most of the tests described above consider brand new rope, with or without knots, subjected to loading in one of three ways: a constant load for a long time (creep), simulating suspended loads; a steadily increasing load (the standard method of measuring breaking strength) simulating overloading a winch; or a shock force simulating a falling body. Most ropes that fail in practice are not new, but have already been subjected to a variety of loads and, quite often, to some amount of abrasion and weathering; the force that actually causes the break is often a small jerk on top of a heavy pull. The laboratory tests can be expected to give only a rough guide to what happens in the field.

There is still a lot to be learnt about the behaviour of rope and knots under various circumstances. There are so many discrepancies between the findings of different workers that it seems to me necessary to get more reliable data before advancing too many or too detailed hypotheses to explain the findings. The data should be obtained from tests in which the experimental conditions are carefully described, so that others may repeat the work exactly; and an adequate number of replicates are made, so that confidence limits may be calculated. The position and nature of the final break should be described. Without all this, the data have minimal meaning.

Acknowledgements

I thank Tim Lamble and Roy Belshaw for their helpful comments on a draft of this paper.

References

1. Alpine Club Special Committee, 'Report on rope, axes and alpenstocks', *Alpine Journal* **1** (1864) 321.
2. Alpine Club Special Committee, 'Report on equipment for mountaineers', *Alpine Journal* **16** (1892) Supplement.
3. Anon. 'Manila rope fastenings', *Engineering Record*, **70** (1914) 706.
4. Arova-Lenzberg AG, 'How old is your rope?' *Off Belay* No. 20 (1975) 16.
5. C. W. Ashley, *The Ashley Book of Knots* (Doubleday Doran, New York, 1944).
6. R. Baillie, 'The Side Figure Eight Knot for rapelling', *Explore Magazine Technical Series*, No. 2 (1980s).
7. S. Barnes, *Anglers' Knots in Gut and Nylon.* (Cornish, Birmingham, 1951).
8. C. Benk and G. Bran, *Edelrid Guide to Mountaineering Ropes* (Edelrid, Isny, 1980).
9. P. W. Blandford, 'A better mousetrap', *Knotting Matters* No. 37 (1991) 4.
10. British Mountaineering Council, *Ropes* (BMC, Manchester, 1991).
11. G. Budworth, *The Knot Book* (Kingswood, Elliott Right Way Books, 1983)
12. N. Cannon, 'Air voyagers', *Climbing* No. 88, (1985) 54.
13. R. Chisnall, (ed) *Rock Climbing Safety Manual* (Ontario Rock Climbing Association, 1985).
14. R. Chisnall, 'Sheetbends"oblique versus direct"', *Knotting Matters* No. 38 (1992) 5.
15. R. Chisnall, 'A few notes on testing knot strength and security', *Knotting Matters* submitted 1994.

16. C. L. Day, *The Art of Knotting and Splicing* (US Naval Institute Press, Annapolis, 1970) 26.
17. J. Dill, 'Are you really on belay?', *Response!* Summer, Fall 1990.
18. Edelrid-werk, *Ropework* (Edelrid, Isny, 1988).
19. H. A. Ewing, 'Evaluation of "Polytie" 3-strand polypropylene rope', *Hydro-Electric Power Commission of Ontario*, Research Report 1973.
20. D. Himmelfarb, *The Technology of Cordage Fibre and Rope* (Leonard Hill, London, 1957).
21. J. Lehman, 'Systematik und Geographischen Verbreitung der Geflechtsarten', *Abh. u. Ber. d. K. Zool. u. Anthr.-Ethn. zu Dresden* **11**, (1907) No. 3, p. 26.
22. D. Lewers, *Fishing Knots and Rigs* (Reed, Sydney, 1981).
23. G. Marbach and J-L. Rocourt, *Techniques de la Spéléologie Alpine* (Techniques Sportives Appliquées, Choranche, 1986).
24. R. P. Mears, 'The climbing rope defined', *Alpine Journal* **57** (1950) 325.
25. R. P. Mears, 'Experience with climbing ropes', *Alpine Journal* **58** (1951) 81.
26. I. Meredith, 'Climbing rope tests', *American Alpine Journal* **12** (1960) 191.
27. H. F. Microys, 'Climbing ropes', *American Alpine Journal* **21** (1977) 130.
28. *Off Belay* No. 50 (1980) 38.
29. A. Padgett and B. Smith, *On Rope* (National Speleological Society, Huntsville, 1987) 35.
30. W. Paul, 'Determing rope safety', *Response!* (Spring, 1983) 9.
31. H. Prohaska, 'Ein Klemmknoten für Bandschlingen', *Die Alpen* (Okt, 1982) 232.
32. H. Prohaska 'The double bight bowline', *Summit*, **34**, No. 3 (1988) 22
33. A. Wexler, 'The theory of belaying', *American Alpine Journal*, **7** No. 4 (1950) (seen in [27]).
34. W. Wheelock, *Ropes, Knots and Slings for Climbers* (La Siesta Press, Glendale, 1967).
35. *Wildsports Catalogue* (Wildsports, Sydney, 1993).
36. C. E. I. Wright and J. E. Magowan, 'Knots for climbers', *Alpine Journal* **40** (1928).

CHAPTER 11

A HISTORY OF TOPOLOGICAL KNOT THEORY

Pieter van de Griend

The subject is a very much more difficult and intricate one than at first sight one is inclined to think.
Peter Guthrie Tait, 1876.

1. Introduction

In this Chapter, *knot theory* will be used as a generic term to signify what mathematicians often distinguish into two separate theories, the one concerned with n-links and the other dealing with braids. To a mathematician a *knot* is a single closed curve that meanders smoothly through Euclidean three-space without intersecting itself. An n-link is composed of n of such components, which may link and intertangle but not intersect each other. This affords a simple and intuitive picture, capturing the most essential aspects of a real-life knotted structure. The mathematical concept of a braid will be treated in a later section.

Any account of the history of mathematical knot theory inevitably will be fragmented over many aspects of the fields across which the subject stretches. Combinatorics, topology and group theory are but a few of these fields. In this exposition I have chosen to give a broad outline which touches upon the main conceptual developments as seen in an historic perspective, in which the more formal theoretical developments feature in the background. Knot theory has now become a subject in its own right, which has grown by leaps and bounds (sometimes in quite unexpected directions) along a multidisciplinary front. This causes it to have a sparkling history, involving a wide diversity of ideas, methods and applications, and linked with the names of many famous mathematicians and scientists.

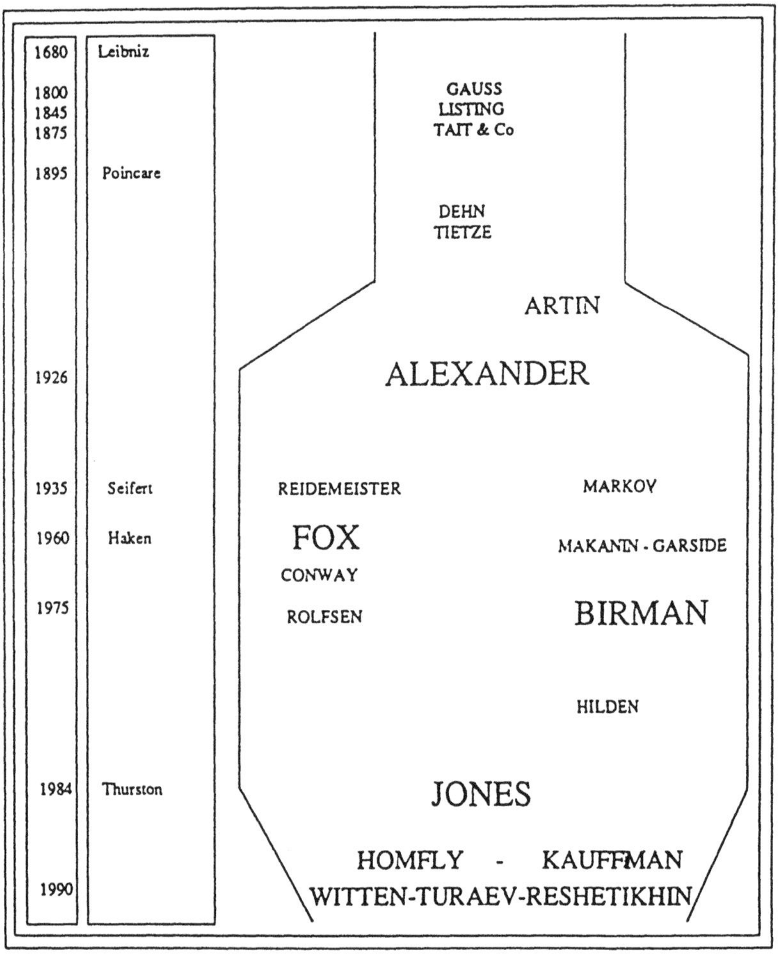

Fig. 1. A Chronology of Topological Knot Theory

From primitive man to present-day scientist, rudimentary knot theory can be traced from the construction of knots and braids for decorative purposes to, what once was held to be, a rather esoteric and deceptively tranquil branch of pure mathematics. However that is a misleading image. From its inception as a proper mathematical discipline, in the second half of the 19th century, knot theory has been associated with many projects on the frontiers of fundamental scientific research. Although much of knot theory is mathematically abstruse, it has found important application in fields as diverse as atomic modelling, quantum physics, theoretical psychology [82] and molecular biology [95].

The table of (Fig. 1) gives an approximate chronology of the chief discoverers and developers of Knot Theory. The column to the far left indicates the time scale, in years. The second column gives some of the great names in topology. The split column to the right is used to indicate the historic placings of the greatest names in Knot Theory, and also to symbolise its dual nature: the left branch covers mathematical knot theory, whilst the right branch caters for the mathematical theory of braids.

2. The Very Early Days

Throughout history many different interpretations of the phenomenom *knot* have been proposed. In this section we shall consider knots to be those structures which can be realized in a knottable medium, such as a length of rope. In order to show how modelling of the simple act of tying a knot progressively translates into amazingly complex mathematical machinery, we shall first be concerned with questions about *the awareness of mathematical problems associated with knots.* In this context *prehistory* is defined to be the interval of time before the recording of data or information in formalized mathematical ways began to take place.

Although knot theory exists nowadays as a highly specialized and concrete set of mathematical ideas, its origins are not easily traced. Yet somewhere in a far and distant past, inklings of those ideas must have been born. Unfortunately, in a mathematical perspective, the theory's tremendously dispersed history prevents them from being localized with certainty. One may be inclined to believe that there was no theorizing about knots in the very early days of Mankind. Surely there was none in any of the stringent ways we now use to model the phenomenon knot. Nevertheless, by making certain assumptions, plausibility for which is provided by results from anthropological and psychological studies concerning knots [26], [82], one is able to make statements about the roots of the theory. Hence some tracking of the subject's gradual evolution is feasible.

The discovery and use of knots would seem to predate those of fire and the wheel by countless aeons. Knots were used long before the practice of mathematization began to influence the thoughts and actions of Mankind. To primitive Man even the simplest of knots would pose vexing and crucial problems. Yet, it is doubtful whether he would analyze them in any degree other than what was required to employ them as practical tools in his struggle for survival. In order to comprehend their workings, his mind would model knotted structures by translating their relevant spatial, topological and mechanical properties into some kind of logical framework of the mind. We could suitably describe the result, achieved through the cognitive processes of transferring the most obvious properties of knots into a mental model, as *intuitive knot theory*. The most important aspects of intuitive knot theory, being structure and its transformation properties, also constitute the main ingredients of contemporary knot theory. How knots and their workings manifest themselves in the real world was of absolute importance to primitive Man. In many instances his life would literally depend on that kind of vital knowledge. Often certain symmetries determine a knot's ability to operate either in a desired manner or utterly to fail. A point in case is provided by the pair consisting of the Reef Knot and Thief Knot (Fig. 2) which are shown below and whose respective behaviour depends on subtle symmetry properties.

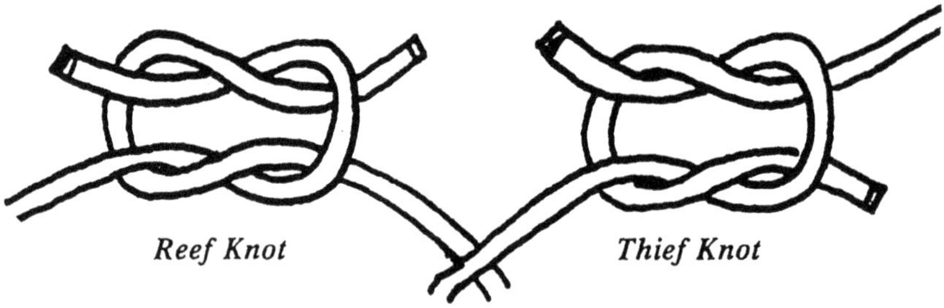

Fig. 2. A Reef Knot and a Thief Knot

To primitive Man, the often incomprehensible and erratic workings of knots were attributable to Divine intervention. The mysterious workings of these topological machines led him to endow them with supernatural powers, placing them into regions of superstition and metaphysics. Such attitudes can still be observed today, with magic knots being found on amulets worn to protect, or bring luck to, the bearer. For similar reasons, knots occasionally were put to decorative uses, in ceremonies for the worshipping of divinities. They were also used for more mundane pastimes, such as in the finger games of cat's cradles, examples of which are to be found in many cultures. The mathematical prehistory terminates at different times in different places. As a rule, this may be said to occur as soon as artefacts, or remnants thereof, for the various cultures come forth.

Knots have been used to represent numbers in different ways. One such numerical application of knots is found in the quipus employed by the Peruvian Incas, a people who used sets of knotted strings for administrative purposes during the better part of a thousand years [11], [53]. The knots themselves functioned only as symbolic and mnemonic devices; but their arrangements on various lengths of string, connected in ordered, meaningful ways, would encourage deeper mathematical thought. The Incas were aware that their quipu knots (Fig. 3) could not transform themselves (without Divine intervention!); and they were so tied and arranged that cheats could not tamper with them, without considerable difficulty. Thus their bookkeeping was assured of consistency and safety.

In order to employ knots in such a fashion, further demands by their accounting systems would relate to problems of structure recognition. The knot-properties they exploited thus concerned structural stability and mutual structural distinctness. Luckily the favoured Overhand Knots possess quite reliable character in both respects.

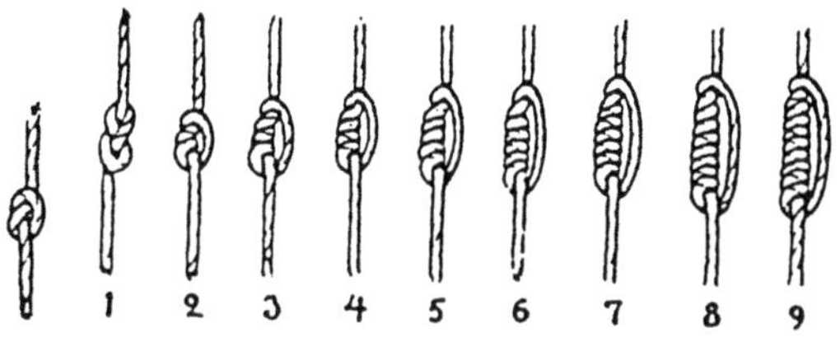

Fig. 3. Quipu Knots

To use knots for decorative purposes, it would be natural to draw pictures of them. This would be an initial step away from intuitive knot theory, as pictures are a first stage of abstraction. Furthermore, drawings entail a process of geometrization, which brings things down to two dimensions. It must be emphasized that this process is not a deliberate attempt to resolve conceptual problems about knots, but arises as a side effect of their application to art. There have been many so-called primitive people that drew fascinatingly complex curves which can be readily recognized as projections of knots. For example, the Bushoong and Tshokwe people in the Zaire-Angola-Zambia region in southwest Africa traced, and their descendants still do trace, complicated and regular figures in the sand. Their unoriented curves, lacking crossings with any distinct parity, are not in any sense knotted and are more akin to graphs. Although their activities have an underlying mathematical base, they seem to be unaware of it [11, p. 34–37].

Evidence of mathematical ideas which tend towards a more Occidental understanding of scientific study concerning knots' planar geometric properties, is discernible in the work of Celtic scribes, work carried out a thousand years ago.

The Celts produced diagrams of knots in which, when following a fixed direction along the curve, the line makes a succession of over-passes and underpasses which alternate regularly throughout the whole curve. We now call the special type of knots which project into such a picture *alternating knots.*

The Celts made extensive use of such pictures (Fig. 4), for decorative and presumably religious purposes. We surmize that their features had to symbolize a number of things. The line would represent time, or possibly life. In the case of a closed curve, the knot's periodicity might relate to the regularity of seasonal changes, with the alternating aspect symbolizing night and day. They seem to have been aware of the non-trivial fact that an alternating knot could be made to correspond with any simple closed planar curve. Their desire to draw such knots posed geometrical problems. This contributed to the process of mathematization of their worldviews, because they had to discover *how* to geometrically create the truly knotted curves and zoömorphics which they employed to adorn surfaces [12], [24].

Fig. 4. Celtic Knotwork

In a sense the foregoing examples of diagrammatic representations of knots and their uses are like an overture. They witness of relatively primitive mathematical thought, and were described in order to illustrate the transition from intuitive knot theory, lacking any apparant formalism, to a vestigial form of the subject. The introduction of a kind of planar geometry was doubtless not directly an attempt to *understand* knots. The geometry of the Celts is of an essentially different kind from Euclidean, but nevertheless it involves elusive properties like transformation and symmetry. The awareness of such problems posed refined demands, requiring the development of new ideas in mathematics.

3. The Birth of Knot Theory

The subject's next steps were related to spirals and closed intertwined curves, and were mainly a German affair. As far back as 1679 Leibniz, in his *Characteristica Geometrica*, tried to formulate basic (geometric) properties of geometrical figures by using special symbols to represent them, and to combine these properties under operations so as to produce other properties. He called his study *Analysis Situs* or *Geometria Situs*, and it comes closest to what we now would call *Combinatorial Topology* [57], the discipline in which geometrical figures are considered as aggregates of smaller building blocks. Leibniz did not go so far as to study knots; but his endeavours at finding a geometry of this kind, different from the only one known at the time, predated other work in this direction by more than half a century.

Although thinkers like Leibnitz recognized the need for different geometries, it was not until 1771 that the birth of knot theory occurred. In that year Alexandre Theophile Vandermonde (1735–1796) wrote a paper [90] (see also [30]), in which he specifically places knots into the arena of the geometry of position. In the opening paragraphs, Vandermonde includes the lines:

> *Whatever the twists and turns of a system of threads in space, one can always obtain an expression for the calculation of its dimensions, but this expression will be of little use in practice. The craftsman who fashions a braid, a net, or some knots will be concerned, not with questions of measurement, but with those of position: what he sees there is the manner in which the threads are interlaced.*

The possibility for a mathematical study of knots was probably first recognized by the truly great mathematician and physicist, Carl Friedrich Gauss (1777–1855), of Göttingen, Germany. One of the oldest notes found amongst his papers after his death was on a sheet of paper dated 1794, which bore the caption *A Collection of Knots*. It contains thirteen sketches of knots with English names written beside them. It is probably an excerpt he copied from an English book. With it are two additional pieces of paper with a few more sketches of knots. One is dated 1819, the other some eight years later [31]. Notes of Gauss referring to the knotting together of closed curves appear in his collected works [38]. During the period of 1823–1827 he was working on *Geometria Situs* about which he later wrote, on 22 January 1833:

> *Eine Hauptaufgabe aus dem Grenzgebiet der Geometria Situs und der Geometria Magnitudinis wird die sein, die Umschlingungen zweier geschlossener oder unendlicher Linien zu zählen.**

His work on electromagnetism had led him to compute inductance in a system of two linked circular wires; and he introduced the concept of winding numbers (or linking numbers), which are now a basic tool in knot theory and other

*[One of the main tasks in the borderland between Geometria Situs and Geometria Magnitudinus will be to count the 'windings around' of two closed or infinite lines.]

branches of topology. One result, which he gave without proof (after the quotation just given) is the following integral:

$$\iint \frac{(x'-x)(dydz'-dzdy')+(y'-y)(dzdx'-dxdz')+(z'-z)(dxdy'-dydx')}{((x'-x)^2+(y'-y)^2+(z'-z)^2)^{\frac{3}{2}}} = 4m\pi$$

where m is the number of 'windings around' (*Umschlingungen*), and the integral extends over both curves.

Another note of special interest, recorded in December 1844, gave numerous forms which closed curves with four knots can exhibit.

These mere snippets represent Gauss' known researches relating to knots; further mention of his knot work may be found in Stäckel [81]. One can only surmize what further thoughts this genius may have had, and what results gained, on the nature and properties of knots.

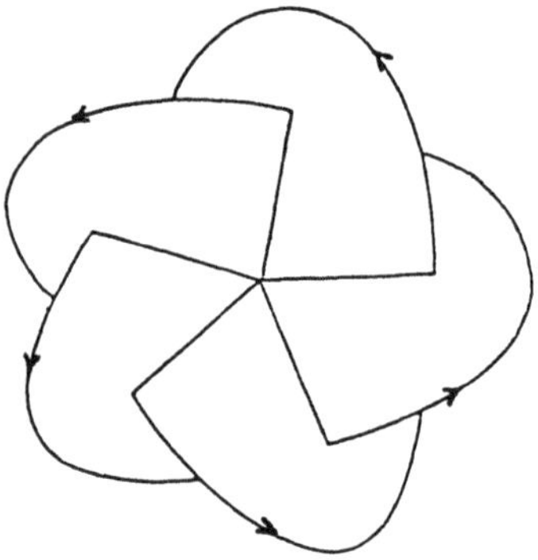

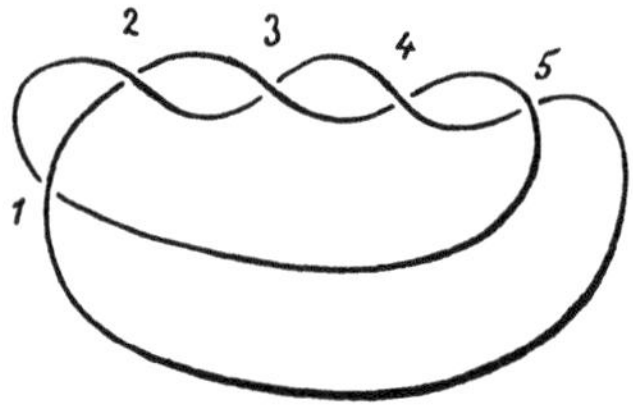

Fig. 5. A Knot by Otto Boeddicker

By contrast with all his other fields of interest, Gauss was not a very active researcher in topology. It has been alleged that Schnürlein, a pupil of Gauss, carried on intensive research with his help, on the application of higher analysis to topology; but no one has been able to verify this. On the other hand, Otto Boeddicker's work from 1876 is with certainty an independent continuation of Gauss' work. In his inaugural dissertation Boeddicker discusses at considerable length the value of the above-mentioned integral [18]. He later expanded this work to illustrate the connection between knots and Riemann surfaces [19, p. 316]. A diagram from one of Boedicker's papers is shown in Fig. 5.

In any case, Gauss' knotting attempts made him (Gauss) conscious of the semantic difficulties continually to be found in topological studies. This placed him among the first to display, and encourage, deep scientific and mathematical interest in knotted structures. Gauss certainly led some of his students to study the intricacies of topology. Fortunately one of them, Johann Listing, was inspired to pursue vigorously the quest for knot knowledge. He thereby secured for himself a name amongst the founders of the subject. Through his work, which we describe next, the roots of the family tree of modern knot theory are firmly anchored in nineteenth century mathematics.

4. Johann Listing's Complexions-Symbol

Johann Benedict Listing (1806–82) was a student of Gauss in 1834 who later became professor of physics at Göttingen. His topological researches eventually led him to publish some of his work on knots in an essay entitled *Vorstudien zür Topologie*, in 1847 [62]. In this work he discussed what he preferred to call *the geometry of position*, but since this term had been reserved for projective geometry by von Staudt, he used the term *topology* instead. This became the collective name for the mathematical disciplines which study the more general concepts of geometric structures. Listing, even though he carried out quite considerable work on the subject, seems to have published a mere fraction of these researches.*

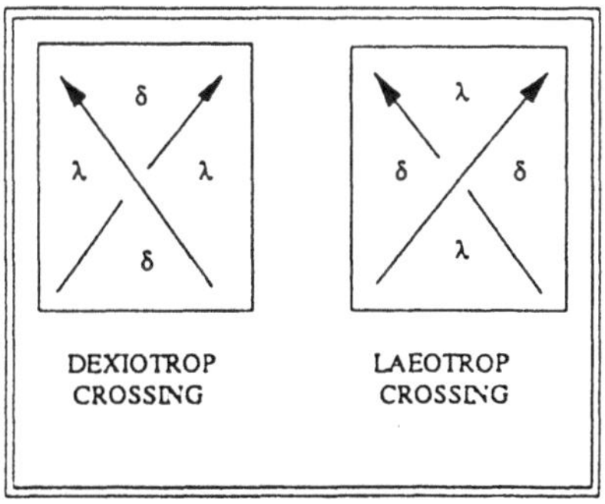

Fig. 6. Handedness of Oriented Crossings

In his 1847 publication he considered the handedness of spirals and discerned between their ability to be either left- or right-handed, which he termed respectively *dexiotropy* and *laeotropy*. Planar projections of these spirals led

*Letter by Listing to the Proc. Roy. Soc. Edinburgh (1877). p. 316.

him to introduce the concept of *handedness for an oriented crossing* (Fig. 6). To each type he assigned a certain symbol distribution consisting of δs and λs. This is illustrated below. The orientation becomes insignificant after the regions have been assigned a type.

From some simple experiments with two- and three-stranded braids and their closures, he came to consider the possibility of listing and classifying all knot projections having fewer than seven crossing points.

Listing was the first to persist in representing *knots* as knotted circles, and obtaining diagrams by projecting these onto a plane. By attaching symbols to the crossings in a diagram, to indicate their types, and considering the resulting symbol distribution in each of the diagram's regions, he was able to propose an 'invariant' for a knot. In general an *invariant* is a mathematical expression (it may be just a number) which carries information about a system, and whose values do not change when the system is transformed in some defined way. An invariant in knot theory is generally an expression derived from a knot diagram which depends solely on the knot or link under consideration, in any of its forms, and not on any particular picture of them. The easiest invariant to visualize, but one which is not very useful for distinguishing between knots, is the number of components in an n-link L. By definition it equals n, and remains so whatever continuous deformations L is subjected to.

Invariants are useful aids in the classification of knots for the following reason. Suppose we compute the value of a particular invariant from two knot diagrams, and obtain two different values. Then we can conclude that the two knot forms from which the diagrams were obtained are different knots. However, the converse is not true; diagrams having the same invariant value may or may not come from the same knot. A perfect knot invariant, which always takes the same value for any particular knot, and a different value for any other knot, has yet to be discovered.

Listing concocted his invariant as follows. He called a region of a knot diagram *monotypical* if all the angles on the region's boundary had been assigned the same type-symbol (that is, were all δ or all λ); in which case, the region was itself given the same type-symbol. If a mixture of δs and λs had occurred, he called the region *amphitypical.* In this way, Listing typified each region, including the unbounded one (which he called the *amplexum*).

He defined a diagram to be *in reduced form* if it had a minimal number of crossings, over all possible diagrams obtainable from the knot. He knew that if one or more regions were amphitypical in a diagram, then the diagram would not necessarily be in reduced form; but he gave no methods for reducing the numbers of crossings in such diagrams.

He proposed an invariant for knots which have a monotypical diagram in reduced form; and he gave it the name *Complexions-Symbol.* Later, we shall give an example which shows that it is not a true invariant.

Briefly, a pair of polynomials are computed from the diagram, one in the 'variable' δ, and the other in the 'variable' λ. The exponents on the terms of these polynomials correspond to the numbers of sides surrounding the regions in the diagram: thus, for example, suppose that there are 5 regions, each having 3 sides and bearing the symbol δ; then the term $5\delta^3$ will appear in the *δ-polynomial* of the Complexions-Symbol.

A full example will clarify the matter. The extraction of the Complexions-Symbol from a diagram of a specific 7-crossing knot is shown below (Fig. 7). Note that all the regions in the diagram are monotypical.

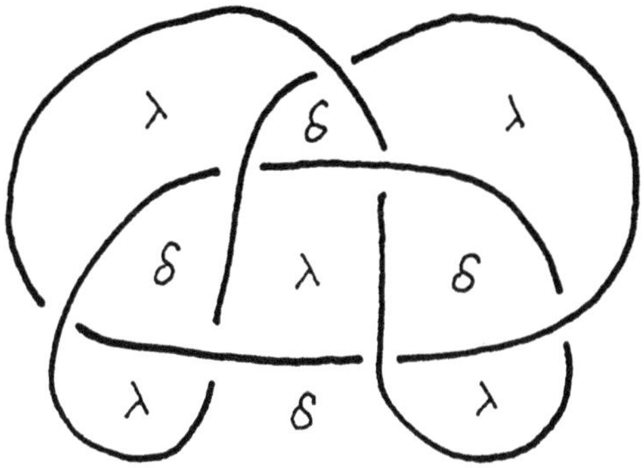

Fig. 7. A 7-crossing knot, with labelled regions

In this case there are four δ-regions, of which three are 3-sided and the unbounded region adjoins five sides. There are five λ-regions, of which two pairs are respectively 2- and 3-sided, while the remaining one is 4-sided. The pair of polynomials for the knot would be shown by Listing thus:

$$\left\{ \begin{array}{c} \delta^5 + 3\delta^3 \\ \lambda^4 + 2\lambda^3 + 2\lambda^2 \end{array} \right\}$$

In general, a Complexions-Symbol has the form

$$\left\{ \begin{array}{c} a_0\delta^n + a_1\delta^{n-1} + \ldots + a_{n-1}\delta^1 \\ b_0\lambda^n + b_1\lambda^{n-1} + \ldots + b_{n-1}\lambda^1 \end{array} \right\}$$

where the coefficients a_i and b_i, with $0 \leq i \leq n-1$, indicate the numbers of the various kinds of region occurring in the diagram.

Listing noted that if a term with exponent unity existed in either of the two polynomials, that term would derive from one or more simple twists in the diagram. These could be removed by 'untwisting'; and so such term(s) could be dropped from the Complexions-Symbol. Note that Euler's polyhedral formula can also be used to check the calculations; the sum of the coefficients in both polynomials is equal to the number of regions in the diagram, which equals $(n+2)$ by Euler's formula.

Listing's Complexions-Symbol has several serious defects. First, it is not defined for the *Unknot*, when projected into a diagram having no crossings. Nor is it defined for non-alternating knots, all of whose diagrams must have at least one amphitypical region; examples of these occur first among knots of eight crossings. The Prussian Heinrich Weith aus Homburg von der Höhe, following up Listing's work, noted this in 1876 [93, pp. 15–16], and gave a diagram of a non-alternating knot to prove the point.

Johann Listing himself noted that occasionally the so-called invariant proved not to be invariant at all! To illustrate how this can happen, we give below an alternative diagram for the 7-crossing knot used above; the Complexions-Symbol derived from *this* diagram is clearly not the same as the one obtained above.

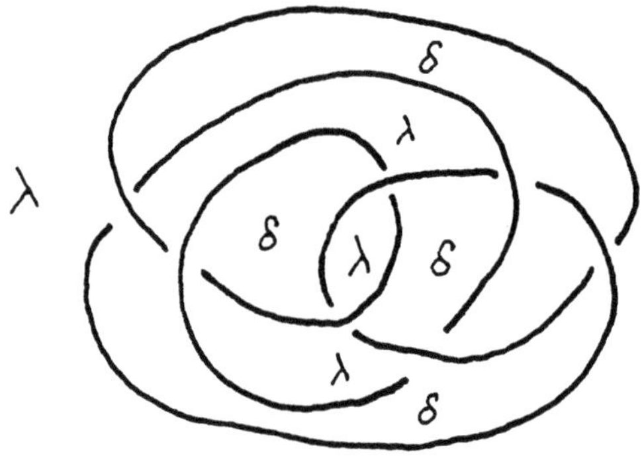

$$\left\{ \begin{matrix} 2\delta^4 + 2\delta^3 \\ \lambda^4 + 2\lambda^3 + 2\lambda^2 \end{matrix} \right\} \neq \left\{ \begin{matrix} \delta^5 + 3\delta^3 \\ \lambda^4 + 2\lambda^3 + 2\lambda^2 \end{matrix} \right\}$$

Thus a single knot can give rise to two different Complexions-Symbols.

Any hopes that Listing might have had that his 'invariant' would be a complete invariant were destroyed by P. G. Tait's finding of two distinct 8-crossing alternating knots which both had the same Complexions-Symbol [84, p. 326]. These two knots, (numbers 8_{11} and 8_{12} from the listings by Reidemeister and Rolfen [74] and [77]), and their Complexions-Symbol, are shown in Fig. 8.

Summarizing the contribution of Listing to Knot Theory, from this brief description of his work, we can see that he established a basis for the mathematical study of knots, working with the natural tool of the diagram of a knot projection. He saw the need for invariants which would help distinguish between different knots; and he proposed one, to be computed from a knot diagram in terms of his crossing-type symbols.

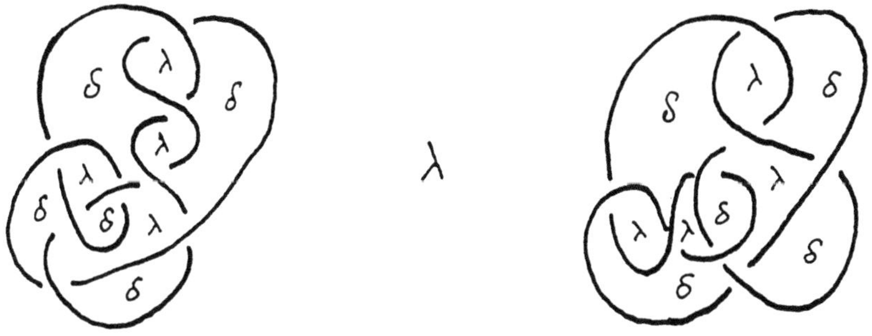

$$\left\{ \begin{array}{c} \delta^5 + \delta^4 + \delta^3 + 2\delta^2 \\ \lambda^5 + \lambda^4 + \lambda^3 + 2\lambda^2 \end{array} \right\}$$

Fig. 8. Tait's two different knots with the same complexions-symbol

Although his Complexions-Symbol had too many serious defects for it to be of much use as a knot invariant, it posed a challenge to other workers coming into the field: namely, to find good invariants. As such, it was perhaps Listing's greatest contribution to Knot Theory. The major quest in twentieth century knot research has been for ever stronger invariants.

Before describing this further work on invariants, however, we must mention other work that was done in the final thirty years of the nineteenth century. In particular, we shall describe the monumental achievements of P. G. Tait and his collaborators; and they demand a section of their own. Passing mention will first be made of several other items related to knot research.

H. Weith, in his inaugural dissertation which elaborated on Listing's work, elegantly showed that there is an infinite number of different knot forms [93].

About that time, there was a curious connection made between knot theory and psychic research. The mathematician Felix Klein appears to have observed that no ordinary knot can exist knotted in a space of four dimensions; one can always use the extra dimension in order to untie it, without, of course, cutting the string [56], [83].

This proposition would be experimentally testable if only we had access to a fourth dimension. Several very reputable and distinguished scientists, including J. C. F. Zöllner from Leipzig, began conducting knot experiments which involved psychic mediums [98]. If the mediums were able to untie closed knots, without cutting the string, it would be a reasonable conclusion that they had some kind of access to a fourth dimension. In the account of Zöllner's investigations, a record is made that this kind of experiment was successfully carried out in December 1877, by Mr. Slade, an American medium.

Some time later, Henry Slade's psychic abilities were proved to be fraudulent. It was shown that the phenomena he produced in the experiments were achieved by trickery. Below [Fig. 9] we reproduce an illustration, by Zöllner, which shows the Overhand Knots produced by Slade's conjuring.

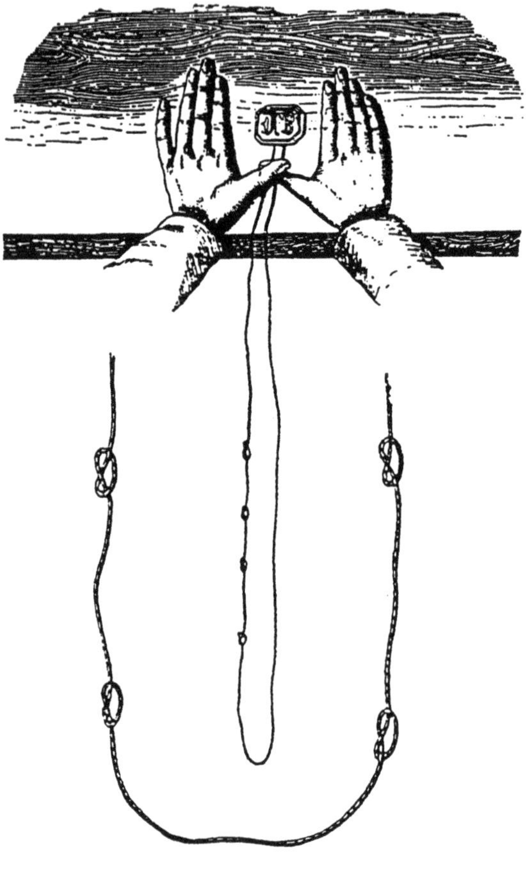

Fig. 9. Overhand knots on Zöllner's sealed cord

There were some publications in which a mathematical explanation was given for 'conjuring' knots from apparently thin air. The first, by Oscar Simony, was based on a prior discovery by Augustin Ferdinand Moebius (1790-1868), that if one makes three or more likewise-handed twists in a flat strip, and pastes the ends together, then on cutting along the centreline of this one-sided surface a knot or knots may occur [80]. In 1890 Friedrich Dingeldey, then a professor at the Darmstadt technical university, published a more rigorous account, which also gives a detailed overview of the early history of topology [30].

In 1897 Hermann Brunn observed that any knot has a projection with a single multiple-point [20]. This proposition became attributed to James Alexander, some 25 years later.

In this period André Hurwitz did some work on Riemann manifolds, which were first steps leading towards theories of braids.

This completes our summary of German work in the field around that time. We must cross the Channel to England to continue our story of the history of knot theory.

5. The Work of Peter Tait

Further substantial progress in knot theory was not made until Sir William Thompson (1824–1907; Baron Kelvin of Largs) announced his model of the atom, his 'vortex theory'. Thompson began writing about this concept in the mid-1860s. He believed that all material matter was caused by motion in the hypothetical ether medium, which he termed *vortex motion* [99]. How did knots relate to these ideas? Scotsman Peter Tait (Dalkeith, 1831–1901) was a close associate of Thompson. In 1867, having been greatly taken by Helmholtz's papers on vortex motion, Tait devised an apparatus for studying vortex smoke rings in which the rings underwent elastic collisions exhibiting interesting modes of vibration. The experiment gave Thompson the idea of a vortex atom. The imaginative picture painted by the theory he developed subsequently was one of particles as tiny topological twists, or knots, in the fabric of space-time. The stability of matter might be explained by the stability of knots; their topological nature prevents them from untwisting. His aim was to achieve a description of chemistry in terms of knots. More specifically, Thompson wanted to produce a kinetic theory of gases, a theory which could explain multiple lines in the emission spectrum of various elements. A swirling vortex tube would absorb and emit energy at certain fundamental frequencies: linked vortex tubes would explain multiple spectral lines. In short, he believed that the variety of chemical elements could be accounted for by the variety of different knots. The main advantage of Thompson's model was that its indivisible bits would be held together by the 'forces' of topology. This construction would avoid the problems inherent in devising forces to hold together an atom made up of little billiard balls.

The theory was taken seriously for quite some time, and even eminent scientists like James Clark Maxwell stated that it satisfied more of the conditions than any other hitherto considered. In fact, in retrospect one could add *transmutation* to its merit-list. The ability of atoms to change into other atoms at high energies could be interpreted as the cutting and recombining of knots.

Even though Thompson's vortex theory of atoms stood for only about two decades, Tait's fascination for knots had been aroused. Thus the study of vortices stands as the starting point of a highly important pioneering study of the topology of knots [87].

The vortex theory led Tait immediately to the problems at the heart of the subject: with insufficiently developed mathematics to help out, knots and links could not be characterized. As already indicated, the accessible work on knot models was very scattered and fragmentary; and results on knots had usually been arrived at almost simultaneously, and independently, by mathematicians ignorant of each other's work. At that time, it was not even clear (to Tait) whether or not there were finitely many knots. Therefore his first self-appointed task, which gradually became his main occupation, was one of enumeration, in which he tried to *find* and *classify* knotted structures. Tait called this *the census problem.* The main thrust of his work was how to find all possible (distinct) knotted structures which can be represented by plane diagrams of continuous curves having n crossing-points. He studied ways by which such diagrams could be 'reduced', a reduction corresponding to a topological change in a knot which led to fewer crossing-points in the plane diagram. He called the minimal number of crossing-points achievable for a diagram of a given knotted structure, the degree of knottiness of that structure. Tait gave several methods for making the reductions. During his attempts to develop these ideas, he followed (roughly) two lines of combinatorial attack. The first involved the development of a method which he called the *Scheme-method*, one which seems to have been known to Gauss [86, p. 13]. The second attempt he came to develop was partially inspired by information gleaned from Johann Listing's work. He termed it the *Partition-method.*

5.1. The Scheme-Method

The central problem, in Tait's case, was to find the combinations of symbol sequences which would encode a connection relation for n points in the plane. He introduced a tool, for which he coined the name *scheme*, which was basically a symbolic shorthand for a connected graph. For ease of demonstration, we shall describe this method in reverse.

Given a diagram D of a 1-link L, with at least one crossing. Choose any crossing from which to start. Call it A. Move from A in either of the two possible directions. Traverse the knot diagram and name the crossings encountered in the odd places respectively B,C,D and so forth until all crossings have been assigned a letter. Traverse the structure over again from any starting-point and jot down the sequence of crossing-points visited. This symbol-string, along with the respective crossing-parities (i.e. 'over' and 'under' crossing symbols), encodes the structure. These symbol-strings, however, were of no significance to Tait. Very much like Listing before him, at first he only considered knots which were represented by alternating diagrams (i.e. diagrams where 'over' and 'under' crossings alternate throughout, in a traverse around the string). He did so because he erroneously believed that all knots, without changing their minimal number of crossing points, could be made to fit such a diagram [13].

An example will illustrate the procedure. For the knot with 5 crossing-points, as displayed below, we find the following scheme.

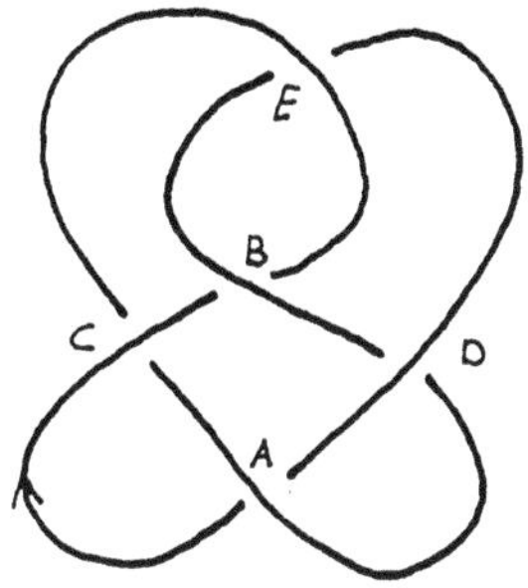

$$ACBECADBED|A$$

The $|A$ indicates the starting point of the sequence. Note that the letters A, B, C, D, E occur at the crossing positions 1,3,5,7,9, respectively as met during the knot traverse.

As said, Tait's procedure was based on a reversal of this process. Thus he tried to find, empirically, all sequences which could yield such schemes. It is easy to see that this task would quickly lead, with increasing values of n, to a massive undertaking. An upper bound for the number of schemes, when n crossings are involved, can be obtained by solving the following combinatorial problem:

> *How many arrangements are there of n letters, when A cannot be in the first or last place, B not in the second or third, and so on, and the sequence must have length 2n?*

With the help of Cayley and Muir it was soon found that the number of combinations rises sharply, into thousands, even with a modest number of crossing points [32]. To complicate matters: many combinations do not even represent knots. The hopelessness of manually finding knotted structures by means of the scheme method was realized. It was abandoned in favour of a more effective one, to be described next.

5.2. The Partition-Method

This was also based on graph properties of knot diagrams. An n crossing-point diagram will have $(n+2)$ regions, which can be denoted by R_i, $0 \leq i \leq (n+1)$. Crucial to the method is the observation that for any region R_i the number of corresponding sides (edges of the graph) s_i has to fulfill the inequality: $2 \leq s_i \leq n$. It is known that the total number of sides in the graph equals $2n$. These facts enable one to partition the number n and, via use of polyhedrons, to arrive at graphs which can be assigned a crossing-coding, thereby yielding representations of knotted structures. This method eventually was made to work well. It was also the method which Tait developed further still, after sharing ideas with the Reverend T. P. Kirkman and C. N. Little, an American professor. They had independently pursued the same lines. Their communications led to a happy collaboration, which resulted in the trio collectively listing virtually all alternating knots up to 11-fold knottiness.

5.3. The Final Results

What did Tait and his collaborators eventually achieve? They found 82 types of knots of 9 or less crossings. An especially remarkable achievement was their work in the class of 10-fold knottiness. A mammoth undertaking which, with their tools, took them 6 years to complete; it resulted in some beautiful tables of 10-crossing knot diagrams. Little continued the struggle, and published the results of his attempts on 10-fold knottiness in 1885 [63]. They were finally able to resolve, too, a large number of the alternating 11-crossing knots. Kirkman provided Little with a manuscript of 1581 polyhedral drawings, from which he distinguished 357 different knot-types.

It is impossible to summarise adequately, in a few paragraphs, the extent of Tait's contributions to the birth of Knot Theory. His researches affected all aspects of the subject. He empirically discovered a great number of useful results, while experimenting with many ideas which future researchers would take up. He worked on the so-called *Gordian number*, which is the minimal number of crossing-point changes required in a knot to produce an unknot. He made some pertinent conjectures which were not resolved for well over a century. He had already considered knots such as Moebius braids, and he both toyed and toiled with problems relating to symmetries such as mirroring. He found a nice little theorem on amphicheirality (a knot is said to be amphicheiral if it can be topologically transformed into its mirror image). He was very much aware of the difficulties which symmetries in mirroring brought along. He introduced a Scottish verb, *to flype*, to denote an operation which can be carried out on certain portions of knots or their diagrams. In fact, one of his foremost contributions was to introduce nomenclature of this kind into knot studies [58].

In order to develop the subject rigorously, however, he needed to discover some form of knot invariant, which would help him to distinguish and identify knot types. He gave no formal proofs that any of his methods actually came to define or to implement one.

The main underlying problem which confronted Tait and his co-workers was deciding when two knotted structures were isotopic, i.e. telling whether either of them could be deformed, by a continuous transformation, into the other. Two knots or links are said to be the *same*, or *isotopy equivalent*, if they can be made to look exactly alike by pushing and pulling, but not cutting, the string(s) in which they are realized. This problem of isotopy became established as the central problem in knot theory, and it became known as the Knot Problem. It was not to be dealt with satisfactorily until the advent of algebraic topology.

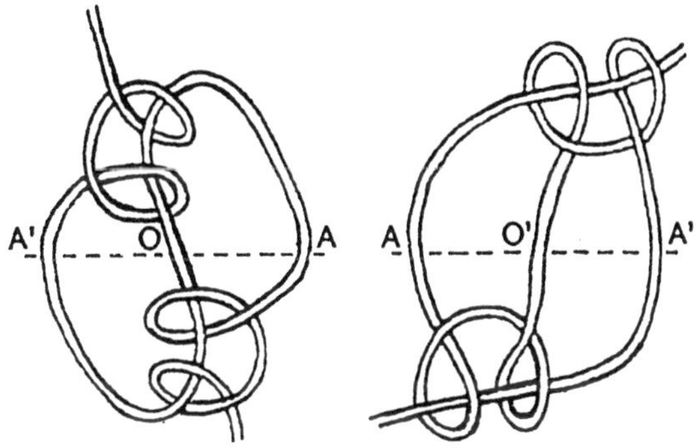

Fig. 10. Knots from Mary Haseman's dissertation

Some thirty years after Tait's endeavours, Mary Gertrude Haseman tackled amphicheiral knots of 12-fold knottiness (Fig. 10). The results of her brave expedition into the then uncharted regions of 12-crossing knot-projections are presented in the charming dissertation [43], which gives a census of amphicheirals in that class of knottiness.

6. The Beginning of the 20th Century

By 1900 there were almost-complete tables available listing knots of up to 11-fold crossings. They represented the fruits of the arduous labor by Tait and his collaborators, and of physicists who had been working in an atmosphere of 'applied mathematics'. Their work had provided sufficient concepts, terminology and knot-diagrams to enable the development of a formalized theory to begin. Tabulating knots had two goals. Completeness of the list was the first. Distinctness of all tabled structures the other. Generally speaking the first goal could be achieved via (cumbersome) combinatorics. The second required methods for dealing with problems involving isotopy; and for that, new mathematical methods were required. Powerful knot invariants had to be discovered.

As the emphasis in the theory of knots turned away from enumeration, under the awareness of the problems due to isotopy, the hunt for good knot invariants began. The ensuing period of transition showed great quantitative and qualitative differences in knot research, as compared with the early and rather empiric work in enumeration. In fact the changes effectively caused the census problem to cease to be the theory's major research topic for the next six decades; although its importance continued to be recognized. Its goals and achievements served as testing grounds for new invariants and other important tools which began to be discovered, in algebraic topology, group theory and other mathematical fields. Much mutual interplay took place between the old and the new approaches.

Knot theory as attempted from the purely topological side became possible only after the development of the required mathematical machinery. This was pioneered by Henri Poincaré (1854-1912) around the turn of the century. Poincaré was a professor at the university of Paris, and a leading mathematician of his day. It has been claimed that he was the last man to possess a universal knowledge of mathematics and its applications. His prime motivation for mathematical research invariably sprang from scientific problems. He was the first person to make a systematic and general attack on the combinatorial theory of a special type of geometric figures called *complexes*. Due to this work he is usually regarded as the founder of combinatorial topology. He decided that a systematic study of the analysis situs of general or n-dimensional figures was not only desirable but also necessary. After some notes which appeared in the *Comptes Rendu* of 1892 and 1893, he published a basic paper in 1895; this was followed by 5 lengthy supplements running in various journals, appearing in the years up to 1904. He did not regard his work on combinatorial topology as a study of topological invariants, but rather a systematic way of studying n-dimensional geometry. However, the influence of his work on subsequent knot theory was to reverse these priorities—the study of topological invariants came to the fore.

Poincaré introduced a number of topological tools, such as the so-called fundamental group of a complex, also known as the Poincaré group in his honour; it was the first in the string of homotopy groups [72]. It came to play a role of utmost importance in topology.

In his efforts to distinguish complexes, Poincaré came to introduce torsion coefficients, and a method for computing Betti numbers of an n-dimensional complex. These concepts are defined as follows. Given a finitely generated Abelian group A, it can be written as the direct product of a free Abelian group F and a family of cyclic groups A/H_i, where each A/H_i is a finite cyclic group of order h_i and such that $h_1|h_2|\ldots|\#A$. The rank of the free Abelian part F and the uniquely defined numbers h_i are invariants of the group A, and completely determine its structure. If A is the homotopy group in dimension, say d, then the rank of F is the d-th Betti number, and the h_i are the torsion coefficients. They are numerical invariants of isomorphism classes of finitely generated Abelian groups. The rank of F is used to calculate the Euler characteristic.

It is interesting to note that Poincaré used only methods of continuous mathematics at the beginning of his series of papers; but by the end he relied heavily on combinatorial techniques. This was not without impact on the newly founded schools that formed to take up and develop his ideas. For the next 30 years researchers concentrated almost exclusively on combinatorial and algebraic methods.

The belief in the power and aptness of combinatorics ran deep. The knot problem's solution demanded a formal definition of a knot, which in true combinatorial spirit became a set of straight arcs making up a closed non self-intersecting polygon in space. Max Dehn and Poul Heegaard in their article [29] in *Encyklopedie der Mathematischen Wissenschaft* in 1907 noted that the knot problem could be formulated entirely in terms of arithmetic, i.e. combinatorics. However this kind of reduction seemed to be of no practical value, nor did it seem to have any theoretical consequences (e.g. for decidability of knot equivalences). There are many natural numerical invariants of knots which may be defined quite easily, such as the already-met number of crossing-points, the Gordian number, the maximal Euler characteristic and so on; but difficulties in computing them by solely combinatorial techniques seem to be inversely proportional to the ease in defining them. There is something general about this matter. There is for instance, to date, still no known algorithm for finding the minimal number of crossing points for an arbitrarily given n-link. In fact there seems to be no hope for finding this number with any tool at all! On the other hand, a recent attack on the Gordian number has yielded good bounds for it (1994). A good account of this work, by William Menasco and Lee Rudolph, can be found in [67].

The first successful algebraic topological invariant attached to a link L was the fundamental group of the 3-manifold, which is constituted by the link's complement in 3-space, namely $(\mathbf{R}^3 - L)$; this invariant is sometimes called *the group of the knot-complement* or, simply, *the knot group*. For an arbitrary link L, and with reference to a basepoint p in L's complement, the knot group is denoted by $\pi_1(\mathbf{R}^3 - L, p)$. This group is one in which the elements are homotopy classes of (unknotted) loops which traverse the complement space of a knotted structure, starting from and terminating at the basepoint p. The binary operation for the group is the composition of two loops, carried out by concatenating them at p. Since this composition is non-commutative, the fundamental group (knot group) is non-Abelian.

The fundamental group expresses in algebraic language some of the topology of the knot-complement, which makes it possible to compare different knots by comparing their algebraic descriptions. A knot's complement, which is three-dimensional, carries a richer topological structure than the knot itself, which is one-dimensional. The topological structure of the complement necessarily contains certain information about the knot. In 1908 Tietze conjectured that it contained *all* such information; an idea that did not become an established fact for 1-links until 1988 [40]. The uncertainties surrounding this conjecture did not prevent this avenue being pursued vigorously; presentations of certain knot groups appeared fairly soon in the literature. General methods for writing down a presentation of the knot group from a knot projection were introduced by Wirtinger, who did not publish them; but he got credit for the idea anyway [65]. Max Dehn, in 1910, also published methods for presenting knot groups.

7. Max Dehn's Work in Knot Theory

It was thought that by considering the knot groups one might be able to classify knotted structures. The initial notions on groups had arisen from 19th-century algebra, analysis and geometry. By the time that Max Dehn began his work on knots, early in the 20th century, group theory had proceeded so far that it was no longer necessary to describe groups by means of their cumbersome Cayley (composition) tables. At the beginning of the 1880's von Dyck had shown how every group is the homomorphic image of a free group, and how one could present such a group by giving so-called *generators* and *defining relations*. Armed with these tools Dehn attacked the knot group.

In his 1910 paper *Über die Topologie des dreidimensionalen Raumes* [27], Dehn discussed a method for extracting a description of the knot group, the so-called *Dehn presentation*. He did so by the following algorithm:

1. List and denote all *bounded* regions of a knot-diagram by $C_1, \ldots, C_n$. These are to be considered the group generators.

2. Every over-crossing yields a relation R_i, $1 \leq i \leq n$ by noting down a relation containing a sequence which is the product of generators as

encountered when traversing clockwise the crossing:

$$R_i = C_{1i}C_{2i}^{-1}C_{3i}C_{4i}^{-1} = 1.$$

In case the crossing includes the unbounded region, the specific generator's place is taken to be unity. Thus a three-generator relation is obtained.

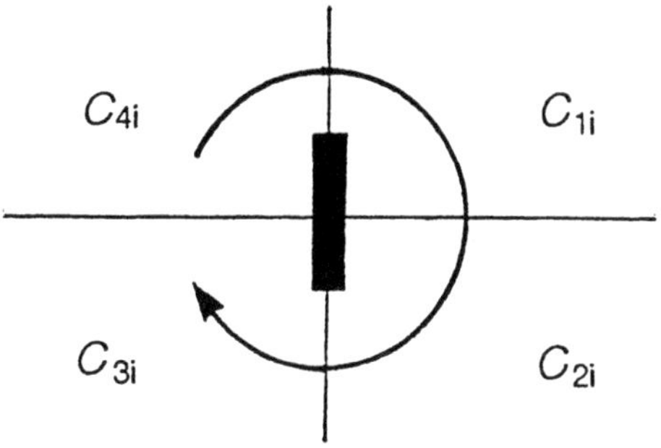

3. The collection of n generators, and n relations, as written below, is a presentation of the knot group.

$$G_k = \begin{cases} C_1, \ldots, C_n \\ R_1 = R_2 = \ldots = R_n = 1 \end{cases}$$

In order to provide an example, we shall apply the algorithm to the right-handed version of the Trefoil Knot, illustrated below.

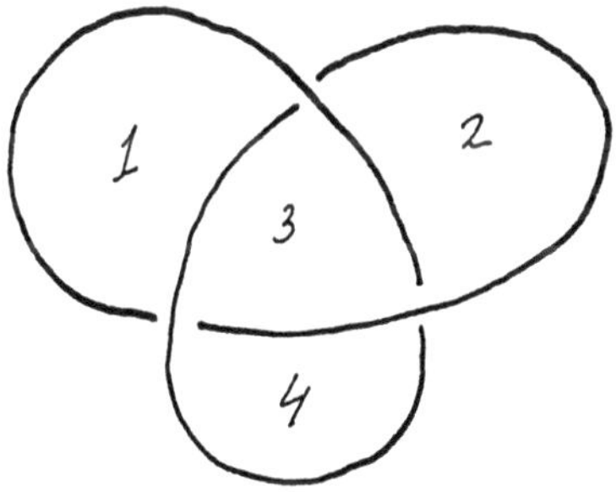

The diagram gives rise to the following three relations. Note that we have omitted the identity **1** in each, as we may.

$$\begin{cases} C_1C_2C_3^{-1} = 1 \\ C_1C_3^{-1}C_4 = 1 \\ C_2C_4C_3^{-1} = 1 \end{cases}$$

These yield the following presentation for the fundamental group of the Trefoil Knot:

$$(C_1, C_2, C_3, C_4 : C_1 C_2 C_3^{-1} = C_1 C_3^{-1} C_4 = C_2 C_4 C_3^{-1} = 1)$$

In fact, this is not the most economical presentation possible: using Tietze operations we can reduce the number of generators to 2, and the number of relations to 1, thus:

$$(C_1, C_2 : C_1 C_2 C_1 = C_2 C_1 C_2)$$

Wirtinger's presentation is derived in similar fashion, but with generators being associated with overpass arcs, rather than with regions (see [25] for details). The end-result is, of course, the same.

Further contributions by Dehn are described in the next section on Alexander's work.

8. James Alexander's Influence

Applications of the fundamental group quickly yielded several breakthroughs. Proofs of the existence of non-trivial knots, via knot groups, had already been given by Tietze as early as 1906. However, the first successes from use of the knot group lay in the verification of the correctness of the knot tables. To achieve this, the group was used with other tools which Henri Poincaré and Enrico Betti had introduced. The proof that Betti numbers and torsion coefficients define combinatorial knot-invariants was first given by James Waddel Alexander (1888-1971), a professor of mathematics at Princeton University and later at the Institute for Advanced Studies. Collaborating with G. B. Briggs and using the torsion numbers, he distinguished all tabled knots up to 8 crossings and all, except three pairs, up to 9 crossings [7]. Alexander also showed that two 3-dimensional manifolds may have the same Betti numbers, torsion coefficients and fundamental group and yet *not* be homeomorphic. His example, of course, involved knot complements. Thus he had shown that a knot contains (at least *a priori*) more information than just its group.

With the tools just introduced, Dehn proved that an arbitrary knot K, its mirror image K^*, and its version with reversed orientation $\tilde{K}$, produce three knot-complement groups which are mutually isomorphic. (Later Reidemeister also proved this, more rigorously.) Using π_1 to denote a knot-complement group, this theorem is stated as follows:

$$\pi_1(\mathbf{R}^3 - K, p) \cong \pi_1(\mathbf{R}^3 - K^*, p) \cong \pi_1(\mathbf{R}^3 - \tilde{K}, p)$$

It was thus realized that the complement alone could not provide complete invariants. The situation was repaired by equipping the complement with an orientation. Let $K \subset \mathbf{S}^3$ be our given smooth knot. By thickening the knot's actual curve to a knotted tube and removing this tube's interior from 3-space we are left with X, the knot's *exterior*. By laying a 'coordinate system' over the tube around the knot, the exterior thus acquires more structure than the complement. The exterior with the coordinate system is called the *peripheral system*. Using additional information from the peripheral system Max Dehn could show by 1914 that neither of the oriented Trefoils (see Figs. 11, 12) is isotopic to its mirror image [28].

Fig. 11. Left-handed Trefoil Knot

Fig. 12. Right-handed Trefoil Knot

He did so by taking one Trefoil, removing it from $\mathbf{S}^3$, reinserting it with opposite orientation, and showing that the result was not homeomorphic to the original knot. This procedure is known as *Dehn surgery*.

The natural question arises as to what extent the peripheral structure is determined by the group alone. It was known at an early date that the Reef Knot and the Granny Knot (see Figs. 13, 14) possess isomorphic groups. Seifert had shown in 1933 that their complements were non-homeomorphic [79]. In 1952, using the peripheral system, R. H. Fox showed that irrespective of the orientations they may have been given, they are two distinct knots [35].

Fig. 13. Reef Knot

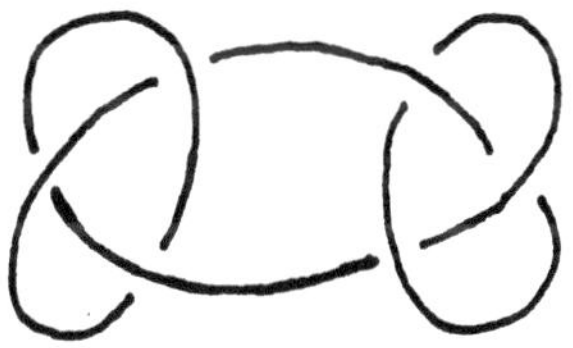

Fig. 14. Granny Knot

The knot group did not immediately fall from grace; but now it was known to be an incomplete knot-invariant. In algebraic topology terminology: the group of a knot determines the knot's complement merely up to homotopy type.

This disturbing example put paid to the generally-held idea that the knot group contained all information about the knot; worse still, it continued to cause trouble over the next few decades. However, despite such examples revealing its shortcomings, the group of a knot was still a powerful invariant. And in the late 1960s the role of the peripheral system was finally clarified; it was shown to be a complete invariant. This demonstration resulted from Waldhausen's work on irreducible, sufficiently large, 3-manifolds, which in turn was based on earlier ideas by Haken [42], [44].

The knot group, even though it was an unwieldy mathematical object, formed the basis for much further research on knots. The new approach via knot groups effectively brought the unknot U into the picture (it was a knot that had been ignored or not taken seriously by early researchers). This knot, a kind of 'limit' in the class of knots, now became an important one in the knot tables, because its group turned out to be a special one, namely the infinite cyclic group with one generator, which is isomorphic to the group of integers under addition. The proof of this was settled in 1956 by the great mathematician Papakyriakopoulos, who also proved that the group of a knot determines the homotopy type of its complement.

Equivalent knots which are projected into distinct diagrams can yield different presentations of their knot group. These presentations must, as theory tells us, relate to isomorphic groups. However, there is no general algorithm which will enable us to decide whether two given representations relate to two isomorphic groups. It is known that no such *general* algorithm is possible. Nevertheless, in working to resolve particular cases, the main efforts in knot research came to concentrate on the problem of finding reduced presentations of knot-groups; in the process, the problem of knot-equivalence was cast into an (algebraic) word-problem mould. The main question became: *When are two presentations of knot groups equivalent?* The complexity of this problem (which is, as already noted, generally unsolvable) led to a quest for simpler invariants, ones more tractable than the knot-group.

This research direction began with a discovery by J. W. Alexander; in 1928 he 'launched' the knot polynomial which was later named after him. It was a totally new idea. He described a method for associating with each knot a polynomial, such that if one form of a knot can be topologically transformed into another form, both will have the same associated polynomial; it quickly proved to be an especially powerful tool in knot theory. For example, the polynomial was able to distinguish 76 knots out of the first 84 in the knot-tables; they were found to have unique Alexander polynomials.

Alexander first obtained his polynomial of a knot K by labelling the regions in the plane bounded by an oriented knot-diagram of K having n crossings. By noting the types of crossing around the knot, in relation to the arc labels, he extracted a certain $n \times n$ matrix (now called the *Alexander matrix*). All of the elements in an Alexander matrix are either 0, or -1, or t, or $1 - t$, where t is a dummy variable or parameter. By removing the last row, and the last column, of the matrix, and taking the determinant of the remaining matrix, a polynomial in t is obtained. This is known as the *Alexander polynomial* of the knot. We may denote it by $\Delta_K(t)$, or simply Δ_K. Alexander was able to show that $\Delta_K(t)$ is an invariant for the knot K (see [6] for full details). In fact, Alexander presented a sequence of polynomials, $\{\Delta_n(t)\}$, with $n = 1, 2, 3, \ldots$, all invariants of the knot K. The first one (case $n = 1$) is the one known as the Alexander polynomial.

Why associate a *polynomial* with a knot diagram? The schemes and partitions which Tait, Kirkman and Little had worked with were unwieldy. Listing's complexions-symbols were not quite what was needed to yield an unambiguous invariant. But why Alexander's polynomial worked the way it did was not clear at the outset. Alexander himself suspected that it was some kind of shorthand for homology groups. A rather reasonable hunch, as later work placed it on a sound homological base.

Alexander's polynomial proved to be a fairly powerful invariant of isotopy in knots. Differently deformed versions of the same knot yield the same polynomial Δ_K. The following comments illustrate a few attractive aspects of the polynomial's behaviour.

Given two prime knots with respective Alexander polynomials, the Alexander polynomial for their knot-composition is given by the multiplication of the two original polynomials. Another aspect almost amounts to a pun: for an alternating knot, Δ_K has coefficients of alternating sign [70]. These, and other more refined pleasant properties, made it knot theory's main tool for almost half a century. The Alexander polynomial's weak points are that it always takes the same value for a knot and its mirror image; and that its power to distinguish between knots terminates for certain example pairs and classes of knots with more than 9 crossings. In 1934, classes of non-trivial knots with trivial Alexander polynomial were discovered [78].

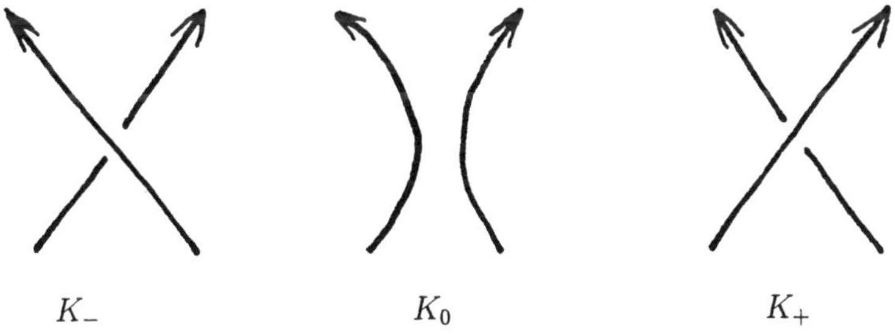

Fig. 15. Set of Alexander crossings

Alexander also discovered a relationship between the polynomials of three oriented knots whose diagrams are identical except within a neighborhood of one fixed crossing where they are as shown in Fig. 15, [6, p. 301]. For further reference the diagrams will be called 'a set of Alexander crossings'.

The relationship is:

$$\Delta_{K_+}(t) - \Delta_{K_-}(t) + (t^{-\frac{1}{2}} - t^{\frac{1}{2}})\Delta_{K_0}(t) = 0, \quad \text{with } \Delta_U = 1$$

Many years later this relation, and others like it (now called *skein relations*), came to have great significance in the development of recursive methods to produce knot invariants. In spite of its early discovery, the literature has shown a remarkable tendency to remain loyal to the calculation of Δ_K by means of determinants. This is rather strange, as this relation bears within it the possibility of calculating $\Delta_K(t)$ recursively by 'untying—or splitting repeatedly—a knot-diagram'.

The idea of unknotting was not born here, though, as Tait had already considered the Gordian number. The relation given above is in a sense deceptively simple. There is no reason *a priori* why it should define any invariant. It may after all depend on things like projections or properties of the plane. Alexander did not find sufficient conditions to give any recursive process by which to obtain his polynomial. He did however prove the well-definedness of his proposed invariant.

The period of Alexander's work can suitably be called one of change and crossroads. The idea that knots could perhaps be understood by studying braids was one of the most promising ones to be pursued at that time. It caused Emil Artin to introduce the braid group, and Alexander to make some fundamental discoveries which bridged the gap between the two theories of knots and braids. We shall discuss these developments later, when we come to focus on braid theory's contribution to the study of knots.

9. Kurt Reidemeister and His Moves

In 1923 Kurt Werner Friedrich Reidemeister (1893–1971) accepted an associate professorship in Vienna where he did research on the foundations of mathematics. In 1925 he obtained a full professorship in Königsberg where his interests went to the foundations of geometry. It is not surprising that he was the person who disposed of many of the basic problems and early difficulties in the field of knot theory. His thorough work covered fundamental treatments of knot enumeration, projections and isotopy. Tait and his collaborators had found many knots, but they had not catalogued them in any workable manner. Reidemeister ordered and numbered them, using a notation which gave their positions in the list and their minimal numbers of crossing points. His notation, and tables of knot diagrams, stood for many years.

In the field of planar knot-projections, Reidemeister studied small, local changes made to a knot and how they corresponded to changes in the diagram obtained by projection of the knot into a plane. He discovered that there were three fundamental changes (and their inverses); they are shown in diagram form below (Fig. 16). The left-hand sketch shows how a loop can be untwisted (removing one crossing); the centre one shows the pulling apart of two flaps (removing two crossings); and the right-hand one shows a portion of string passing over a crossing point (leaving the number of crossings unchanged). These three types of change, together with their inverse changes, are known as the *Reidemeister moves.* It should be evident that none of these changes relates to a change in the topological nature of the underlying knot.

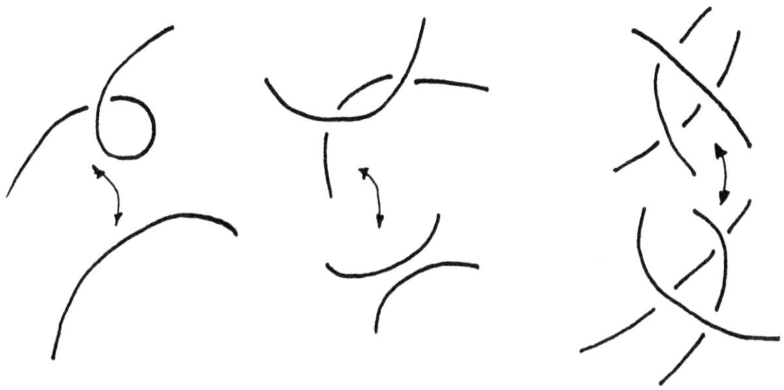

Fig. 16. The Reidemeister moves

The importance of the Reidemeister moves in combinatorial knot theory is embodied in the following key theorem:

> *If two knots (or links) are topologically equivalent, their diagrams can be transformed one to the other by some (finite) sequence of Reidemeister moves.*

It should be noted that in any given case, there are many (indeed an infinite number of) different sequences of the three Reidemeister moves and their inverses which effect a transformation from one diagram to the other.

Reidemeister published the first book*on knot theory, in German, in 1932: an English edition of this book was published in 1983 [74].

Midway during the 20th century the history of knot theory, like much else, was temporarily disrupted by a seemingly global desire to practice politics

*J. B. Listing wrote the first book [62] on topology in 1847; it was dedicated primarily to knot theory. Bernhard Riemann was a student of Listing, and he learned about knots from Listing's book.

with violent means. The influence of combinatorial topology on knot theory declined markedly during this period.

10. The Fifties and Sixties

Ralph Hartzler Fox (1913–1973) was a mathematician who fostered an impressive mathematical environment around his person. Since Alexander's time, Princeton University had been the great name in knot theory's geography; and Fox's extensive publications on the subject made it even greater.

After the interruption in efforts caused by World War II, research came to focus mainly on the knot group, its subgroups and the principal ideals of its group ring. The way to represent a knot group by means of generators and defining relations led Fox to discover a free differential calculus. The ideas behind this calculus caused the Alexander polynomial to emerge as a determinant value of a matrix in which the entries are 'partial derivatives' (in Fox's calculus) of the knot group's relators with respect to its generators. The calculus came to be christened Fox's, and it led to the discovery of links between hitherto unrelated other fields in mathematics. In knot theory itself, it showed that the knot polynomial is determined by the group of the knot, and provided a link between the combinatorial and geometric definitions of the Alexander polynomial. On the practical side, the calculus supplied one more method for calculating Alexander polynomials. Specifically, it became one of the most important tools for studying knot groups defined by generators and relations.

As a person Ralph Fox has left quite an impression. After his death former students dedicated a 350-page book of their research papers to his memory [100]. From his school came people like Joan Birman, whom we shall meet later, and Lee Neuwirth working in knot groups; and Elvira Strasser Rappaport who studied 'knot-like' groups, addressing the question of which groups are knot groups.

Many of the developments in topology during the 1950–1980 period came to affect ideas about knots. Typifying the general development of knot theory and its techniques is that the concept of 'knot', so far treated as a *polygonal non-intersecting curve in 3-space (i.e.* $\mathbf{R}^3$*) upon which certain moves were permitted* became modernized to 'knot' being an *equivalence class of embeddings of the unit circle* $\mathbf{S}^1$ *in* $\mathbf{S}^3$. Topological studies had made it clear that $\mathbf{R}^3$ should be replaced by $\mathbf{S}^3$, in view of compactification properties of the latter. At the end of the fifties this led mathematicians such as André Haefliger and Christopher Zeeman to elaborate upon a theme, traceable back to Emil Artin's work, which considered mappings $\mathbf{S}^n \to \mathbf{S}^m$, for which $m - n = 2$ [41], [97]. These mappings 'tied the n-dimensional unit sphere into a knot' in the m-dimensional unit sphere. The objectives of this higher-dimensional, generalized knot theory included classification of knots with respect to isotopy. A difficulty was that the construction of these knots could not be visualized by simple-minded drawings of knot projections. Classification, and hence finding invariants, had therefore to be coupled to construction methods, showing how the invariants were realizable.

The more formal demands on smoothness of mappings brought in the notions of *tame* and *wild* knots. A knot is *tame* if it is equivalent to some polygonal knot; otherwise it is *wild*. The distinction was of vital importance; the principal invariants of knot type, namely the elementary ideals and the knot polynomials, were not necessarily defined for a wild knot. Knot theory was largely confined to the study of polygonal knots, and it was natural to ask what kinds of knot other than these were tame. An early theorem, and partial answer to this question was: *If a knot parametrized by arc length is continuously differentiable, then it is tame.*

There are infinite classes of wild knots, and their study forms a field of its own within the topological theory of knots.

11. John Conway's Tangling

As we have seen above, the problem of distinguishing knot-types for given numbers n of crossings, and tabulating them, was first tackled by the three men Tait, Kirkman and Little, in the final fifteen years of the 19th century. They succeeded in resolving the problem, by largely empirical methods, for $n = 3, \ldots, 10$ and for most of the alternating knots on 11 crossings.

There the matter rested for some seventy years, until John Horton Conway devised entirely new methods for studying knots, based on a construct called a *tangle*. Essentially, a *tangle* is a portion of a knot-diagram from which a number, usually four, free-end strands emanate; an example (Fig. 17) is given below.

Fig. 17. An example of a Conway tangle

Conway gave a notation for describing knots in terms of their construction from tangles: using this notation, he was able to give rules for determining equivalences between knots. His methods were much simpler than previous ones, and lent themselves to programming for computer analysis. In a 1970 paper [23], Conway presented these ideas, and also listed knot-types in his notation for the following: all the prime alternating and nonalternating knots with crossing numbers $n = 3, \ldots, 11$; the 2-links up to $n = 8$; the 2-, 3-, and 4-links for $n = 9$; and all links for $n = 10$. In addition, for most of the knots in his tables, Conway gave values for several classical, and also new, invariants, obtained by his methods.

Thus, in a very short time*, using his newly devised methods of tangles, Conway had checked and extended the tables of Tait, Kirkman and Little, produced so laboriously about eighty years previously.

Certain manipulations on the Conway tangles gave rise to polynomials. Sample calculations with these were made, and they revealed certain algebraic relations between the polynomials (which in fact obviated the use of a computer for calculating his tables). It was natural to think of tangles as elements in a vector space, in which certain identities became linear relations. There were many natural questions to be asked about these spaces, and study of these led Conway to his discovery of a polynomial knot-invariant. Initially, Conway only wanted to further the enumeration and tabulation of knot-types, which task had been at a standstill for the past six decades when he began his attack upon it. But his contribution turned out to be a major one, in the hunt for knot invariants. Even though his find, in a sense, was Alexander's polynomial disguised in a normalized form, it was obtained by totally new methods. It became known as *the Conway polynomial*, often denoted by ∇_K. It was also a polynomial which could be calculated directly from a diagram by means of a recursive method, not requiring the evaluation of any determinant.

John Conway wanted to call the relationship between three links whose diagrams differ only in a set of Alexander crossings a *potential function*; but instead, this relationship went on to lead its own life in knot research, acquiring the name of *skein relation*. In Conway's original work this potential function had the form:

$$\nabla_{K_+}(t) - \nabla_{K_-}(t) = t\nabla_{K_0}(t) \text{ , with } \nabla_{\mathbf{U}} = 1$$

It relates to the Alexander polynomial Δ_K via the equation:

$$\Delta_K(t) = \nabla_K(t^{\frac{1}{2}} - t^{-\frac{1}{2}})$$

The important idea, which set new trends in knot research, was that the skein

*In [23] Conway claims that he could check in a mere afternoon much of the work that Tait and Co. took six years to complete!

relation became the invariant's definition. Its well-definedness could be proved by showing its invariance under the Reidemeister moves.

More so than with Alexander polynomials, which require definition and computations of certain determinants, the preferred way for defining polynomial invariants obtained via skein relations is to proceed from knot-diagrams. A polynomial is computed recursively, by a kind of 'unknotting process' when one systematically obtains diagrams on reducing numbers of crossings, making use of a given skein relation. When diagrams with already known polynomials are arrived at, the process can be retraced, and the polynomial for the original knot is arrived at.

Success with, and increased use of, this procedure caused knot-diagrams to become notational devices at the same level as other symbols in mathematical writings.

Incidentally, as we noted above, Conway did expand the knot-tables; and his work was later continued by Thistlethwaite and Perko [86], [71]. The latter completed the census problem for 10-fold knottiness in 1974, and detected some errors in Little's 1885 table of 11-fold crossing knots. Now we have complete listings of knots with up to 13 crossing-points [86]. And researchers are working to enumerate knots on 14 and 15 crossings [8]. There is an estimate that there exist over 150 000 different prime alternating knots on 15 crossings.

The following table shows the totals of prime alternating knots which have from 3 to 13 crossing-points. The top row gives the numbers of crossing-points, and the bottom row the corresponding frequencies of knots.

n	3	4	5	6	7	8	9	10	11	12	13
f	1	1	2	3	7	21	49	165	552	2176	9988

In the early 80s, John Turner [89] studied knot-graphs, and experimented with operations similar to Conway's. He obtained various knot invariants, working from both non-oriented and oriented diagrams. One idea he pursued was to take an alternating knot-diagram, and reduce it systematically by a process he called *twinning*. Crossings were 'deleted', one at a time, and two new knots (the 'twins', each with one fewer crossing than the original knot) were formed at each 'deletion' (see the example in Fig. 18). He continued this twinning, producing a binary tree of knots, and stopping the deletion process whenever a twist knot was arrived at.

The ultimate result, from any given starting knot, was a collection of twist knots (situated at the tree leaf-nodes), each of which had well-defined twist-senses, labelled plus or minus. He saw these as being fundamental building units of the original knot, and was able to prove, subject to one of Tait's many conjectures being true, that the end-collection of twists was independent of the order of reduction by 'deletions' of crossings, and that it was a knot invariant.

He assigned the symbol u^n to each n-twist, where n (positive, zero or negative) was the 'sum' of the senses of crossings in the twist (e.g. see Fig. 18).

Collecting all the symbols together, he arrived at a polynomial which he called *the twist spectrum* of the starting knot*. This, then, was a polynomial knot-invariant. In [89], Turner gave tables of knot twist-spectra for all alternating prime knots with $n = 3, \dots, 9$ crossings, and all alternating 2-links with $n = 2, \dots, 8$ crossings. The twist spectrum distinguished all these knots. He conjectured that it would distinguish all alternating knots with fewer than 15 crossings. In that sense it clearly outperformed the Alexander polynomial. Moreover, it could be used to provide a simple test for nonamphicheirality in a knot; for if a knot is amphicheiral its twist spectrum is symmetric about the constant term (the converse of this was conjectured, but unproven).

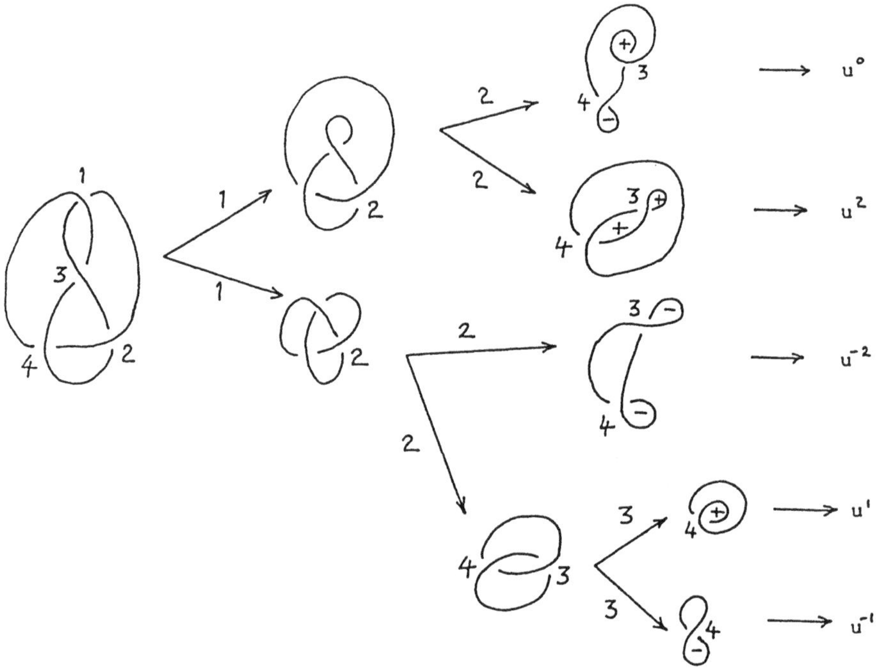

Spectrum: $T(u) = u^{-2} + u^{-1} + u^0 + u^1 + u^2$
Vector of coefficients: $(1, 1, \underline{1}, 1, 1)$

Fig. 18. Computing the Twist Spectrum of Listing's Knot (amphicheiral)

*This was a precursor of Kauffman's bracket polynomial, to be described later. Kauffman used the same deletion process, but continued beyond the twists, until no crossings at all remained. If his process were stopped at n-twists, his polynomial would be the same as the twist spectrum.

Figure 18 demonstrates the above process, producing the twist spectrum for the 4_1 knot (Listing's). The final twists, with their orientations and their corresponding polynomial terms, are shown on the right of the diagram.

An interesting connection between the Alexander polynomial $\Delta(t)$ of a knot, and the twist spectrum $T(u)$, is that the so-called determinant of the knot, given by $|\Delta(-1)|$, is equal to the *torsion coefficient* value $T(1)$. Also like the Alexander polynomial, the twist spectrum of a composition of two knots is equal to the product of the twist spectra of the two knots. Further, this time like the Jones' polynomial, $T_{K^*}(u) = T_K(u^{-1})$ if K^* and K are mirror images. So, for example, the trefoil and its mirror image are distinguished, since their twist spectra coefficients-vectors are $(1, \underline{0}, 1, 1)$ and $(1, 1, \underline{0}, 1)$.

Very soon after the twist spectrum was discovered, Vaughan Jones' great knot polynomial discovery was announced [43]. As we shall see below, this triggered an explosion of discoveries of polynomial invariants, and markedly changed the face of topological knot theory, both pure and applied. Before going on to describe these developments, it is necessary for us to review the history and achievements of braid theory.

12. Researches in Braid Theory

The beginning of the 1920s witnessed an impasse in the theory of knots. With the omnipotence of the knot group fatally punctured, and presentations of knot groups stuck in generally unsolvable word problems, it was not strange that knot theorists should seek new ways for achieving progress. The problems of those days attracted some of the most prominent algebraists and topologists. Minds like Seifert's pursued further research via Riemannian manifolds, while the actions of others appeared more desperate. Reasoning that knots consist of *bits of knotted patterns*, they broke them into smaller pieces which fulfilled certain conditions and called these objects *braids*. Braids were not a new idea when they entered the scene in the 1920s. Listing and Tait had already studied procedures which generated simple knots after plaiting samples with two and three strands. On the other hand the *breaking up* was something entirely new. Emil Artin (1898-1962), with the help of Otto Schreier, formalized the ideas and provided tools to carry on the halted quest. His landmark paper on them [9], *Theorie der Zöpfe*, appeared in 1925.

Basically, he provided an entirely new algebraic environment for knot studies by introducing the so-called (algebraic) *braid group*, denoted by B_n. This is the set of all braids on n strings satisfying certain conditions, together with a binary operation which consists of the simple process of concatenation of two braids, joining the lower ends of one to the upper ends of the other. Examples of 3- and 4-string braids appear in Fig. 19.

The geometric picture of a braid in $\mathbf{R}^3$ is easy to envisage. Consider n

parallel strings in a plane, all hanging vertically from a line drawn on, for example, the ceiling, and dropping down to a line on the floor. There are thus $2n$ string endpoints, n on the ceiling and n on the floor. In a given braid, the endpoints are all to be regarded as fixed. This first configuration, with all strings parallel, is the *null braid*; it acts as the unit element of the *geometric braid group* $B(n)$. If now the lower endpoints are removed from the floor, the strings interwoven in some manner, and finally the endpoints are refixed to the floor in some order, then a new n-string braid will be achieved; such are the members of $B(n)$. With this geometric picture in mind, it is easy to imagine the concatenation operation, which joins two braids 'one on top of the other'.

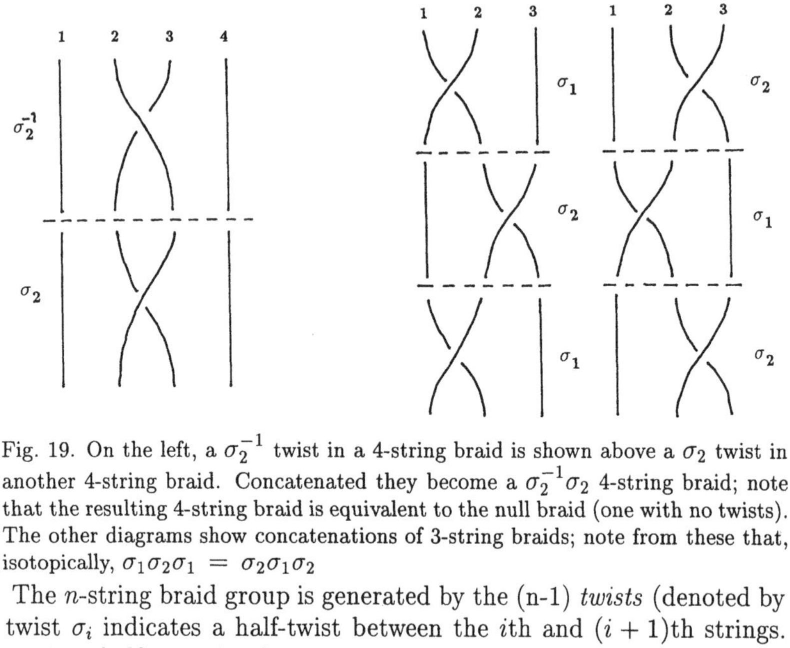

Fig. 19. On the left, a σ_2^{-1} twist in a 4-string braid is shown above a σ_2 twist in another 4-string braid. Concatenated they become a $\sigma_2^{-1}\sigma_2$ 4-string braid; note that the resulting 4-string braid is equivalent to the null braid (one with no twists). The other diagrams show concatenations of 3-string braids; note from these that, isotopically, $\sigma_1\sigma_2\sigma_1 = \sigma_2\sigma_1\sigma_2$

The n-string braid group is generated by the (n-1) *twists* (denoted by σ_i): the twist σ_i indicates a half-twist between the ith and $(i+1)$th strings. Its inverse is a half-twist in the opposite sense. The diagrams above illustrate this, with the 4-string braids.

Artin showed that $B(n) \simeq B_n$, and that they permit a presentation in terms of generators and relators given by

$$(\sigma_1, \ldots, \sigma_{n-1} : r_1, r_2), \qquad \text{in which}$$

$$\begin{cases} r_1 : \sigma_i\sigma_j = \sigma_j\sigma_i\,, & |i-j| \geq 2,\ 1 \leq i,j \leq n-1 \\ r_2 : \sigma_i\sigma_{i+1}\sigma_i = \sigma_{i+1}\sigma_i\sigma_{i+1}\,, & 1 \leq i \leq n-2 \end{cases}$$

Note how the two relators catch the principal features of a braid with three or more strands.

The braid group elements are *open* patches of weaving. They can be closed by linking up the respective ends. The most obvious way this can be done for an arbitrary $\beta \in B_n$ is by pairing the ends at the ceiling one to one with the corresponding ends at the floor. This yields the *closed* braid denoted by $\hat{\beta}$.

The main evidence that braids would be useful in studying knots and links was the existence of an $(n-1)$-dimensional representation of B_n, discovered by Werner Burau in 1936 [21]. This representation is expressed with a parameter t; and from it one can extract the so-called *Burau Matrix*. The following relationship connects Burau's braid group representation with Alexander's knot polynomial:

$$\det(\mathrm{Id} - \text{Burau Matrix}(\beta)) = \pm t^n \Delta_{\hat{\beta}}(t) \qquad \text{for some } n$$

Burau showed this by connecting the Burau Matrix of β with a known way of calculating Δ from a presentation of the knot group.

The first firm connection between knot theory on the one hand and braid theory on the other was made by Alexander in 1923 with a theorem which showed that any n-link is equivalent to some closed braid. He also gave a simple algorithm [5] for converting an n-link into a closed braid. The existence of such an algorithm was already noted by Herman Brunn in 1897 [20]. However, troubles with this algorithm are twofold. It may cause one and the same n-link to become a closed braid which, upon cutting can belong to two distinct braid groups B_n and $B_{n'}$, $n \neq n'$. If the algorithm consistently causes the n-link to yield a closure of a braid belonging to just one braid group, then there may be many words representing it.

How does this result affect classification and knot isotopy? Closed braids have two very interesting properties, which are caught by the so-called Markov moves.

1. There is a particularly diabolical way of making a knot from an open braid. If one takes two braids $\alpha, \beta \in B_n$ and constructs the closure: $(\widehat{\alpha\beta\alpha^{-1}})$ then it will equal $\hat{\beta}$. The closing operation causes α to be cannibalized by its inverse. This is worded by saying that conjugate elements in B_n yield equivalent links.

2. Imagine a closed braid on a spar. Adding any number of strands by means of simple twists cast in any of its outer bights over the spar does not change the type of link.

The essence of these two properties was captured by A.A. Markov in his theorem of 1936 which states that closed braids $\hat{\xi}$ and $\hat{\zeta}$ are equivalent as links if and only if they can be connected by a finite sequence of elementary moves, which are precisely those described by the two properties of closed braids above. The braids are then said to be Markov equivalent. This is an equivalence relation on B_∞, the disjoint union of all braid groups.

Markov's theorem is of particular interest because it allows one to restate the knot problem as a purely algebraic problem known as the *Algebraic Link Problem*, which is the classification of Markov classes in B_∞. This comes down to finding a well-behaving class function on the Markov classes. It is worthy of remark that Markov never proved the theorem which got named after him. It had to wait until 1974 when the doyen of Braid Theory, Joan Birman, published a complete proof, achieved by cleverly combining the results of many other workers. It was no mean feat.

Another classical result is due to Artin. His original paper already contained the fundamental isomorphism between braids and automorphisms on B_n. It establishes that these mappings may be used to obtain a presentation for the knot group of any tame n-link [9]. The proof, though, was too intuitive for his liking. In 1947 he published a new paper on braids. This time he gave rigorous definitions and proofs including normal forms of a braid, which may be used to give a complete characterization of knot groups [10]. When Artin first suggested braid theory as an approach to the study of knots and links, he conjectured that the chief obstacle in the approach would be the conjugacy problem in B_n. This is the problem of deciding whether there exists a $\gamma \in B_n$ such that for two braids $\alpha, \beta \in B_n$ the equation $\alpha = (\gamma\beta\gamma^{-1})$ holds. In that case α and β are said to be *conjugated.* There have been many attempts to resolve this conjecture since Artin's 1925 paper. Partial solutions were attained, such as Frölich's in 1936, but it was not until Makanin and Garside around 1968–69 completely solved the problem. Garside [37] invented an ingenious, though rather complicated and hard to prove algorithm, by which he could decide whether or not two braids are conjugate.

Since we have available an algorithmic solution to the conjugacy problem it is natural to ask whether this might lead to a general solution of the knot problem? The answer is no, since there is trouble with the Markov moves. An arbitrary sequence of them applied to the closure of $\alpha \in B_n$ may either increase or decrease the number of strands, but if the final closed braid ultimately returns to B_n then there is no guarantee that one has not replaced $\hat{\alpha}$ with a conjugate of itself. This is not so bad as it may sound. Birman succeeded in finding some relaxed conditions for the knot problem which resulted in simplifying Garside's solution a little. Her work can easily be explained by introducing some nomenclature. A *positive word* denotes an open braid in which all $\sigma_i (1 \leq i \leq (n-1))$ have non-negative exponents. If such a braid is closed then it yields a *positive link*. Birman found that the knot problem on positive links reduces to the conjugacy problem. This implied that such links can be classified.

13. Problems in Paradise

Until this time, the beginning of the 1980s, the overall picture of the knot theoretical arena was one of relative tranquility. Purists insisted that knots and braids were different things. The former lived in the world of topologists, whilst braids belonged to the algebraists. This view catered for a state of peaceful coexistence, the result of an evolution in which both camps more or less went on minding their own business. This changed in a radical manner when a new knot polynomial erupted onto the scene. In the 1980s the New Zealander Vaughan Jones, through work in von Neumann Algebras associated with certain physics problems in statistical mechanics, had found a new way into knot polynomials. In order to study those aspects of theoretical physics he had constructed an algebra J_n given by:

$$(a_1, \ldots, a_{n-1} : r_1, r_2, r_3) \text{ where } \begin{cases} r_1 : a_i^2 = a_i \\ r_2 : a_i a_{i\pm 1} a_i = \tau a_i \\ r_3 : a_i a_j = a_j a_i \, , \ |i-j| \geq 2 \end{cases}$$

What happened is perhaps best expressed in his own words [48, p. 53]:

> *In my work I had been astonished to discover expressions that bore strong resemblance to the algebraic expression of certain topological relations among braids.*

He was so struck by the resemblance between the definition of his algebra and that of the braid group B_n, that in May 1984 he journeyed to Columbia University to meet Joan Birman, to discuss his ideas with her. Their initial deliberations were discouraging. But Jones found, very soon after, that a representation of B_n could be transformed into one of J_n; which in turn possesses a trace map. Since trace maps are natural class functions, and Jones' trace also supported the second of the Markov moves, he thus had effectively constructed a link invariant!

General acknowledgement of the importance of this discovery was not immediate; but eventually his ideas came to have immense impact on much mathematical research in topology. In fact, he was later awarded the prestigious Fields' Medal for his contribution to the advancement of mathematics. Not only did Jones' work link knots to statistical mechanics, but also it sparked an interaction between knot-theory and braid-theory, the like of which had not been seen since the times when Artin's and Alexander's ideas became enjoined.

The new Jones knot invariant became known as *the Jones polynomial*; it is denoted by V. Jones himself had published its skein relation in his first article [47] which documents his ideas. The relation has the form:

$$t^{-1}V_{K_+}(t) - tV_{K_-}(t) + (t^{-\frac{1}{2}} - t^{\frac{1}{2}})V_{K_0}(t) = 0 \, , \text{ with } V_{\mathrm{U}} = 1$$

At first it seemed to be yet another polynomial invariant in the long line of such objects which had been proposed previously. Some of them were truly exotic in their definitions, while others were plain monsters in their time complexities; but V did something different. Knot theorists had begun to understand that symmetries, such as handedness and orientation, which are not caught by the Alexander/Conway polynomial, belong to entirely different regions of the mathematical universe. The reason that Alexander's polynomial does not cater for invariance under mirroring is due to the fact that $\Delta_{K^*}(t) = \Delta_K(\pm t^n)$. A key property of V is that (like the twist spectrum):

$$V_{K^*}(t) = V_K(t^{-1})$$

Simple examples show that $V_K(t)$ need not be invariant under $t \to t^{-1}$, so that it can sometimes distinguish knots from their mirror image. For example, in the case of the two Trefoils T_l and T_r we have:

$$\begin{cases} V_{T_r^*}(t) = t + t^3 - t^4 = V_{T_l}(t^{-1}) \\ V_{T_l^*}(t) = t^{-1} + t^{-3} - t^{-4} = V_{T_r}(t^{-1}) \end{cases}$$

V is pretty good at detecting this sort of symmetry, though not infallible. Its merits for this are first dashed with knot 9_{42} from the tables by Reidemeister and Rolfsen [74], [77]. But even so, this property shows that V is not a knot-group invariant like Δ or ∇.

It is a remarkable fact that ∇'s appearance on the knot scene, some 40 years after Δ's, did not trigger any generalizing activity in the mathematical community. Whereas the appearance of Jones' polynomial seemed to present an explicit invitation to do so. It immediately led to an outburst of discoveries of knot polynomials with more than one variable. Moreover, time since then has shown that V was to be generalized in two quite distinct ways.

The first general polynomial to appear was to be known as *Homfly*; it is also called *Homfly-PT*, and also *Thomflyp*. The names are acronyms which are derived from the initials (of 6) of the 8 people who discovered it [36], independently and almost simultaneously. Four of them happened to submit an article on their work to one and the same major mathematical journal on virtually the same day! Although they reached their results via different routes, the main idea was inspired by generalizing the coefficients in the skein relation. They were invited to pool their ideas, and present a single paper, jointly under all their names; this they did. Their two-variable polynomial, denoted by P, has the following skein relation:

$$\ell P_{K_+}(\ell, m) + \ell^{-1} P_{K_-}(\ell, m) + m P_{K_0}(\ell, m) = 0\,, \text{ with } P_{\mathbf{U}} = 1$$

For the sake of comparison all four skein relations are listed below:

$$\begin{cases} \Delta_{K_+}(t) - \Delta_{K_-}(t) + (t^{-\frac{1}{2}} - t^{\frac{1}{2}})\Delta_{K_0}(t) = 0\,, \text{ with } \Delta_{\mathbf{U}} = 1 \\ \nabla_{K_+}(t) - \nabla_{K_-}(t) - t\nabla_{K_0}(t) = 0\,, \text{ with } \nabla_{\mathbf{U}} = 1 \\ t^{-1}V_{K_+} - tV_{K_-} + (t^{-\frac{1}{2}} - t^{\frac{1}{2}})V_{K_0}(t) = 0\,, \text{ with } V_{\mathbf{U}} = 1 \\ \ell P_{K_+}(\ell, m) + \ell^{-1}P_{K_-}(\ell, m) + mP_{K_0}(\ell, m) = 0\,, \text{ with } P_{\mathbf{U}} = 1 \end{cases}$$

However, very soon after the public announcement of P's discovery, examples of indistinguishable pairs of mirror image knots emerged. Of course, one counter-example is sufficient to torpedo a conjecture, but in 1986 Taizo Kanenobu produced infinitely many classes of, in turn, infinitely many *distinct* knots with the same P-polynomial. Using the second elementary ideal of the Alexander module, he showed [52] that for the knots $K_{p,q}$ (see Fig. 20):

$$P(K_{p,q}) = P(K_{p',q'}) \quad \text{if and only if} \quad p + q = p' + q'.$$

P contains the information of Δ, ∇, V and more; but there the similarity ends. And to-date all attempts to interpret V in the same topological framework as Δ have failed.

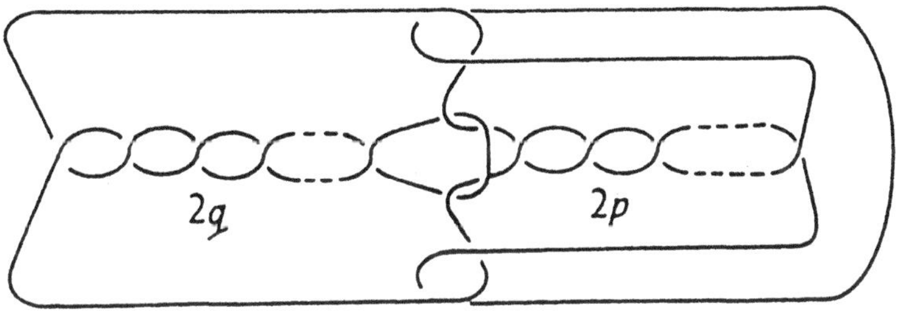

Fig. 20. The knot $K_{p,q}$ used by Taizo Kanenobu

Louis Kauffman's F-polynomial was the second distinct generalization of V to appear. He obtained his polynomial by cutting up a knot in a special way [54]. The skein relations we have seen so far are recursive definitions on knot diagrams which differ in the set of crossings which was shown in Fig. 18. Kauffman's polynomial is based on the very imaginative idea of a *state model*, which sees an unoriented knot diagram as a *state* σ, and which provides information carried by the diagram. Given a diagram (of a knot K, say), he proposed splitting a crossing in two ways, thereby obtaining two new states (the same operation as used to obtain Turner's twist spectrum). These were assigned a 'weight' A or A^{-1} according to their type of split (if a new state included a circle, a more complex weight, a function of A and A^{-1}, had to be assigned). He continued this process until no crossings were left; and then took the weighted sum over all the resulting states, obtaining what is now called the *Bracket polynomial* of K. This is denoted by:

$$< K > = \sum_{\sigma} < \sigma | K > (-A^2 - A^{-2})^{||\sigma|| - 1}$$

The Bracket can be generalized to yield a two-variable polynomial invariant, named the *L-polynomial*. By using L, divided by a factor involving one of the variables raised to the sum of the knot's crossing parities (the so-called *writhe* $\omega(K)$), one can construct yet another polynomial invariant, which is customarily denoted by F, and is known as the *Kauffman polynomial*. Thus:

$$F_K(\alpha, z) = \alpha^{-\omega(K)} L_K(\alpha, z)$$

In [60] Lickorish shows that the Jones and Kauffman polynomials, V and F, are related by the following equation:

$$V_K(t) = F_K(-t^{-\frac{3}{4}}, t^{-\frac{1}{4}} + t^{\frac{1}{4}})$$

The F-polynomial is quite good at detecting chirality. It is probably a little better for this purpose than P, because it originates from four instead of three terms; but it has its shortcomings too. However the Bracket polynomial is a truly amazing construct. For instance, it was used to confirm the old conjecture (made by Tait) that *the number of crossing points in a connected, reduced, alternating projection of a link is a topological invariant.* Several other long-standing problems were also dealt with quickly by means of this and the other new knot polynomials [69], [85].

In spite of its more powerful generalizations, $V_K(t)$ has retained its interest and value in knot research and applications. Ironically, one reason for this is, simply, that it has only one variable; which makes it easier to work with!

Amongst all the questions asked after the discovery of the new knot polynomials, the dominating one was: *Granting its existence, how may these polynomials be placed in a sound mathematical setting*? Many researchers have tried to shed light on this question, in the past decade, working from a variety of viewpoints, moving in uncharted territories of topology, algebra and statistical mechanics. The schematic diagram on the next page indicates the need to underpin satisfactorily the array of new knot invariants with a unifying theoretical base. And the next section summarizes some of the work that has been done in this 'terra incognita' since the invariants appeared.

14. Charting Terra Incognita

Jones' discovery implied that statistical mechanics must hold clues for an understanding of his polynomial. Therefore the discovery of V, F and

P triggered off even more ambitious research, which came from roughly two (interacting) directions. One of these was from physicists churning out link invariants. The other was from the mathematical camp of knot-theorists, trying hard to understand them.

It became obvious that via Markov's and Alexander's Theorem, one should be able to relate algebraic interpretations for link invariants to the braid groups. Algebraically, the link problem translates into domesticating a class function on the Markov classes. However, a head-on attack on Markov equivalence in B_∞ is hopelessly difficult. Luckily, representation theory's richness provides plenty of room for finding invariants. As the endpoints of braids define a permutation in a natural way, the symmetric group S_n thus exists as a quotient in B_n. It is natural, then, to study how representations of S_n and B_n are related. It turns out that *every* irreducible representation of S_n transforms to a parametrised family of irreducible representations of B_n. In fact, S_n transforms to an algebra $H_n(q)$, the so-called Hecke algebra, when $q \to 1$.

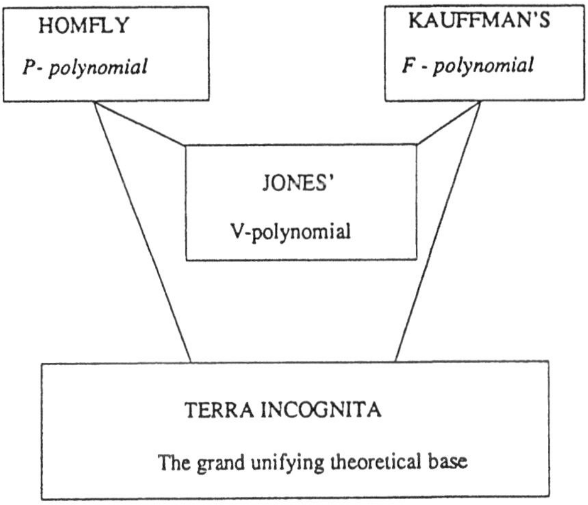

Adrian Ocneanu discovered Homfly as a trace function on the algebras H_∞, which supported a Markov trace as a weighted sum of matrix traces on their irreducible summands. The quadratic defining relation of a Hecke algebra afforded an explanation for its skein relation. The Japanese research team of Akutsu and Wadati found new link invariants by interpreting further statistical mechanical concepts [1], [2], [3], [4]. Their invariants turned out to be P again, but now in terms of 'cubic' Hecke relations, and supporting skein relations for triplets of links which are equivalent except in one crossing point, where they may have left- or right-handed spiralling segments instead of a single crossing. This was an idea that Conway had already explored when he discovered ∇ [23]. It is called *cabling.*

Unlike V, and to a certain extent P, the F-polynomial was discovered by purely combinatorial techniques, and it seemed at first glance to be completely unrelated to braids. However, so-called BWM algebras were constructed by Joan Birman, Hans Wenzl and also, independently, by Jun Murakami [17], [68]. Geometrically speaking, they extended the braid groups with U-turns, making them into (braid) monoids. The BWM algebras are quotients of the complex group algebra CB_n, and they support a 2-parameter family of Markov traces whose associated link invariant is the Kauffman polynomial. Each of these algebras contains H_n as a direct summand, and the Markov trace that associates to Homfly is the restriction to H_n of the Markov trace that defines F.

Each of the algebras just described supported a Markov trace, and so determined a link-type invariant. In this way a uniform picture of the old and new link invariants gradually emerged, with the representation theory of B_n being an important central part of the picture.

However, the various generalizations of link polynomials have been subsumed under an even more general and unifying procedure, via the so-called *Yang-Baxter Equation* (YBE). A Yang-Baxter operator on a vector space V is a linear isomorphism $R : V \otimes V \to V \otimes V$ such that the following hexagon commutes:

$$\begin{array}{ccccccc} & & V\otimes V\otimes V & \overset{R\otimes 1}{\longrightarrow} & V\otimes V\otimes V & & \\ & {}^{1\otimes R}\nearrow & & & & \searrow^{1\otimes R} & \\ V\otimes V\otimes V & & & & & & V\otimes V\otimes V \\ & {}_{R\otimes 1}\searrow & & & & \nearrow_{R\otimes 1} & \\ & & V\otimes V\otimes V & \overset{1\otimes R}{\longrightarrow} & V\otimes V\otimes V & & \end{array}$$

This is equivalent to requiring that:

$$(R \otimes Id_V) \circ (Id_V \otimes R) \circ (R \otimes Id_V) = (Id_V \otimes R) \circ (R \otimes Id_V) \circ (Id_V \otimes R)$$

This equation is the Yang-Baxter Equation (YBE), introduced by C. N. Yang in 1967 in the context of the 1-dimensional quantum n-body problem as a factorization condition on the S-matrix. Later it was used by Rodney Baxter to obtain explicit formulae for the partition function of the 8-vertex model by the transfer matrix method. The YBE plays a fundamental role in two physical theories: namely the theory of exactly solvable models in statistical mechanics, and the theory of completely integrable systems.

Driven by an idea of Vaughan Jones [49], and returning to the mathematics behind statistical mechanics, Vladimir Turaev [88] showed that R-matrices (i.e. YBE solutions) yield a mechanism to produce new representations for B_n, each of which lead to new polynomial invariants. Turaev imposed conditions which ensured that the representations so obtained supported a Markov trace. Thus was born a machine, ready to produce further link invariants provided it was fed with R-matrices. Since then the history of the YBE has intertwined and interacted with developments in knot theory. For example, Hans Wenzl discovered how, by cabling braids one can find new representations for B_n, which in turn yield new solutions to the YBE [94].

The YBE can be written in several forms. There are general methods which allow one to construct R-matrices for the various versions [59], [75]. Originally, it was not written in the form given above, but as an equation in a Lie algebraical setting and called the classical YBE (the CYBE). This is easily explained. Let $\mathcal{L}$ be a Lie algebra, and let $r : \mathcal{L}^{\otimes 2} \to \mathcal{L}^{\otimes 2}$ be an isomorphism. Let r_{12} be $r \otimes 1 : \mathcal{L}^{\otimes 3} \to \mathcal{L}^{\otimes 3}$ and r_{ij} the image of r_{12} under the automorphism of $\mathcal{L}^{\otimes 3}$ induced by the permutation $(1,i)(2,j)$. The isomorphism r is said to be a solution for the CYBE if:

$$[r_{12}, r_{23}] + [r_{13}, r_{23}] + [r_{12}, r_{13}] = 0$$

The CYBE has been well studied. To any finite-dimensional representation of a simple complex Lie algebra on a vector space V endowed with an automorphism of the Dynkin diagram, there is a matrix $R \in \text{End}(V \otimes V)$ which satisfies the CYBE [50], [59]. It was found that CYBE solutions satisfied Turaev's extra conditions. Michio Jimbo gives the R-matrices corresponding to the fundamental vectorial representations of the non-exceptional Lie algebras of the series $A_n^1, B_n^1, C_n^1, D_n^1, A_n^2$ and D_n^2, where the upper index denotes the order of the automorphism of the Dynkin diagram [45]. Jimbo showed that the R-matrices of the series A_n^1 result in operator invariants of ordered tangles and the Homfly polynomial of a link. The R-matrices of the series B_n^1, C_n^1, D_n^1 and A_n^2 result in operator invariants of framed tangles and the Kauffman polynomial of a link. These are 'constant' YBE-solutions. A daunting idea is to parametrise them and the YBE, much like a Hecke algebra is a parametrised version of S_n. A quantised universal enveloping algebra is such a transformation of a classical Lie algebra that depends on a transformation parameter q and recovers the classical algebra in the limit as $q \to 1$. Furthermore, it is endowed with a comultiplication, as well as an antipode and a co-unit, which gives it the structure of a Hopf algebra. These objects are better known as quantum groups, due to their intimate relationship with the quantum inverse scattering method, and in that connection first studied by the St. Petersburg school of L. D. Faddeev. Quantum groups arose in 1982 as algebraic structures describing symmetry properties encountered in solvable statistical models. However, they have not much to do with quantum mechanics, and are not groups either! Drinfel'd, who introduced the term *quantum group*, defined the structure as a Hopf algebra, essentially a bialgebra with an antipode. Quantum groups constitute an exciting generalization of the concept of symmetry. In this context, the parametrised YBE becomes the Quantum YBE (i.e. QYBE).

The invariants from the quantum group setting are polynomials. They are gathered under the heading quantum invariants, also called generalized Jones invariants. Since transfer matrices satisfy the YBE if and only if they are a representation for B_n, every quantum invariant is obtained from a trace function on an R-matrix representation for B_n.

The renaissance in the interactions between between physics and knot theory (recall Gauss's use of properties of knots in the solution of an electromagnetic problem, and Thompson's plans to describe chemistry in terms of knots) was due to statistical mechanics studies. But nowadays there are at least three other, different, ways in which physics and knot theory are related. Not only is there topological quantum field theory, but the theory of quantum invariants has also proved to be closely related to conformal field theories. In this connection one should mention Edward Witten's papers [96], where it is shown that Jones' polynomial and its generalizations are related to the topological Chern Simons actions.

So, order is emerging from chaos, and new results are being achieved continuously. The order appears to be part of an even larger order, which involves the physics of conformal field theory, and leads to further invariants, now in arbitrary 3-manifolds. As we have seen, the theory of R-matrices gives a systematic description for the quantum invariants. It has been known for a long time that any compact orientable 3-manifold arises (up to PL-homeomorphism or diffeomorphism, depending on the choice of category) as the boundary of a 4-manifold M, where M is obtained by attaching 2-handles to the 4-ball, along some framed link in S^3, i.e. by surgery along framed links. Lickorish and Wallace proved this at the beginning of the 1960s [61], [92]. One can think of framing as the thickening of the knot into a ribbon-like object. Any closed oriented 3-manifold may thus be obtained by performing surgery along *different* framed links in the 3-sphere. This yields an equivalence relation on framed links. Robion Kirby proposed a set of moves which generated this equivalence relation [55]. Roger Fenn and Colin Rourke simplified them. Their moves may be described by means of tangle generators [33]. By using Kirby, Fenn and Rourke calculus, Nicolai Reshetikhin and Vladimir Turaev in 1991 defined 3-manifold invariants using the theory of quantum groups. They produced new 3-manifold invariants, which can be defined from any simple Lie algebra, provided the associated quantum groups have the structure of a finite dimensional modular Hopf algebra [76].

The theory of quantum invariants had led to the discovery of tangle categories. By looking at the most generalized form of 'knot', which is a graph with 'knotted ribboned parts', one can construct so-called ribbon categories. The finite dimensional representations for quantum groups comprise such a ribbon category. This connection could thus be used to find the new invariants for 3-manifolds. Other results made Reidemeister's theorem a covariant functor between the categories of links and diagrams, while quantum invariants became covariant functors from tangle categories to categories of modules. On the whole, category theory has been able to cut a lot of cake, as the language was quite effective to formulate and extend very general ideas about the central link polynomials.

So far, the story has been one of 'simple' hierarchical generalizations. An exciting change of perspective comes from V. A. Vassiliev [91]. He considers the so-called *knot space* $\mathcal{M}$, which is the space of all embeddings $\gamma : \mathbf{S}^1 \hookrightarrow \mathbf{S}^3$. This allows one to study more than just a single knot and the ways in which distinct knots relate to each other. The object of utmost interest is the natural stratification of $\mathcal{M}$. Deforming knots to the level where we permit self-intersections of the cord in which they are realised leads to the notion of *chambers* in $\mathcal{M}$. The discriminant $\mathcal{C}$ of $\mathcal{M}$ is defined to be the set of mappings which are not embeddings. This is a singular hypersurface in $\mathcal{M}$. The components of $\mathcal{M} - \mathcal{C}$ are clearly in one-to-one correspondence with the knot types. Thinking of a numerical knot invariant as a function on the components of $\mathcal{M} - \mathcal{C}$ one is led to study the cohomology of $\mathcal{M}$. Vassiliev introduced a system of subgroups of $\tilde{H}^0(\mathcal{M} - \mathcal{C})$:

$$0 = G_1 \subset G_2 \subset G_3 \subset \ldots \subset \tilde{H}^0(\mathcal{M} - \mathcal{C})$$

where $\tilde{H}^0$ is reduced cohomology with integer coefficients and G_i is free abelian of finite rank. The evaluation of an element in G_i/G_{i-1} on the component of $\mathcal{M} - \mathcal{C}$ corresponding to an oriented knot type K yields a rational number $u_i(K)$ associated to K. This is a Vassiliev invariant of order i. Like the Jones invariants, one computes Vassiliev invariants from a diagram by changing crossings. However, the combinatorics of the computation are very much more difficult.

Vassiliev's invariants are rational numbers, while generalized Jones invariants are Laurent polynomials over $\mathbf{Z}[q, q^{-1}]$. Joan Birman and Xiao-Song Lin showed that there is a relation between them [16]. Let a *knot* K have Jones polynomial $V_K(t)$. Set $U_x(K) = V_K(e^x)$ and express it as a power series in x:

$$U_x(K) = \sum_{i=0}^{\infty} u_i(K) x^i$$

then $u_0(K) = 1$ and each $u_i(K)$, for $i \geq 1$, is a Vassiliev invariant of order i. Although this scheme so far works only for knots, the Vassiliev invariants seem to offer at least part of the topological framework we seek for the quantum invariants. Hence there are speculations abounding. For instance, it is well-known that quantum groups do not detect invertibility of knots, but Vassiliev's invariants just might. This would make them stronger than quantum invariants. The research with ribbon categories and Vassiliev invariants have caused the singular braid monoid to place itself permanently on the scene. Moreover it threatens to become just as fundamental a mathematical tool as the braid groups. However, amongst all of these recent developments, mainly involving the braid groups, the knot complement group is far from forgotten. David Joyce, Colin Rourke and Roger Fenn have concocted an algebraic structure, which dates back to John Conway and van Brieskorn in the 1960s, and is now called a *rack*. It generalizes the knot group, but also captures aspects of the peripheral system. This completely classifying invariant seems to be a promising and exciting new part of the overall picture [34], [51].

15. The Future?

In the foregoing sections we have seen how vague, intuitive notions about knotted structures, beginning with the work of Listing, Gauss, Kirkman, Tait and Little of last century, were gradually developed until they reached, by the last decade of this century, extremely high levels of abstraction and complexity. Concerning the future of this process, one can only speculate on how far, and in which directions, the current attempts to solve a variety of outstanding problems will take us.

The prime knots with up to 13 crossings have been distinguished and tabled; and these knots are relatively simple objects. Attacks on the classification of 14- and 15-fold crossing knots are in progress; there are very many more of such knots, and no doubt it will require combinations of several of the available invariants to distinguish them all. It will be difficult, and perhaps not sensible, to produce diagrams for these vast numbers of knots; most will be 'known' only by their corresponding invariant values, arranged in classes and stored in some digitised form. The baffling problem of finding a single, complete, knot invariant (if one exists) still remains.

Future research will certainly be affected by the amazing developments stemming from Jones' discovery. They continue unrelentingly; yet many simple questions remain unanswered. It is still not known whether a non-trivial link can have the same V-, P- or F-polynomial as has the trivial link of the same number of components; we know this can happen for Δ and ∇. Resolution of this question would lead to significant conceptual progress. No generalization of it to knots and links of higher dimension has yet been achieved.

From the point of view of contemporary topological knot theory, the chief problem is to find an interpretation of the new invariants in terms of classical algebraic topology (homology theory, homotopy theory and such) or differential geometry (differential forms, connections, etc.). Here considerable speculation has produced little of note. Perhaps there can be no such interpretation, and states-model theories from statistical mechanics must be incorporated into topology. Perhaps we shall witness the emergence of other, exciting and entirely new theories, such as quantum mathematics, enriching mathematics. Who can tell? The recent interactions between knot theory and the rest of mathematics are really *quite* bewildering. They indicate that there is much still to be done.

Knot theory as it stands today represents a significant stream of ideas, flowing from the challenging difficulties of describing and understanding the phenomenon *knot* and its observable properties. The abstract heights it has reached, and the applications it has so surprisingly found in the wake of Jones' discoveries, give eloquent support to the often-mentioned notion of: *'the seemingly inevitable utility of mathematics conceived symbolically without reference to the real world.'*

It has been said that knots are more numerous than the stars, and are equally mysterious and beautiful. Like the stars seen at night, knots pervade our senses and challenge us to understand them. This happens now, not only in our everyday working world but also, as we learn from the quantum physicists, in our deeper philosophical efforts to explain the mysteries of fundamental physical and biological phenomena. The needs to understand these mysteries will continue to give impetus to the currently widening spread of research into theories and applications of knot theory.

Bibliographic Notes

Knot theory has a substantial literature, albeit very scattered; literature on the history of the subject is also scattered, fragmentary and sporadic. The earliest works, before the turn of this century, tend to mention many interesting sources; but as a rule authors on knot theory after 1900 are rather sparing with their historical information. Luckily there are a few exceptions such as Dehn/Heegaard [29]. From a mathematician's point of view, undoubtedly the most impressive accounts of knot theory's history may be found in Gordon [39] and Thistlethwaite [86]. The encyclopaedic work by Burde/Zieschang [22] evaluates and records the state of the field immediately before the spectacular discovery of the Jones polynomial. Their book supplies fragmentary historical data; but their bibliographic listing has over 1000 entries to compensate. Wilhelm Magnus has written about the early history of braid theory in [64]. Józef Przytycki has described parts of the modern history of knot theory in

[73]; and Joan Birman has written a speculative and exciting overview article [15] of the very latest developments.

After the discovery of the Alexander polynomial, knot literature came in a steadily increasing flow. Nowadays one may speak of an explosive growth of papers in the field. Yet comprehensive books, both monographs and textbooks, are still few and far between. Kurt Reidemeister's pioneering work *Knoten Theorie* of 1932 appeared in English translation [74] in 1983. Its approach, of course, follows the combinatorial spirit of its times, and so only supplies a historical introduction to the subject. Another book still of much value is *Introduction to Knot Theory*, by Richard Crowell and Ralph Fox (1963) [25]. This gives a beautiful introduction to the subject from the classical algebraic topological point of view, and is a fine tribute to the developments which emerged in the post World War II period. *Knots and Links* by Dale Rolfsen (1976) is remarkable for a number of reasons [77]. It is a giant leap into (geometric) topology, and introduces all developments up to the mid-70s. An excellent introduction to the theory of braids is *Braids, Links and Mapping Class Groups* by Joan Birman [14].

Following the explosion of activity in applied knot studies in the late 80s, a stream of books on the topic has been published. For example, Louis Kauffman's *Knots and Physics* [54], hard on the heels of books such as *Braid Group, Knot Theory and Statistical Mechanics* (edrs. C. N. Yang and M. L. Ge, 1989) and *New Developments in the Theory of Knots* (edr. Toshitake Kohno, 1990); these last two are volumes 9 and 11 in World Scientific's *Advanced Series in Mathematical Physics*. This present book is volume 11 in World Scientific's *Series on Knots and Everything*. And in January 1992 the first edition of *Journal of Knot Theory and its Ramifications* appeared, also published by World Scientific; the subject has, at last, its own Journal.

It is inevitable that the new ideas and theories about knots will gradually be introduced into syllabuses for graduate and undergraduate mathematicians and physicists. Textbooks for teaching the subject will come forth. An excellent recent example is *Knot Theory*, by Charles Livingston (Mathematical Association of America, 1993); he covers much of the classical theory, and continues through to high-dimensional knots and the combinatorial techniques of various of the new polynomial invariants. He includes many exercises suitable for undergraduates, to whet their appetites and help them come to grips with this exciting but demanding subject.

References

1. Y. Akutsu and M. Wadati, 'Knots, Links, Braids and Exactly Solvable Models in Statistical Mechanics', *Communications in Mathematical Physics*, **117**, (1988). 243–259.

2. Y. Akatsu and M. Wadati, 'Knot Invariants and the Critical Statistical Systems', *Journal of the Physical Society of Japan*, **56** No. 3, (March 1987). 839–842.
3. Y. Akutsu and M. Wadati, 'Exactly Solvable Models and New Link Polynomials', *Journal of the Physical Society of Japan*, **56** No. 9, (Sep. 1987). 3039–3051.
4. Y. Akatsu, T. Deguchi and M. Wadati, 'Exactly Solvable Models and New Link Invariants. II. Link Polynomials for Closed 3-braids', *Journal of the Physical Society of Japan*, **56** No. 10, (Oct. 1987). 3464–3479.
5. J. W. Alexander, 'A Lemma on Systems of Knotted Curves', *Proc. Nat. Acad. Sci.*, **9** (1923). 93–95.
6. J. W. Alexander, 'Topological Invariants for Knots and Links', *Transactions of the American Mathematical Society*, **30** (1928). 275–306.
7. J. W. Alexander and G. B. Briggs, 'On Types of Knotted Curves', *Annals of Mathematics*, series 2, **28**, (1928), 562–586.
8. B. Arnold, M. Au, C. Candy, K Erdener, J. Fan, R. Flynn, R.J. Muir, D. Wu, J. Hoste, 'Tabulating Alternating Knots through 14 Crossings', *Journal of Knot Theory and its Ramifications*, **3**, No.4, (1994), 433-438.
9. E. Artin, 'Theorie der Zöpfe', *Abh. Mat. Sem. Univ. Hamburg*, **4** (1925). 47–72.
10. E. Artin, 'Theory of Braids', *Annals of Mathematics* **48**, No. 1 (January, 1947).
11. M. Ascher, *Ethnomathematics*, (Pacific Grove 1991).
12. I. Bain, *Celtic Knotwork*, (London, 1986).
13. C. Bankwitz, 'Über die Torsionzahlen der Alternierende Knoten', *Mathematische Analen*, **103**, (1930). 145–161.
14. J. S. Birman, *Braids, Links and Mapping Class Groups*, Annals of Mathematics Studies No.82, Princeton University Press 1975.
15. J. S. Birman, 'New Points of View in Knot Theory', *Bulletin of the American Mathematical Society*, **28**, (April 1993). pp. 253–287.
16. J. S. Birman and X-S Lin, 'Knot Polynomials and Vassiliev's Invariant', *Inventiones Mathematicae*, (1993). 225–270.
17. J. S. Birman and H. Wenzl, 'Braids, Link Polynomials and a New Algebra', *Transactions of the American Mathematical Society*, **313** No. 1, (May 1989). 249–273.
18. O. Boeddicker, *Beitrag zür Theorien des Winkels*, Dissertation, Göttingen (1876).
19. O. Boeddicker, *Erweiterung der Gauss'schen Theorie der Verschlingungen mit Anwenddungen in der Elektrodynamik*, Stuttgart 1876.
20. H. Brunn, 'Über verknotete Kurven', *Verhandlungen des Internationalen Matematiker-Kongress Zürich* (1897). 256–259.
21. W. Burau, 'Über Zöpfgruppen und gleichsinnig verdrillte Verkettun-

gen', *Abh. Mat. Sem. Univ. Hamburg*, **11** (1936). 179–186.
22. G. Burde and H. Zieschang, *Knots*, (New York, 1985).
23. J. H. Conway, 'An Enumeration of Knots, Links and some of their algebraic Properties', *Computational Problems in Abstract Algebra*, (Pergamon Press, 1970). 329-357.
24. P. R. Cromwell, 'Celtic Knotwork: Mathematical Art', *Mathematical Intelligencer* **15** No. 1, (1993) 36–47.
25. R. H. Crowell and R. H. Fox, *Introduction to Knot Theory*, (Springer Verlag, 1963).
26. C. L. Day, *Quipus and Witches' Knots*, (University of Kansas Press, Lawrence, 1967).
27. M. Dehn, *Über die Topologie des dreidimensionalen Raumes*, Mathematische Annalen, **69**, (1910), 137–168.
28. M. Dehn, *Die beiden Kleeblattschlingen*, Mathematische Annalen, **102**, (1914), 402–413.
29. M. Dehn and P. Heegaard, *Encyklopedie der Mathematischen Wissenschaft,* **III** (Leipzig, 1907-1910). 207–213.
30. F. Dingeldey, *Topologische Studien über die aus ringförmig geschlossenen Bändern durch gewisse Schnitte erzeugbaren Gebilde* (G. B. Teubner, Leipzig, 1890).
31. G. W. Dunnington, *Carl Friedrich Gauss: Titan of Science*, (New York, 1955).
32. C. Ernst and D. W. Sumners, 'The Growth of the Number of Prime Knots', *Math. Proc. Camb. Phil. Soc.*, **102** (1987). 303–315.
33. R. Fenn and C. Rourke, 'Kirby's Calculus for Links', *Topology*, **18**, (1979). 1–15.
34. R. Fenn and C. Rourke, 'Racks and Links in Codimension Two', *Journal of Knot Theory and its Ramifications*, **1**, (1992). 343–406.
35. R. H. Fox, 'On the Complementary Domains of a Certain Pair of Inequivalent Knots', *Indag. Math.*, **14**, (1952). pp. 37–40.
36. P. Freyd, D. Yetter, J. Hoste, W. B. R. Lickorish, K. C. Millet and A. Ocneanu, 'A New Polynomial Invariant of Knots and Links', *Bulletin of the American Mathematical Society*, **12** No. 2, (April 1985). 239–246.
37. F. A. Garside, 'The Braid Group and other Groups', *Quarterly Journal of Mathematics*, **20**, No. 2, (Oxford 1969). 235–254.
38. C. F. Gauss, *Werke*, **V**, 605, **VIII**, 271–286.
39. C. McA. Gordon, 'Some Aspects of Classical Knot Theory', *Knot Theory* (Lecture Notes in Mathematics No. 685, Springer Verlag, 1977). 1–60.
40. C. McA. Gordon and J. Luecke, 'Knots are determined by Their Complements', *Bulletin of the American Mathematical Society*, **20**, No. 1, (1989). 83–88.

41. A. Haefliger, 'Knotted (4k–1) Spheres in 6k-space', *Annals of Mathematics*, **75** No. 3, (May 1962). 452–466.
42. W. Haken, 'Über das Homoömorphie Problem der 3-Mannigfaltigkeit I. *Math. Zeit.*, **80**, (1962). 92–120.
43. M. G. Haseman, 'On knots, with a census of amphicheirals with 12 crossings', *Trans. Roy. Soc. Edinburgh*, **52**, (1918). 235–255.
44. G. Hemion, *The Classification of Knots and 3-dimensional Spaces* (Oxford 1992).
45. M. Jimbo, 'Quantum R Matrix for the Generalized Toda System', *Communications in Mathematical Physics*, **102**, (1988). 537–547.
46. V. F. R. Jones, 'A Polynomial Invariant for Knots via von Neumann Algebras', *Bulletin of the American Mathematical Society*, **12**, No. 1 (January, 1985).
47. V. F. R. Jones, 'A New Knot Polynomial and von Neumann Algebras', *Notices of the American Mathematical Society* **33**, No. 2, (1986). 219–225.
48. V. F. R. Jones, 'Knot Theory and Statistical Mechanics', *Scientific American*, (November 1990). 52–57.
49. V. F. R. Jones, 'On Knot Invariants Related to some Statistical Mechanical Models', *Pacific Journal of Mathematics*, **137** No. 2, (1989). 311–334.
50. V. F. R. Jones, *Subfactors and Knots* (CBMS No. 8, American Mathematical Society, 1991).
51. D. Joyce, 'A Classifying Invariant of Knots, the Knot Quandle', *Journal of Pure and Applied Algebra*, **23**, (1982). 37–65.
52. T. Kanenobu, 'Infinitely Many Knots with the same Polynomial Invariant', *Proc. of the American Mathematical Society*, **97** No. 1, (1986). 158–162.
53. A. Kaselowsky, *The Peruvian Quipu* (Essay in the History of Mathematics, Århus University, 1993). (See chapter 5 of this book).
54. L. H. Kauffman, *Knots and Physics*, (WSP Singapore 1991).
55. R. Kirby, 'A Calculus for Framed Links in S^3', *Inventiones Mathematicae*, **45**, (1978). 35–56.
56. F. Klein, *Annals of Mathematics* **IX**, 478 (1876).
57. M. Kline, *Mathematical Thought from Ancient to Modern Times* (Oxford University Press, 1972).
58. C. G. Knott, *Life and Scientific Work of Peter Guthrie Tait* (Cambridge University Press, 1911).
59. P. P. Kulish, N. Yu. Reshetikhin and E. K. Sklyanin, 'Yang-Baxter Equation and Representation Theory I, *Letters of Mathematical Physics*, **5**, (1981). 393–403.
60. W. B. R Lickorish, 'A relationship between link polynomials', *Math.*

Proc. Camb. Phil. Soc., **100**, (1986), 109-112.
61. W. B. R. Lickorish, 'A Representation of Orientable Combinatorial 3-Manifolds', *Annals of Mathematics*, **76** No. 3, (November 1962). 531–540.
62. J. B. Listing, *Vorstudien zür Topologie* (Göttingen, 1847). 857–866.
63. C. N. Little, 'On Knots, with a Census for Order 10', *Trans. Connecticut Ac. Arts. and Sci.*, **18**, (1885). 374–378.
64. W. Magnus, 'Braid Groups', *Proc. Sec. Int. Conf. on Theory of Groups* (Canberra, 1973). 463–487.
65. W. Magnus, 'Max Dehn', *The Mathematical Intelligencer*, **0–2**, (1977). 132–143.
66. W. Menasco and M. Thistlethwaite, 'A Geometric Proof that Alternating Knots are Non-trivial', *Math. Proc. Camb. Phil. Soc.*, **109** (1991). 425–431.
67. W. Menasco and L. Rudolph, 'How Hard Is It to Untie a Knot?', *American Scientist*, (Jan 1995), 38–49.
68. J. Murakami, ' The Kauffman Polynomial of Links and Representation Theory', *Osaka Journal of Mathematics*, **24**, (1987). 745–758.
69. K. Murasugi, 'Jones Polynomials and Classical Conjectures in Knot Theory', II. *Math. Proc. Phil. Soc.*, **102**, (1987). 317–318.
70. K. Murasugi, 'On the Alexander Polynomial of the Alternating Knot', *Osaka Journal of Mathematics*, **10**, (1958). 181–189.
71. K. A. Perko, 'On the Classification of Knots', *Proc. of Amer. Math. Soc.*, **45** No. 2, (August 1974). 262–266.
72. H. Poincaré, Analysis Situs Journal, École Polytech, (1895). 1–121.
73. J. H. Przytycki, 'History of Knot Theory from Vandermonde to Jones', *Proc. of the Mexican National Congress, November 1991* (to appear).
74. K. Reidemeister, *Knot Theory*, (BCS Associates; Moscow, Idaho/USA 1983).
75. N. Y. Reshetikhin, 'Quantized Universal Enveloping Algebras, the Yang-Baxter Equation and Invariants of Links', (Pre-print No. E.4–87, Leningrad Branch, Steklov Inst., 1987).
76. N. Y. Reshetikhin and V. G. Turaev, 'Invariants of 3-Manifolds via Link Polynomials and Quantum Groups', *Inventiones Mathematicae*, **103** Fas. 3, (1991). 547–598.
77. D. Rolfsen, *Knots and Links*, (Publish or Perish, Inc., 1976).
78. H. Seifert, 'Über das Geschlecht von Knoten', *Mathematische Annalen*, **110**, (1934). 571–592.
79. H. Seifert, 'Verschlingungsinvarianten', *Sitzungsber. Preussen Akad. Wiss. Berlin*, (1933), 811-828.
80. O. Simony, *Lösung der Aufgabe: In ein ringförmig geschlossenes Band*

einen Knoten zu machen (Gerold & Comp., Vienna, 1881).
81. P. Stäckel, Gauss als geometer, *Carl Friedrich Gauss Werke* **10**, pt. 2, sec. 4. Gesellschaft der wissenschaften in Göttingen (B. G. Teubner, Leipzig, 1923). Also in *Materialien für eine wissenschaftliche biographie von Gauss*, compiled by F. Klein, M. Brendel and L. Schlesinger. Nachrichten der K. Gesellschaft der Wissenschaft zu Göttingen (1917).
82. C. Strohecker, *Why Knot?* (Doctoral thesis, MIT 1991).
83. P.G. Tait, 'On Knots; I, II and III.' *Scientific Papers* (I. London 1877-85, 273–347: C. U. P. 1898).
84. P. G. Tait, 'On Knots' Transactions of the Royal Society of Edinburgh, **IX**, 237–246, 289–298, 306–317, 321–332, 338–342 (on links); also Cayley, 363–366, 383–391; and Muir, 391, 392, 403, 405. (1876–1877).
85. M. B. Thistlethwaite, 'Kauffman's Polynomial and Alternating Links'; *Topology*, **27** No. 3, (1988). 311–318.
86. M. B. Thistlethwaite, 'Knot Tabulations and Related Topics', *Aspects of Topology*, London Math. Soc. Lecture Notes Series, No. 93 (Cambridge Universities Press 1985).
87. W. Thompson, *Philosophical Magazine* **34**, (July, 1867). pp. 15–24.
88. V. G. Turaev, 'The Yang-Baxter Equation and Invariants of Links', *Inventiones Mathematicae*, **92**, (1988). pp. 527–553.
89. J. C. Turner, *A Study of Knot-Graphs*, D.Phil thesis, Department of Mathematics, University of Waikato, New Zealand (March, 1984).
90. A. T. Vandermonde, 'Remarques sur les problèmes de situation', *Mémoires de l'Académie Royal Des Science* (Paris, 1771). 566–574.
91. V. A. Vassiliev, 'Cohomology of Knot Spaces', *Theory of Singularities and its Applications*, Advances in Soviet Mathematics **1**, ed. V. I. Arnold, American Math. Soc. (1990).
92. A. H. Wallace, 'Modifications and Cobounding Manifolds', *Canadian Journal of Mathematics*, **12**, (1960). 503–528.
93. H. Weith, *Topologische Untersuchung der Kurven-Verschlingung*, Inaugural Dissertation, Zürich (1876).
94. H. Wenzl, 'Representations of Braid Groups and the Quantum Yang-Baxter Equation', *Pacific Journal of Mathematics*, **145** No. 1, (1990). 153–180.
95. J. H. White, K. C. Millet and N. R. Cozzarelli, 'Description of the Topological Entanglement of DNA Catenanes and Knots by a Powerful Method involving Strand Passage and Recombination.' *Journal of Molecular Biology*, 1987, 585–603.
96. E. Witten, 'Quantum Field Theory and the Jones Polynomial', *Communications in Mathematical Physics*, **121** No. 3, (1989). 351–399.
97. E. C. Zeeman, 'Unknotting Spheres', *Annals of Mathematics*, **72** No. 2, (September 1960). 350–361.

98. J. C. F. Zöllner, *Transcendental Physics*, (Harrison, London 1880).
99. *Knots and Applications*, (ed) L. H. Kauffman, Series on Knots and Everything, **6**, (WSP, Singapore, 1995).
100. *Knots, Groups and 3-manifolds*, (ed) L.P. Neuwirth, Annals of Mathematics Studies **84**, Princeton University Press, (1975).

CHAPTER 12

ON THEORIES OF KNOTS

John Turner

" 'And everybody praised the Duke,
Who this great fight did win.'
'But what good came of it at last?'
Quoth little Peterkin.
'Why, that I cannot tell,' said he,
'But 'twas a famous victory!' "

[On G. T. Fechner turning psychology into an exact science; quoted in *The World of Mathematics*, J. R. Newman, p. 1165.]

1. Is Knot Theory Topology?

The earliest scientific paper we know in which a mathematician discusses the problem of constructing a mathematical theory of knots, contains the following paragraph:

> *Whatever the twists and turns of a system of threads in space, one can always obtain an expression for the calculation of its dimensions, but this expression will be of little use in practice. The craftsman who fashions a braid, a net, or some knots will be concerned, not with questions of measurement, but with those of position: what he sees there is the manner in which the threads are interlaced.*
>
> Alexandre Theophile Vandermonde (1735–1796)

A few decades later, Carl Friedrich Gauss (1777–1855), one of the greatest mathematicians and physicists of all time, made sketches of knots*and began to think of their properties. It is clear from notes found in his papers after he died that during his working life he gave much thought to the problem of capturing the essence of knots in mathematical terms. However, to our knowledge he published only one paper which referred to knots; this paper dealt with a problem in the theory of electrodynamics, and it involved the linking number of two wires winding together in space. In 1847 his student (and later colleague, at Göttingen University) Johann Benedict Listing (1806–1902) published the first book on Topology [11]. The book was devoted primarily to knot theory; and we may surmise that Listing was influenced by Gauss when developing his ideas on this virgin subject.

The story of how topological knot theories have developed since 1847 is told in the previous chapter of this book. It is a tale which deals almost exclusively with the hunt for topological invariants of knots; that is, with the search for mathematically derived numbers or expressions, obtained from studies of diagrams of knots, which enable one knot to be distinguished from another with mathematical certainty. Knots are truly mysterious objects; it is hard to say unambiguously what they are, and even harder to describe their properties. Before seeking a topological invariant for them, one first has to find a way of defining a knot, in mathematical language. Then one has to state very carefully what one means by saying 'That knot is the same as this knot.' Indeed, it turns out that there are many different, useful ways to define 'sameness', each being with reference to a different set of properties of a knot. In short, one has to define equivalence relations for knots. Then, and only then, can one attempt to classify knots, and to study their properties, using mathematical models and/or experimental methods.

The kind of knot theory we have spoken about so far is 'classical' or 'topological' knot theory. Since the publication [7] of the Jones polynomial invariant in 1986, with its applications in genetics and quantum physics, development of this theory has become a burgeoning industry in Mathematical Science. We submit, however, that it is only one part of the Science of Knots.

There is a great deal to learn about knots which does not hinge upon purely topological notions. Indeed, we assert that the current topological knot theory says nothing at all of interest or use to "the craftsman who fashions a braid, a net or some knots" (see the above Vandermonde quotation).

The craft of braiding is perhaps as old as Humankind. It is highly likely that Eve discovered how to braid her own tresses in the Garden of Eden. Moreover, it is still, today, a highly useful and decorative craft practised extensively in virtually every society. Is it not strange, then, that until recently no useful

*A page from his notebooks, bearing the date 1794, is shown in the Preface to this book.

mathematical theories have been developed to help the braider with his craft? Especially as the process of braiding is a regular, repetitive sort of task, with string or strings being interwoven into patterns which always have symmetries which are capable of geometric or other sorts of mathematical description.

In this Chapter we shall describe various attempts made in this century by knot craftsmen (artisans rather than mathematicians) to list and classify knots by means of diagrams and reference to their uses. Then we shall outline the work done in the last fifteen years by a New Zealand mathematician, Georg Schaake, to provide a strong mathematical theory of braiding processes. We shall call his work 'Braiding Theory', to distinguish it from 'Topological Knot Theory'. It must not be confused with Emil Artin's Braid Theory, which was introduced in 1925 and is part of topological knot theory.

We wish to show that the statement, often heard on the lips of mathematicians, that 'Knot Theory is a branch of Topology', is not only misguided but also false. There are other theories of knots, which are not topological. Moreover, topological knot theory deals with but a small subset of the mysteries of knot lore: to imply that it covers all of Knot Science is rather like saying that Medical Science is the study of diseases of the foot.

2. What is a Science?

The question: What is a Science? is a large one. Indeed the philosophy of science is a subject in its own right. However, simple and general notions of what constitutes a science, leaving aside any mathematical requirements, can be culled from any dictionary. For our starting point we shall take two definitions from a recent Collins Concise Dictionary.

The first definition, which is too weak to be of much value (it makes just about any subject into a science) is: *A Science is any body of knowledge organised in a systematic manner.*

The second states that: *Science is the systematic study of the nature and behaviour of the material and physical universe (or any particular branch of the knowledge gained) based on observation, experiment, and measurement.*

Before a scientific theory of anything, still less a mathematical science, can be developed, one has to settle upon a class (or universe) of objects to study. Then one has to attempt to define objects in that universe with something akin to mathematical precision. Thus the first step towards a Science is essentially descriptive; it involves observing, recording and naming the objects in a chosen universe.

Within and following the initial step, observations are made which determine relationships between the objects. These enable the universe to be partitioned in useful or interesting ways. In other words, one specifies equivalence relations, which then place the objects into different classes according to special, singled-out properties. Some of these properties will grab the attention of investigators—excite them in a way which directs their efforts to understand the properties more clearly, more deeply. Often, with growing knowledge of the objects, they begin to gain some kinds of control over the objects. They learn how to manipulate them in ways which have interesting, maybe useful, consequences. They make hypotheses about them, such as: 'If I do this to these objects, and that to those ones, the results will be such and such.' Then they work out experimental methods for testing their hypotheses.

Testing of hypotheses proceeds essentially in one of two ways:

(i) Experiments are made with the actual objects. Results of the experiments will then suggest answers to the queries, or settle the hypotheses one way or another. Confidence in one's conclusions will depend on many factors, such as experimental design, sample sizes, accuracy of the observations, and so on.

(ii) If mathematical descriptions of the objects and their relationships are sufficiently well-advanced, mathematical models of the objects can be manipulated in ways suggested by the hypotheses. Other kinds of model, iconic or analogue, may be used similarly. Results obtained by observing changes in the models can then be used to confirm or deny hypotheses.

When using mathematical or other types of models, their suitability must always be kept in question. Experiments must be made with the objects themselves, to test whether the real processes correspond closely with the changes and processes occurring with the models. If they do, then confidence will be gained in the validity of the modelling: and answers to hypotheses, gleaned from the models, will be considered to be new knowledge about the objects, with a fair degree of justification. If they don't, the validity of the models will be in doubt, and they must be improved before further use.

When much of the above is done, or at least well-started and moving along the indicated lines, it is permissible, in our view, to claim the title 'Science' to apply to this universe and its methods of study.

To summarise, it is generally agreed that the key elements involved in developing a science are making observations on a set of objects, establishing descriptions and definitions of key characteristics, classifying the objects, modelling processes and relationships involving the objects, and making and testing of hypotheses about them. And revising the models when necessary.

As to the objects of study, the second dictionary definition would seem to restrict the field to the material and physical universe, although, in parentheses, it extends this to include any particular branch of the knowledge gained. We find this restriction puzzling, old-fashioned maybe, and unacceptable. Does the Dictionary writer find Biology hard to classify as a Science? What of the Social Sciences, such as Economics and Psychology? Can the study of money markets, or labour movements, or social deviants be said to be science? And what about Linguistic Science—can there be one?

In the vast new world of Computers, is it only the study of hardware systems that constitutes Computer Science? Do the theories of the software that drive them not qualify also to be under that umbrella? In the 1960s and 1970s a continuous debate raged as to whether the theory and practice of Computing could be called Science. Now most Universities in the world have Computer Science Departments, even Schools of Computer Science.

Is Mathematics a Science? That seems harder to decide; if not, what about part of it, such as Number Theory? This has been called 'the most exact Science'; but does it qualify* to be called Number Science? Some would have us lump together mathematics, computer science, parts of philosophy and the theoretical parts of science and engineering, and place them in a box labelled *the logical sciences*; because, they would claim, they use the method of logical deduction from axioms. That is patently wrong; or, at best a half-truth. I claim that most new mathematics is arrived at empirically, by mathematicians 'playing with' their materials. They seek and find patterns, which they then formulate as theorems (conjectures). *Only after* this experimental process do they turn to the (often difficult, sometimes impossible) task of proving the theorems, linking them to axioms via previously proved theorems. Anyway, not all branches of mathematics, or the other subjects listed, have been axiomatised.

It is evident that there are grey areas, between subjects which are universally labelled 'Science' (such as Chemistry and Physics) and those which are clearly 'non-Science' ('Arts' subjects such as Greek and Latin).

Where does the study of Knots fit into this spectrum? Our next section examines this question.

3. Is there a Science of Knots?

Under the first definition from the Collins Dictionary, it is impossible to deny that the study of knots is a Science. There is a considerable body of knowledge on the subject; and that has been organised systematically in a number of ways, as we shall show below.

Under the second definition, however, the evidence has to be looked at more closely before arriving at a decision. There is no doubt (in this author's mind) that knots are phenomena which make up part of the material and physical universe—even though, in order for them to exist, they usually are first created by man or woman (they do occur naturally; for example, consider

*A distinguished mathematician of the last half-century, Freeman Dyson, is on record [3] as believing that number theory is applied mathematics. He says: 'You are not creating ideas; you're just applying methods and using numbers as your experimental material.'

the coiled and tangled DNA in all our cell nucleii). And there is ample evidence that those who construct and study knots make observations, carry out experiments, and take measurements of one kind or another. These activities may not correspond in kind, physically, with those of Chemists experimenting with chemicals* in their laboratories; however, they do embody the spirit and practices of 'scientific method'.

The evidence to be presented in this chapter, together with that of most other chapters in this book, convince the present author that the Study of Knots can be regarded as a Science under the second definition too. Moreover, he is satisfied that it qualifies as a Mathematical Science, with at least two types and areas of mathematical modelling now available for enlarging its study.

Before relating the above notions on what a Science is to the current states of knowledge and theories about knots, we shall briefly trace the history and contents of one well-known Science, namely Biology. Comparisons between this example and the development and current state of 'knot science' will then be possible and useful.

Biology is often known as the Science of Life. It treats generally of the life of animals and plants, including their morphology, physiology, origin, development and distribution. It attempts to survey all the phenomena manifested by living matter. It would seem that its 'universe' is quite clear; and yet, at bottom, it is notoriously difficult to distinguish the limit at which an object may be said to be 'living', rather than 'inanimate'. Classification of living objects into animals and plants is not always easy, either.

Granted these shaky foundations of object definition, the saga of the development of the biological sciences is a long and exciting one. The study of living beings has proceeded unchecked since prehistoric times. Beginning with empirical folk-knowledge, collected and handed down orally over many hundreds of years, much basic classification would be done as the observations were collected. Leaving prehistory behind, and focussing on the development of animal science, we can trace knowledge of human and other animal forms, and of functions of visceral organs in them, in the records of the Babylonian and Egyptian civilizations. In Greece, we find in the fifth century B.C. the earliest attempts to organize such knowledge in systematic form. Hippocrates (460–377 B.C.) discarded magical theories of disease, and Aristotle (384–322 B.C.) originated scientific classification. And so on, and on through the many centuries up to the present-day, the organised knowledge on life forms has accumulated, the literature on it bespattered with great names such as Harvey,

*Chemistry now has its knot-scientists. In 1989, French chemists synthesised the first knotted compound ever made, a 124-atom molecule in the form of a trefoil. More complex knotted molecules are now being produced.

de Buffon, Linnæus, Wallace, Darwin and Mendel.

Such is the vastness of this study, that splintering into sub-branches of knowledge has perforce taken place. For example, there are now sciences (or sub-sciences) of Physiology, Morphology, Anthropology, Embryology, Evolution and Genetics, and so on; and as each of these grows in size, it forms its own branches, sub-branches, off-shoots.

As with Biology, so it is with the other great Sciences, such as Chemistry, Physics and Medicine. They are each traceable from prehistoric times, and have developed from myths and customs, from gradual accumulation of knowledge and practices handed down from tribal times. The knowledge slowly became more codified, written down, organised; and the practices became more efficient, more likely to be correct, more useful as times passed. And every Science has had its great discoveries, its great men and women, its moments in history when big advances have become possible with a new invention or discovery. The telescope and spectroscope in Astronomy, the microscope in Biology, Dalton's atomic theory and its effects on Chemistry and Physics—these are but a few examples.

With all the subjects which we now unhesitatingly label as sciences, believing them to deal sensibly with their subject matter, it is well to remember that until quite recently (within three centuries of now) they all were little more than arts and crafts themselves, propelled by folk lore, magic and religious cant. For example, it was not until the middle of the 17th century that Newton's work, building upon Galileo's discoveries, moved Astronomy and Physics forward into the scientific paths they travel today. Chemistry grew out of Alchemy; and even as late as the mid-18th century we did not know anything sensible about what constitutes the air we breathe. A glance at the history of medical 'science' will show how primitive that was, right up to this century; even today many of its 'sub-sciences' are primitive, subject to fashion and loosely formed opinion.

How, then, does the study of knots and the practice of knotting stand in relation to the above examples?

It has to be said, first, that if knot theory is a science then it is a hidden one—in the sense that very few people are aware of it. Whilst everyone beyond the age of three has tied a knot, almost no-one knows that knots play a large and quite vital role in their lives. They are taken completely for granted. They may be seen (or, more likely, not seen) everywhere. They are used daily by sailors, surgeons, soldiers, firemen, fishermen, building construction workers and so on, in a wide variety of forms. They are made in many materials. Rope making is a trade that exists today, and has existed for many thousands of years. And, in a wider sense, any form of woven or knitted structure, and any inter-weaving process, can be said to fall within the field of Knot Science.

Thus, although it does not stand in any limelight nowadays, except perhaps within the coteries rejoicing in the recent discoveries of Jones et al., the development of knot practice and knowledge can be compared favourably with any of the other main sciences. Every chapter in this book corroborates the immense history of knots, the ubiquity of their existence throughout societies, and the range of their applications in the service of mankind. It must be said, however, that the writing down of knot lore, and the codification of its objects, methods and applications, is much more sporadic, sparse, and less traceable through the centuries than is that of any of the main Sciences mentioned above. This fault, if so it can be called, is being remedied quite rapidly now. In the last 150 years or so, many people have expended great efforts to catalogue knots and to study their uses and properties. And the list of people whose names will 'live' in future writings on knots is growing longer by the decade.

We now turn to a discussion of a few important attempts to classify knots, made within this century.

4. Organised Knowledge of Knot Forms

As we have said, any Science begins with a class of objects, which are observed closely, given names, and then arranged into classes and subclasses according to some criteria.

There are many problems attached to these early processes. Finding common names for the objects is one of them. Each worker who studies the objects will bring his or her own special viewpoint (and language) to the naming task, and to placing them into classes. Many objects will acquire several names; and different kinds of classification will be made. No Science, however, is immune from these problems; and the differing classifications are to be welcomed, since each adds to the general knowledge about the objects—they enrich the emerging Science.

Some level of agreement has to be achieved, though, when the subject comes to be written about. Attempts have to be made to standardise names and definitions. In the practical world of knotting, the following basic definitions are usually given first when knot lore is being written down.

The word ***knot*** appears to derive from ***canute*** or the Anglo-Saxon word ***cnotta*** [4]. In its most general sense it refers to any kind of fastening made by interweaving of cordage. Virtually every language has a word for this kind of object. For example, the Germans say 'Knoten' and the French say 'nœud'; the Italian says 'nodo', and the Swahili says 'fundo'.

A well-recognized basic classification of knots (though one with many blurred distinctions and overlaps) separates the field into ***Hitches***, ***Bends*** and ***Knots***, thus:

(1) **Hitch**—A knot which secures a rope to an object (such as a spar, post or ring bolt) or to the standing part of another rope.

(2) **Bend**—A knot which joins two ropes together, end to end, and typically in line.

(3) **Knot**—A method of interweaving used to form a knob or a stopper in a rope; or to enclose an object; or to form a noose.

Although there is general acceptance of these terms, it is recognised that they leave much to be desired when attempting to sort knots into classes. Discussion on this may be found in [2], [4].

There is no universally agreed system for naming knots; this also leaves much scope for confusion in the literature. Warner [27] gives the following example of this: 'What I call the Fishermans Knot is also known, somewhere or other, as the Anglers Knot, English Knot, Englishmens Knot or Bend, Halibut Knot, Leader Knot, True Lovers Knot or Bend, Water Knot or Bend, or Watermans Knot.' A little further on, Warner says: 'These confusions in the naming of Knots raise few problems with localised groups of people in the one trade or craft, who are mostly very comfortable with their own traditional names, but they do make difficulties for anyone interested in knots in general.'

A handful of people, in this century, have tried to overcome problems of defining and classifying of knots by producing encyclopedias in which they have gathered together large numbers of drawings or photographs of knots, placed them in some kind of logical order, and given them numbers and names. Their collective work has been of immense value to those who study and practice knot-tying. Perhaps we might reasonably compare their labours with those of the great collectors and cataloguers in the Biological Sciences—an outstanding 18th century example in Botany being Carl Linnæus (1707–78).

An example of a classifying study of knots is one by Dr. J. Lehmann, published in German [10] in 1907. It is hardly encyclopedic; but by means of 166 Figures, 3 Tables and accompanying text it catalogues a large number of knots, nets, basket weaves, etc. from many countries of the world. Its examples are drawn from museum specimens; and various systematic classifications are made, by numbers, letters and names. It is a notable attempt to describe knots with reference to their uses and geographical distribution.

A second example is the important and widely-quoted work of Cyrus L. Day, entitled *The Art of Knotting and Splicing* [2]. It was first published in 1947 by Dodd, Mead and Company, Inc., New York; and it appeared in a second, expanded edition in 1955, under the imprint of The United States Naval Institutes, Annapolis, Maryland. Cyrus Day gathered much historical information about knots. He presents this in the Introduction to the book, and then discusses characteristics of rope, basic knot terminology, and strength and security of knots. With the aid of over 1000 photographs, he then lists, describes and gives names and numbers to over 200 knots and operations with knots and ropes.

Two further scholarly contributions by C. L. Day on the history of knots might well be mentioned here. They are: *Knots and Knot Lore: Quipus and Other Mnemonic Knots* (Western Folklore, 1957), and *Quipus and Witches' Knots* (Lawrence, 1967).

We shall next review four books of wider scope, written in the English language since the 1930s. There is, of course, much overlap; but between them they describe some 4000–5000 different knots. Each gives diagrams, names, textual descriptions, and instructions for tying of the knots. All discuss the various uses of knots; and their classification methods generally depend in some way on the functional uses of knots. Thus their contents constitute a basis for the descriptive part of a practical Knot Science.

4.1. 'Encyclopedia of Knots and Fancy Rope Work'

This extensive work [4] on knots was first published by Cornell Maritime Press, Maryland, U.S.A. in 1939. It was first written by Raoul Graumont and John Hensel; and reached a fourth edition in 1952, when it was completely revised and enlarged by Raoul Graumont. The Encyclopedia is still in print.

The fourth edition is a book of 690 pages. Generally, each right-hand page contains photographs of up to 25 different knots, each of which bears a number; and each corresponding left-hand page supplies a name and brief description for each of these knots. We estimate that over 3500 knots are treated in this way.

In general the photographs are clear; and most 'small' knots could be made up from them and their descriptions. But for the very many designs of, for example celtic weaves, macramé knots, belts, handbags and leashes, it would be very difficult, if not impossible, to make the objects up without further instruction.

The scope of the Encyclopedia, and its main divisions, are best seen from the headings in the Contents thus:

> Notes on the History of Knots and Rope Making;
> I Elementary Rope Work; II Simple Knotting; III End Rope Knots; IV Rope Splicing; V Coxcombing; VI Turk's Heads; VII Sennit Braiding; VIII Ornamental Knotting; IX Macramé Tatting, Fringe, and Needle-Work; X Useful Rope Designs; XI Miscellaneous Knotting; XII Splicing Wire Rope.

Quoting from the Preface, the authors' aims emerge thus:

'Insofar as the authors of this work have any knowledge, there is no complete and comprehensive literature covering the entire subject of tying knots and the making of ornamental rope designs. Most of the knowledge on these subjects was never published, but was handed down from man to man, and from generation to generation, through actual contact with those who were familiar with the work ...'

'... It is therefore the purpose of this book to include, insofar as it is possible, all of the known kinds, types, forms and designs of knots, macramé, lace, tatting, braiding, and ornamental rope work, in order that the art may be preserved as an historical record for future generations.'

4.2. 'The Ashley Book of Knots'

On the front cover of this encyclopedia [1] are the claims that the book includes:

'Every practical knot—what it looks like, who uses it, where it comes from, and how to tie it ... with 7000 drawings representing over 3900 knots.'

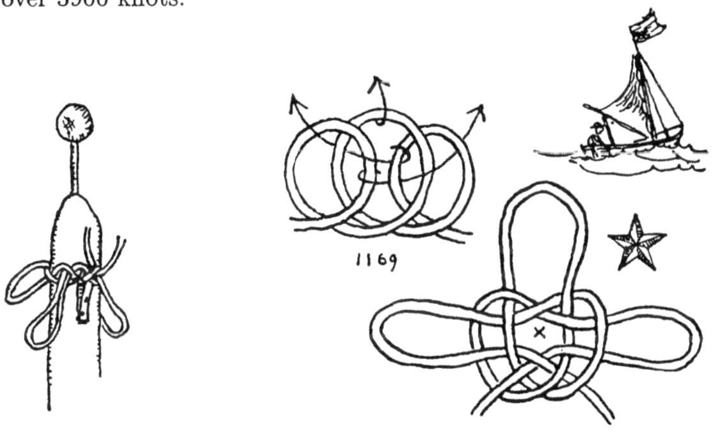

Fig. 1. Tying the Jury Mast Knot, #1169; an example from p. 212 of Ashley's book

This beautiful book of 620 pages is truly a celebration of knot lore, covering all its practical facets with charming drawings and text. It was written by Clifford W. Ashley, and first published in 1944 by Doubleday & Company, Inc., New York. It is still in print, and will no doubt remain so for very many years: it is the 'classic' in its field.

Mr Ashley was born in 1881, in the whaling port of New Bedford, Massachusetts. He had two ruling passions all his life, namely marine painting and knot tying. His knot encyclopedia resulted from forty years of looking for, trying out, and thinking up new knots; and he devoted eleven years to the writing of it. The 7000 drawings are from his own hand, and he enhances the actual knot-diagrams with tiny sketches of people, ships, ladders, landscapes etc. To illustrate, we give in Fig. 1 his drawings from p. 212 which show how to tie a *Jury Mast Knot*. (N.B. This is one of the knots on the front cover of this (our) book.)

The following summary of the contents of the book tells how Ashley classified knots into 41 chapters.

On Knots, chapter 1, introduces history and general methods of construction of knots. Then *Occupational Knots* attempts a classification by 'occupation': thus he covers knots as used by the Archer, the Artilleryman, the Artist, the Angler, and so on, through to the Weaver, the Well Digger, the Whaleman, the Whipper and the Yachtsman.

The remaining 39 chapters deal with knots in topics as follows: Knob Knots, from single-strand stoppers to multi-strand buttons; Single- Double- and Multiple-Loop Knots; The Noose; Knots tied in the Bight; Clove Hitch and other Crossing Knots; Binding Knots; The Turk's Head; Bends; Shroud Knots (multi-strand Bends); Belaying and Making Fast; Hitches of various kinds; Hooks, Beckets, and Toggles; Miscellaneous Holdfasts; Occasional Knots; Lashings and Slings; The Monkey's Fist and other Knot Coverings; Flat or Two-dimensional Knots; Fancy Knots; Square Knotting; Tricks and Puzzles; Long and Short Splices (multi-strand Bends); Eye and Odd Splices; Sinnets—Chain and Crown, Plat, Solid; Marlingspike Seamanship.

This description of the contents of *Ashley's Book of Knots* serves to show the extent of service that knots give to mankind, and brings home the enormous variety of their forms, constructions and functions.

4.3. 'Encyclopedia of Rawhide and Leather Braiding'

This book, written by Bruce Grant and published by Cornell Maritime Press, Maryland in 1972 (still in print) claims to be 'the most definitive work on the subject [of leather braiding] in the English language.' In 528 pages, Grant gives diagrams (on right-hand pages) and text (on left-hand pages) which catalogues and give excellent instruction for the tying of a very large number of braids and leather-braided objects. The diagrams are largely hand-drawn (see the example in Fig. 2), but there are also many photographs.

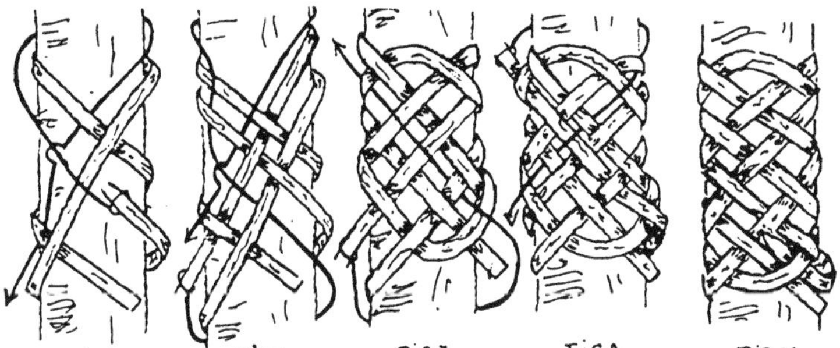

Fig. 2. Constructing a 9P/4B Turk's Head braid; from p. 384 of Grant's book

Grant organises his material under the following headings:

> The Romance of Leather Braiding; Rawhiding in the West; Making and Working Rawhide—Leather-braiding tools; Twist Braids; Slit Braids; Flat Braids; Round Braids; Crocodile Ridge Braids; Square, Rectangular, Spiral, Twist Spiral, Oval and Triangular Braid; Edge-Braiding; Buckle and Ring Coverings; Other Braided Coverings; Handle Coverings; Hobbles; Turk's-Heads; Braided Knots; Lazy-Man Button and Pampas Button Knots; The Braiding Detective; Braided Appliqués; Braid on Braid; Additional Projects and Examples.

Grant tells us in the Foreword that the book was the result of almost a lifetime of work. It arose 'as the inevitable consequence of two prior books, as well as numerous magazine articles on braiding ...' It certainly makes good his statement that braidwork takes many forms, with applications that are practical as well as decorative; combinations of beauty and utility lend themselves to a large array of items of use to mankind.

4.4. 'A Fresh Approach To Knotting and Ropework'

This recent book [27] was written by Charles Warner, of Australia; and published by him in 1992. It is a well-presented, comprehensive work of 272 pages, supplying nearly 1000 hand-drawn illustrations of knots and their applications. A name and a number (linked to its illustration) is given to each knot, and alongside there is some discussion on its construction, properties and uses. The author states in the Preface:

> 'Just about all knots are useful, even those we think of as primarily decorative, or use in toys and puzzles. This book gives emphasis to the uses and functions of knots... Now, there is a logic in the construction of knots, and I have tried to follow that logic in the

writing of this book. There has been too little attention in the past to the structure of knots as a basis for the organisation of knotting books or training courses. This tends to make learning about knots a series of isolated exercises in memorising apparently unrelated facts and manipulations. Many knots of diverse functions share common structures. Calling attention to these similarities, as I do here, should help you understand something of the relationships amongst knots, which in turn should help you learn more about them, from how to tie them to what to use them for.'

This emphasis on structure leads Warner to develop a system of classification which is made clear by the Plan of the book, which is as follows:

Part 1: supplies background information 'to help you get going' on ROPE (Manufacture, Preparation and Care) and KNOTS (Tying, Testing and Classifying).

Parts 2 and 3: These constitute the core of the book. They demonstrate the method of arranging the knots by their structure, the way they get their grip. About 500 knots are thus classified. The main, two-way, division is into **Knots with distributed nip** and **Knots with concentrated nip**. The first group is partitioned under the headings 'Twists', 'Overhands', 'Friction', and 'Figure Eights'. The second group divides into 'Hitches' and 'Interweaves'.

Part 4: Deals with designer knots—the search for different knots; designing, tying and testing novel knots, for various uses.

Part 5: Covers Field Engineering uses of rope and knots, and deals with 'Surveying', 'Lashings', 'Anchorages', 'Tackles'.

Part 6: Provides 'a whole set of ways of locating what you want in this book'—a variety of lists, indexes, definitions, etc.

4.5. Summary of the Books

The last four books described above have the common aim of capturing and presenting, in some systematic way, a great deal of information about knots and their uses. They catalogue them, classify them, give instructions on how to tie them, and make the reader aware of their uses.

Perhaps the most widely known of the four (it is often referred to as the Knotter's Bible) is the Ashley Book of Knots. Its numbering system—recall that it gives diagrams and numbers for nearly 3900 knots—is now frequently used by writers of articles on knots, or scientific authors of articles in which knots occur, to identify their knots. For example, if one reads in the I.G.K.T. Newsletter of a use of knot #412, one can turn to page 66 of a copy of 'Ashley' and discover that the reference is to *The Handcuff Knot*, and study the clear diagram of the knot given there. This is evidently an extremely useful function for the Encyclopedia to give to the world.

However, the kinds of diagram given in the encyclopedias described above are not suitable for topological studies of knots. In the next subsection we describe diagrams that have been developed for that purpose.

4.6. Topological Knot Diagrams

Topological Knot Theorists use different tables of diagrams, when they wish to refer to a particular knot. When one approaches a theory of knots, one has to begin by drawing diagrams of them; usually these are essentially two-dimensional (projections into a plane) with some kind of markings indicating where the crossings occur and whether they are 'overs' or 'unders'. For historical interest, we give below (Fig. 3) diagrams of a few ten-crossing knots, as first published by P. G. Tait in *Transactions of The Royal Society of Edinburgh, 1885*.

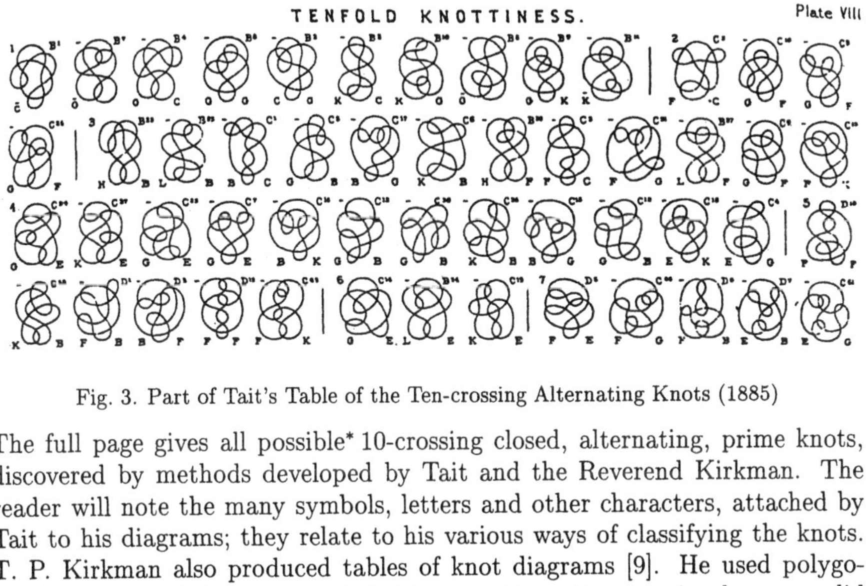

Fig. 3. Part of Tait's Table of the Ten-crossing Alternating Knots (1885)

The full page gives all possible* 10-crossing closed, alternating, prime knots, discovered by methods developed by Tait and the Reverend Kirkman. The reader will note the many symbols, letters and other characters, attached by Tait to his diagrams; they relate to his various ways of classifying the knots. T. P. Kirkman also produced tables of knot diagrams [9]. He used polygonal (straight-line) figures, which didn't resemble ordinary string knots, as did Tait's; further, he used Greek symbols and phrases to denote his classifying concepts—making it very difficult for persons without a classical education to

*The page gives diagrams for 166 such knots. In 1979, nearly a century later, K. A. Perko showed [12], using 'modern' topological invariants, that two of these knots are the same under topological transformation; it is now known that there are just 165 prime alternating knots with 10 crossings. So Tait wasn't far out with his catalogue!

relate to and remember.

The knot diagrams which are now in general use by topological knot theorists are as given by D. Rolfsen [14]; his system (it was begun by Reidemeister [13]) for numbering them is simply n_i, where n is the number of crossings, and the subscript i gives the order in the list of such knots. When $n \leq 8$ the ordering given by i is, for given n, the same as the order of the values of their Alexander determinants. As an example, the 10-crossing closed knot shown on the cover of this book is the knot 10_{20} given in Rolfsen's tables. Alongside are topological knot diagrams for the unknot 0_1, and all single-string knots with n-crossings, for $n = 3, 4, 5, 6, 7$.

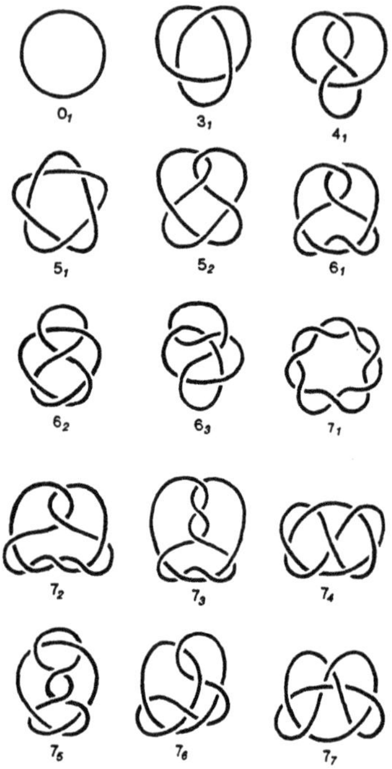

As is explained in the previous Chapter, early knot theorists at the end of the last century were not satisfied with the methods of Tait and his collaborators. Their diagrams were all very well, but was anything really proved by them?

In order to answer this question, they developed a model of a knot as a topological object in three-dimensions (an $\mathbf{S}^1$ embedded in $\mathbf{S}^3$), and began to discover further mathematical objects such as the group of the Knot Complement and the invariants it gave them. Out of this work grew the topological knot theory that we have today.

In the next section, we turn to the work of Georg Schaake, a New Zealand mathematician. Some twenty years ago, he asked himself the question: Does topological knot theory tell us anything about real knots? The kinds of knot that the Encyclopedias catalogue? His answer was firmly in the negative; and remains so. Let us read about what he has done to remedy the matter.

5. The Work of Georg Shaake

Albertus Georg Schaake was born in Holland in 1933. He studied Engineering at Delft Technical University, before emigrating to New Zealand in 1962. His career has been spent first as a Public Works Engineer, and then as a Tutor in Engineering subjects at the Waikato Polytechnic in Hamilton.

He now works at home, extending his researches into braiding theory, and publishing books and pamphlets on his findings. Since 1988 some 3000 pages of published work have flowed from his pen (or, in this age, from his word processor). This is a considerable body of work; and he has at least as much again, of results stored in notebooks, waiting to be put in book or pamphlet form. Virtually single-handedly he has laid the foundations for a 'theory of real knots', as distinct from what he calls 'the classical theory of hypothetical knots'. He makes this distinction quite clear in his pamphlet *Knots—Facts and Fallacies* [15, Pamphlet 8]; in this work (which we shall quote from extensively, below) he sets out his beliefs as to what knot theory should be about, expanding on the theme that 'classical knot theory is concerned with imaginary *closed* knots only, and hence is not able to address the vast majority of real knots.' He supports this statement not by dubious verbal argument, but by making detailed analyses, using his own methods, of several fundamental knots. The types of problem he addresses are quite outside the experiences of 'classical mathematical knot theorists'.

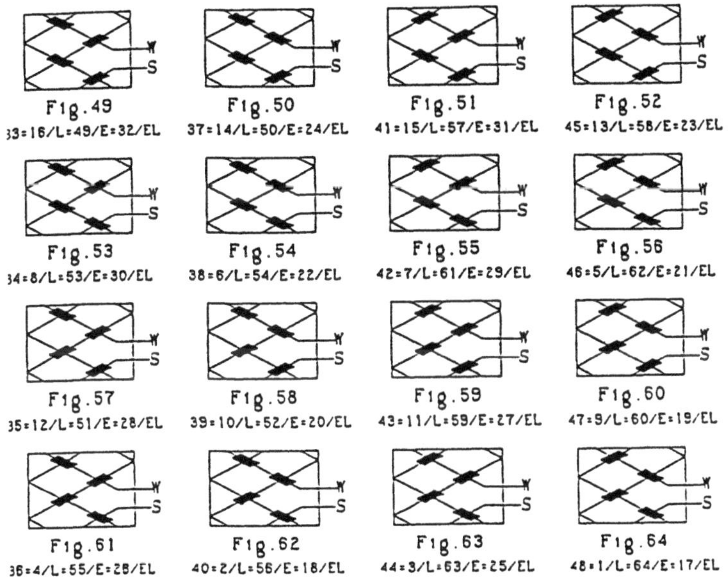

Fig. 4. 16 grid diagrams of 4-crossing knot forms, from 112 diagrams given in [15, Pamphlet 8]

To give just one example here, he discusses all possible forms of the four-crossing knot which are of interest to a braider. A classical knot theorist is generally satisfied that there is only one such form, namely 4_1, or Listing's knot. Schaake gives 112 *grid diagrams** of 4-crossing knot forms! The above Figure (see Fig. 4) shows 16 of these. It will be noted that they are open-ended forms, with **S, W** referring respectively to the 'standing-end' and the 'working-end' of the knot string. Schaake defines five different equivalence relations on these forms, namely *Lateral*, *Evert*, *Evert-Lateral*, *Transpose*, and *Convert*, and discusses classifications of the 112 forms in terms of these relations. We cannot go into details of these relations here, for space reasons. We shall merely quote one of Schaake's comments, with regard to Convert string manipulations, which points up fundamental differences between his kind of knot theory and that of the topologists:

> 'We shall see later, with the aid of examples, that with string manipulations associated with *Convert equivalency*, the results for 'closed' knots are generally quite different from those obtained for open-ended knots of similar structure. This is, of course, not surprising, since the result obtained with the described string manipulations (in finite space) on a knot 'closed' in finite space is not the same as that with the same knot closed at infinity (which is equivalent to the continuation of the string-ends to infinity).'

Having said a little of his methods, with reference to the 4-crossing knots, let us return to our description of the man and his work.

Georg Schaake is a superlative craftsman in many fields, from bone carving to glass engraving, from leather braiding to decorative knotting. In all his endeavours, he brings an original mind to the task of understanding their principles and improving their practices.

He has been particularly successful in developing mathematical theories which describe braiding processes. When he took up leather braiding in the early 80s, and read the literature on the subject, he was immediately struck by the fact that there were no satisfactory definitions of knots, and no mathematical rules by which they could be tied and studied further. He wondered why the whole treatment of the formation of knots had not been placed (and long since) on a surer footing than the graphical, wholly empirical ones given in the encyclopedias of Ashley [1] or Grant [5]. Their works are monumental, of great beauty, worth and scholarship; but they espouse no systematic method for producing knots and braids, no theoretical treatment.

Among the very few references to mathematics in these works, the following two quotations are revealing. The first (from [5, p. 440]) occurs within a two-page section headed 'The Braiding Detective'. It gives the only clues to the reader that in the wonderful and infinitely extensive world of braiding

*The 'grid diagram' is Schaake's basic tool for displaying and analysing knots; through its use, his theory becomes both geometric and analytic.

there lies the possibility of a theory of braiding, and of methods by which the artisan can develop and analyze his own creations of decorative braids.

> '**Duplicating Strange Braids and Creating New Ones.** The braided knot, or even the Turk's-head, is a mathematical marvel. It goes round and round and comes out perfect—the working end of a braided knot returns to its point of origin.
>
> It is interesting to analyze, more absorbing to create, and definitely a challenge. Distinct variations are limitless and all of them are adjusted to a mathematical scale of easy deduction.
>
> Dr Almanzor Marrero y Galíndez, in the prologue of his book, *Cromohipologiá*, said of his Argentine father: 'During his last years almost each day he made an original *criollo* button, which he first posed as a problem, resolving it mathematically, and later executing ther formulae he had worked with success. Some of these buttons were of such complexity that he had to use several strings in the same knot, whose total length reached several metres.'
>
> Alas, his mathematical techniques in planning his buttons have been lost to posterity. Many serious scientists and mathematicians have given profound thought to the Turk's-head, and as an independent result of the studies of Clifford W. Ashley of New England, and George H. Taber of Pittsburgh, the 'Law of the Common Divisor' was discovered.'

The Law of the Common Divisor for single-string Turk's-head knots, as enunciated by Ashley in [1, p. 233], is as follows:

> 'A knot of one string is impossible in which the number of parts and the number of bights have a common divisor [greater than 1].'

This would seem to be the extent of any known theory of braiding processes, before Georg Schaake began his studies. Moreover, it seems that the above law was arrived at empirically; no proof was given for its general truth. We were unable to trace any further of Ashley's or Taber's writings on the matter. Schaake and Turner give a concise proof of the Law in [20], and a discursive one in [26].

There appear to be no extant mathematical theories, certainly no general ones, before Schaake's, which enable algorithms (step-by-step methods) to be calculated and written down so that artisans can produce braids of given patterns. Without such a theory, of course, fundamental questions about the universe of braids cannot even begin to be answered. For example, we might ask: 'Suppose we wish to produce a braid with a certain herringbone pattern of string-crossings, using only one string and having 95 parts and 36 bights. Can it be done? If so, *how* may it be done? Give us a method to do it.'

None of the encyclopedias described earlier give any ways of answering such questions. Nor do any of the papers or books in the literature of topological knot theory contain clues which might lead to answers. Indeed, it must be said (and we shall elaborate on this below, using largely Schaake's own words) that topological knot theorists are not at all interested in this kind of problem, vital though it is in the real, physical world of braiding. A well-known remark is that a topologist is a person who cannot distinguish between a doughnut and a coffee-cup. Similarly, a topologist is a person who cannot distinguish between a Reef Knot and a Lark's-head Knot. In the knotter's world, these are different knots, with different properties and different ways of tying them.

In the early 1980s, Georg Schaake became aware of this gap, namely the almost total lack* of general mathematical theories to model braiding processes; and he decided to do something about it. Within a space of ten years he discovered many beautiful relationships between numbers of parts (p), numbers of bights (b) and numbers of strings (s) in a braid; and these relationships formed the basis of a broad general theory of braiding.

This theory is clearly a branch of number theory, with modular arithmetic and solution of Diophantine equations under boundary conditions being main elements of it. There is much appeal to polygonal grid diagrams, which are precise drawings of the string-runs of braids; as a consequence, one can say that Schaake's braiding theory is geometric and arithmetic; it is not topological. It is a different kind of knot theory, different in goals and methods from the mainline branch which has stemmed from Listing's and Gauss' work, and the classification work of Tait, Kirkman and Little in the 80s and 90s of the last century.

Since 1987 Schaake has been carrying out a project to publish his discoveries, collaborating with other authors to produce a stream of books, pamphlets, articles and research papers on a variety of topics in braiding theory. Items [15]–[24] of the bibliography of this Chapter indicate the extent of his published output in a period of six or seven years. It ranges from the books on Regular Knots, Fiador Knots and Herringbone Knots through to extensive pamphlets on the braiding of Wheelknots. And it includes 'spin-off' work in Number Theory. For example, in 1987 a study of Schaake's evolution tree for Regular Knots enabled Schaake to discover a new way of solving Pythagoras' equation for integer triples (for sides of right triangles); all of them, including their multiples, were given by this method, and they were classified in entirely new ways, using tree graphs. In view of the ancient nature of this problem, this was a remarkable feat. The work resulted in the book (with J. C. Turner

*One or two other authors have taken small steps to fill this gap in the past two decades; for example, van de Griend [6] has written on *Knots and Rope Problems* and Karner [8] has given tables for the development of Turk's Heads. But their work is insignificant beside Schaake's.

[24]) entitled *A New Chapter for Pythagorean Triples*, and a research report [22] on new methods for solving quadratic Diophantine equations.

This publishing project is continuing. Schaake has tackled many more problems and knot classes than appear in the works Referenced in this book. His further results are steadily being written up and published in one form or another.

We shall end this essay by giving brief notes on the philosophy, basic ingredients and methods of Schaake's braiding theory. The final remarks will draw heavily upon Schaake's own statements, made in Pamphlet No. 8, where he gives his views upon *Knots—Facts and Fallacies.*

6. Basic Ingredients and Philosophy of Schaake's Theory of Braiding

In order to begin studying Schaake's theories, it is necessary to know what is meant by the following terms: *parts, bights, string runs, coding, grid diagram, algorithm table.* They are used in reference to a braid*which can be constructed by passing a string from left-to-right, right-to-left, and so on, round and around a cylinder, interweaving the string with itself by making a succession of under- and over-crossings as the passes are made.

The 'grid diagram' is a geometric picture which represents the braid (and the braiding process) completely. The 'algorithm' gives a list of instructions for carrying out the interweaving as the string passes round and around the cylinder. The 'coding' is the particular pattern of string crossings which appears on the grid diagram. A 'string run' is a rectilineal path which the 'string' takes as it traverses the grid diagram, from the left boundary to the right boundary or vice versa. Such grid diagrams are best drawn on *isometric grid* graph paper. We shall illustrate these terms and ideas in relation to the production of the very simple and familiar 3-crossing knot, namely the *trefoil knot.*

6.1. Diagrams and Production Steps for the Trefoil Braid

The diagrams (a) and (b) in Fig. 5 are well-known representations of the right-handed 'trefoil knot' and braid respectively. Note that (a) is formed from a single closed string; whereas (b) is formed from two interwoven strings, one running from point P_1 to Q_2, and the other from P_2 to Q_1.

Next note that if a circular cylinder be placed *behind* the braid (b) with its straight-line generators parallel to P_1P_2 and Q_1Q_2, then the braid can be wrapped backwards around the cylinder until P_1 meets Q_1 and P_2 meets Q_2.

Let us assume this done, and that we have joined the ends P_1 and Q_1,

*'No distinction is to be drawn between 'braid' and 'knot', since any knot is in fact a braided structure.'

and also P_2 and Q_2. We observe that we now have the one-string '2 part–3 bight' knot (trefoil knot) tied around the cylinder. If we slip it off the cylinder, moving carefully in the direction $P_1 \longrightarrow P_2$, we can then arrange it on a flat plane in the manner of the lower diagram in (a).

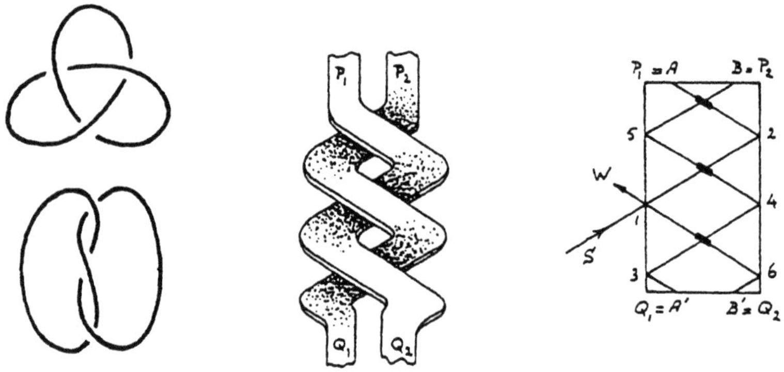

trefoil knot — *2 string—3 bight u/o braid* — *regular knot, 2P/3B Turk's Head Knot*

(a) *(b)* *(c)*

Fig. 5. Diagrams of a Three-crossing Knot and Braid

Now examining diagram (c), we observe that it provides a clear picture of the '2 part–3 bight' Turk's Head Knot (trefoil knot) tied around a cylinder. The thick short-line segments indicate where, and how, one string crosses another (a thick short-line represents an over-crossing): the set of these thick lines prescribes the entire weaving pattern, which is called the *coding* of the knot.

In fact, the diagram tells us much more: it provides full information as to how the *process* of tying the knot on the cylinder must be carried out if a '2 part–3 bight' braid is to be achieved. To be specific, *diagram (c) is a complete visual algorithm for constructing the right-handed trefoil knot.* It is an example of what Schaake calls *the grid diagram* of a knot. For his theories, the grid diagram is the starting point of all studies of the knot which it represents.

It is the basic tool in Schaake's Braiding Theory. It must be noted that it is not a topological diagram (as is diagram (a), which contains less information); it is geometric, although neither the scale nor the fixed angle at which the string half-cycles are laid down on the cylinder matters in the ensuing braiding theory.

By *half-cycle* we mean a complete passage from Left→Right (or vice versa). For example, the first half-cycle is the pass (from the Standing end, labelled S) of the string from point 1 to point 2. There are six half-cycles needed to complete the braid, which is completed by the last half-cycle from point 6 to point 1, when the Working end, labelled W, arrives at point 1.

Note that every half-cycle is bisected by a string crossing; that is why the term '2 part' is applied to the knot. And note also that at the right-hand 'edge' of the knot, the string changes direction at points 2, 4 and 6; we call these *bight points*; that is why the term '3 bight' is applied to the knot. The pair of integers (2,3), serves to specify the 'whole string run' of this Turk's Head Knot; this number pair, together with the given interweave coding, completely specifies the Knot.

Using the diagram (c), we can trace the six half-cycles needed to complete the Knot, and write down the following Algorithm, which specifies the braiding process.

Algorithm

Step 1: L→R upwards; run 1 to 2; no crossings.
Step 2: R→L (right to left); run 2 to 3; no crossings.
Step 3: L→R (left to right); run 3 to 4; no crossings.
Step 4: R→L; run 4 to 5; cross *over* 1–2.
Step 5: L→R; run 5 to 6; cross *under* 2–3.
Step 6: R→L; run 6 to 1; cross *over* 3–4.

If, finally, we join the working end (W) to the standing end (S) at 1, then we shall have completed the braiding of a trefoil knot.

7. On Regular Knots

With the concepts of grid diagram and associated measures explained, we can turn to definition of classes of braids. Schaake's first book deals with The Regular Knot class. He defines this class (see [17]) as follows:

Definition (i)

A *regular flat braid* is a flat braid which has all its left-hand bights on a single straight line, and all its right-hand bights on another, parallel straight line. It consists of two sets of string runs, with all the members of a set being parallel and running from one bight boundary to the other.

Figures 6 and 7 show two regular 7 part/5 bight flat braids, with grid diagrams.

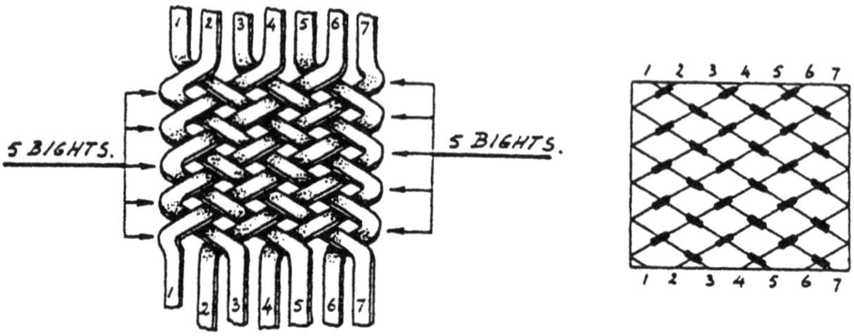

Fig. 6. The (7P, 5B) regular flat braid, with Turk's Head coding

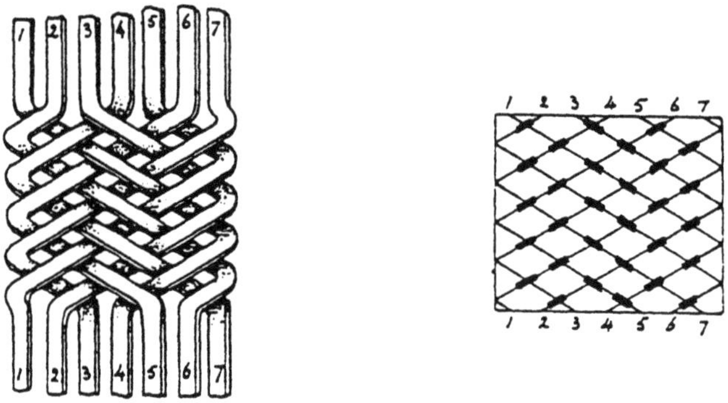

Fig. 7. The (7P, 5B) regular flat braid, with Two Pass Headhunter's coding. Figs. 6 and 7 demonstrate two different braids with the same whole string run

If we wrap a regular flat braid backwards around a cylinder, and join the corresponding string ends when they meet, then we obtain a regular cylindrical braid knot, which has one or more strings in its composition.

Definition (ii)

A *Regular Knot* is a regular cylindrical braid knot composed of a single string.

In the two Figures 6 and 7, if the grid diagrams be 'wrapped round a cylinder' and the string runs traced, it will be found that only one string is involved. It will be readily observed that the two diagrams have different codings. So the two grid diagrams now represent two different Regular Knots having the same whole string run. The first one has a coding which indicates that the string passes alternately 'over' and 'under' at the crossing points, throughout. Such a knot is called a Turk's Head Knot, by Schaake's definition:

Definition (iii)
A *Turk's Head Knot* is a Regular Knot whose super-imposed coding alternates overs and unders throughout.

The description 'Turk's Head' is used freely in the literature, for many kinds of knot, without precise definition. For example, in Chapter 17 of Ashley's Book of Knots [1], a Turk's Head Knot is said to be a 'tubular knot that is usually made around a cylindrical object, such as a rope, a stanchion, or a rail.' Ashley goes on to say: 'It is one of the varieties of the BINDING KNOT and serves a great diversity of practical purposes, but it is perhaps even more often used for decoration only; for which reason it is usually classed with 'fancy knots'.' In his chapter he describes the construction of over a hundred such knots, many of which would not be called 'Turk's Head' by Schaake (either because they are not Regular Knots, or because they do not have the strictly alternating coding pattern).

Schaake knows, of course, the necessity to define one's terms precisely, as one builds up a consistent mathematical theory. So, in his first book on Braiding [17], he lays down the definitions of *grid diagram, string run, coding, Regular Knot, Turk's Head Knot.* All of his theories are grounded upon the first three terms; then 'Regular Knot' defines a vast (infinite) class of single string cylindrical knots with reference to their string run grid diagrams only. Superimposing the alternating coding upon these gives the infinite class of Turk's Head Knots.

8. Development of Schaake's Braiding Theory

8.1. Aims and Achievements

The principal concerns of Schaake's Theory, for any given class of Knot which he defines, are first to discover geometrical and algebraic relationships between the parameters (e.g. the numbers p and b, representing numbers of parts and bights in a knot) involved in the string run grid diagrams for members of the class. Then, if a coding (weaving pattern) is given, the concern is to find an algorithm which enables a braider to construct a braid with that coding. Schaake solves these problems as generally as possible—not for particular braids, but for infinite classes of them. Recurrence equations with boundary conditions are determined for a general knot within a class, and methods, diagrams and tables are supplied which enable any braider to formulate the algorithm by which he or she may arrive at the desired braid.

A third concern is to determine how braids within a given class are interrelated. What does it mean to say 'This braid is an enlargement of that one', or 'This braid *evolves* from that one, but not from this other one'? Again, suppose that knots in a certain class are determined by an integer vector

$\mathbf{P} = (p_1, p_2, \ldots, p_n)$ of parameters. The question can be asked 'Which integer vectors will correspond to knots in that class?'*

It is the pursuit of answers to this kind of question which has caused the greatest fascination and pleasure to this author. The solutions often provide an *evolution tree*, which can take the form of a *number tree*[25] which consists of a tree graph displaying the parameter vectors on its nodes; the paths between nodes indicate relationships between knots in the given class. It takes little imagination to view these trees as new objects in number theory. Every study of their properties constitutes (so this author claims) new investigations into properties of numbers. To give just one example of the charm and power of such studies, we have already mentioned how, in 1987, we discovered new ways for solving second-order Diophantine equations [22], by studying the Regular Knot evolution tree.

The production of a step-by-step plan† for obtaining a braid, having a given string run and coding pattern, is traditionally done by tracing the braid pattern on paper and following it around carefully from a starting point, noting down the crossing-types as they are to occur along the string.

Exactly the same procedure can be followed, though often very tediously, using Schaake's grid diagram for the desired braid. The *algorithm* for producing the braid by passing string back and forth over a cylinder can be similarly determined. We showed how to do this in the simple trefoil example, given above, following the string run and observing the interweaving that had to take place within each half-cycle.

In principle, this 'pencil-and-paper' approach can always be used; even when the braid has many hundreds of crossing-points. However, it is desirable to have mathematical methods to compute the algorithm in more complex cases. Such mathematical methods not only provide the algorithm but also deepen one's understanding of the processes involved; and they can be generalized.

This might be said to be the point where Schaake's theory of braiding becomes a mathematical science. The grid diagrams of braids (any braid can be represented by one) are the basic mathematical objects of that science. They are objects which are essentially geometric (although some topological deformations can be made on a diagram, without destroying or corrupting the information it carries about its braid). The types of mathematics which are used to analyse grid diagrams are those of geometry, modular arithmetic, and systems of recurrence equations with many boundary conditions.

*An easy example is the following. In the class of Regular Knots, the whole string run of a knot is determined by an integer pair (p, b). It is easy to show that any ordered pair (m, n) of whole numbers corresponds uniquely to a Regular Knot string run provided that $g.c.d.(m, n) = 1$.

†Compare the use of a knitting pattern to produce a jersey.

Georg Schaake has discovered the methods necessary to obtain algorithms for braiding of knots in many classes—for different sets of conditions on their string runs and codings. It is not possible to give meaningful summaries, in this short chapter, of any of the formulae or mathematical methods involved in this work. The list of his achievements is long. To name a few of the classes he has dealt with, we begin with the general class called Regular Knots, which includes many well-known types of Knot such as Turk's Head, Head Hunter's, Gaucho, Botón Oriental, Fan, Ring, Slow and Fast Helix and Basket Weave—all knots well-known to experienced braiders. The list of known knots in this class could go on. We emphasise that he has dealt with these classes (or subclasses) in great generality; in principle, from his formulae braiding algorithms can be computed for any given numbers of crossings, for each of the named types. Indeed, his collaborators have produced computer programmes from his formulae, which produce the required algorithms whenever particular parameter values are input. In the Knot Encyclopedias described earlier, just special cases of these braids were treated, with drawings and graphical algorithms given for each. It was descriptive work only.

To continue the list of his achievements, he has studied many general types of Regular Knots—for example those with Row-Coding, with Column-Coding, or with Casa-Coding (see [15, Pamphlets 6, 7, 9, 10]).

Having dealt with single-string braids, one can turn one's attention to multi-string braids. The possibilities for extending the types and complexities of braids for making, studying and classifying grow exponentially. For example, if one relaxes the condition that all the bights of the braid must lie on two parallel edges, the possibilities become truly vast. Of course, Schaake first turned his attention to those knots which are of most interest to artisan braiders—and the types of knot which they wish to braid are generally both functional and decorative (that is, artistically pleasing). Beauty is their guide, when they select coding patterns for their braided objects. If further considerations are added, such as colours of strings (they can be all the same, or be selected from a range), then again, the numbers of possible braid-types within each weave-type class goes up rapidly.

Some of these more complex types, the problems of which Schaake has solved in broad generality, include the Standard Herringbone Knots, the Regular Fiador Knots, the Standard Herringbone Knots, and Wheelknots. The last-named Knots are different from all the earlier, cylindrical type braids, in that they involve braiding in the form of a torus; moreover, holes have to be left in the weaving, at regular intervals around the torus, for wheel-spokes to pass through. To imagine this, think of a steering wheel of a yacht, with perhaps five spokes; the braiding process, using one string only, has to pass around the circumferential rim of the wheel, constructing the required coding pattern as it goes and avoiding the spokes neatly. In [15, Pamphlet 12] Schaake gives formulae, algorithms and tables which solve many Wheelknot problems. He has also discovered an evolution graph for Wheelknots which is 'tree-like' but contains cycles (i.e. it is not strictly a tree graph).

For Schaake's work to constitute a mathematical theory, it is necessary to supply proofs of all the derived formulae and algorithms. In 1988, Schaake and I began a program for writing out the proofs, in a Research Report Series, doing this at the same time as producing the books and pamphlets on algorithms and tables. The first two Reports, [20] and [21], written in 1988, covered the work done on Regular Knots. At that point, several exciting mathematical projects arose out of the evolution theory of Regular knots; in quick succession we wrote books on number properties discovered from the Regular Knot Tree ([22], [24], and [23]). The project for grinding out proofs of braiding algorithms fell into abeyance; and it has not yet been restarted. We felt it more sensible to carry on writing up pamphlets explaining the algorithms themselves. Schaake has most of the proofs for his discoveries; but it would take a further 2000 pages of mathematical writing to present them all. He says 'Let the mathematicians who follow in my wake'—and he is sure that others will eventually do so—'supply all the proofs. Keep them busy for a while!'

To end this subsection, we present several figures, selected from Schaake's books. We give them without comment, allowing them to speak for themselves. They illustrate some of the ideas and points we have tried to make in words; and they show some of the braids that Schaake has studied and 'solved'.

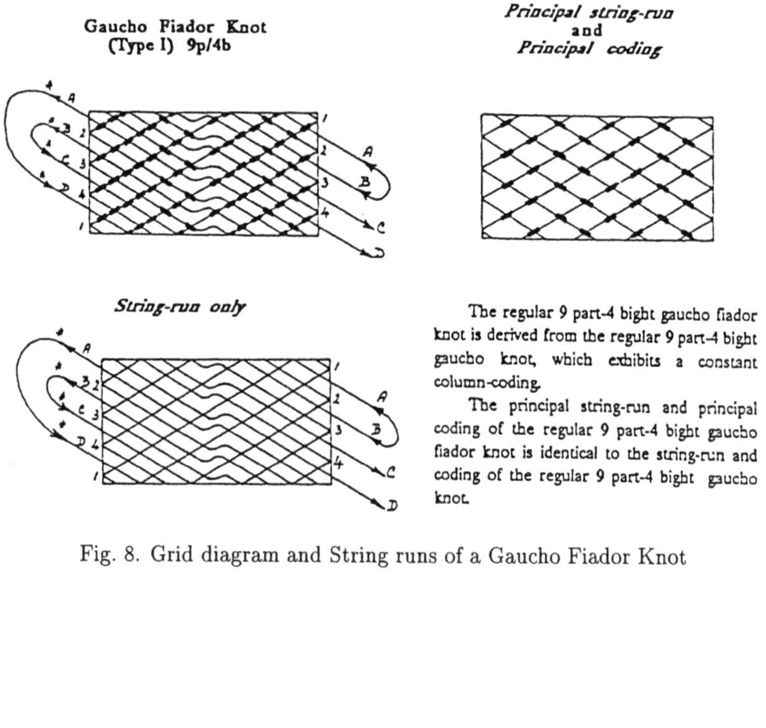

Fig. 8. Grid diagram and String runs of a Gaucho Fiador Knot

Fig. 9. Grid diagram and Braid diagram of a 3-string Standard Herringbone Knot

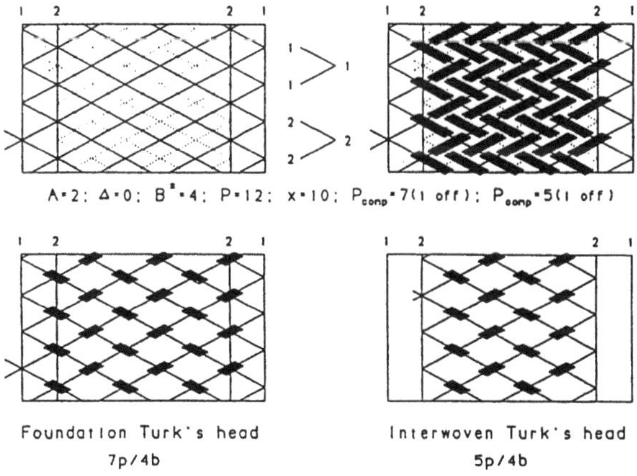

Fig. 10. Diagrams for a Standard Herringbone Pineapple Knot, which is obtained by interbraiding two Turk's Head Knots, one a 7 Part/4 Bight and the other a 5 Part/4 Bight

$$\Delta = 0 \tag{1}$$

$$B^{*} \text{ and } P_{component} \text{ are coprime} \tag{2}$$

$$x = cA + \delta, \qquad \text{wherein} \tag{3}$$

$$\delta = |2(l_i + r_i)|_A \qquad \text{and} \tag{4}$$

$$c = (2m - 3) + \frac{2(l_i + r_i) - \delta}{A} \tag{5}$$

$$P_{component} = \frac{x + 4A - 2(l_i + r_i)}{A} \tag{6}$$

$$P = \sum P_{component} = x + 2A - 2 \tag{7}$$

Fig. 11. Characterizing relationships for Standard Herringbone Pineapple Knots; conditions (1), (2) and (5) are necessary and sufficient for this type of Knot

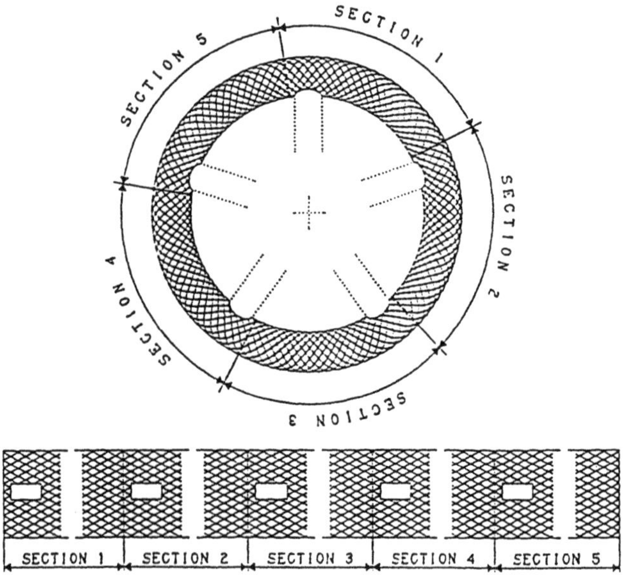

Fig. 12. A 5-Section Wheelknot (torus knot 'with holes'), and its sectioned string run diagram

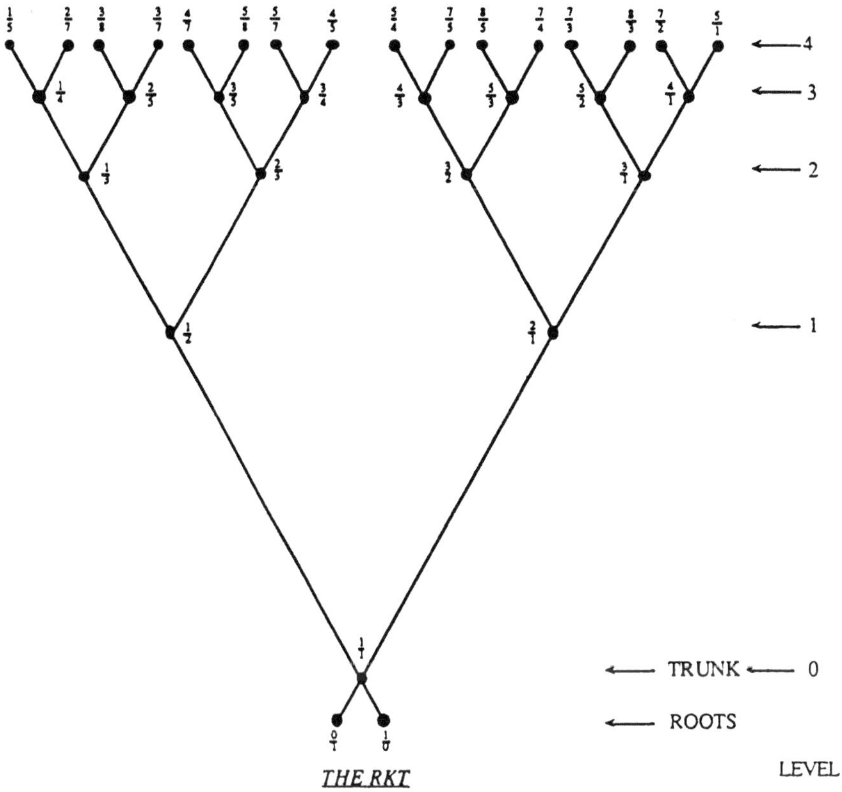

Fig. 13. The Evolution Tree for String Runs of Regular Knots [25], [15, Pamphlet 6]. It shows the way string-runs can evolve upwards, from the trunk. For example, to construct by cylindrical braiding a 5 Part/8 Bight Turk's Head, one has to, in effect, pass through the 1P/1B, 1P/2B, 2P/3B, and 3P/5B forms first. Note that the whole tree grows from the 'forms' 0/1, 1/0 and 1/1, using mediant additions.

8.2. Quotations from 'Knots—Facts and Fallacies'

In 1992 Schaake published, privately, a pamphlet [15, Pamphlet 8] called *Knots—Facts and Fallacies*, in which he set out many of his views on what a 'real knot theory' should consist of. He made many scathing remarks about what he called the 'classical knot theory of hypothetical knots'—and even included inflammatory opinions about mathematicians. We will pass over the latter; but we feel it is instructive to include here some of the former kind of remarks, together with other of his views on knots, drawn from that pamphlet. As with the Figures above, we shall pass them on without further comment. They illustrate his strong beliefs, based on his fifteen years work with his own knot theory, as to what should be the main thrust of knot research.

'This Pamphlet is mainly intended for the general reader interested in knots and braids. It is also intended for those who have always wondered why the academic world has not progressed beyond their "classical knot theory" and consequently has never yet been able to produce a sound knot theory of practical value ... '

...

'It is regrettable that any association which their emerging "knot theory" might have had in the past with physical knots, has been lost without recognizing this.' [Schaake is here referring to the fact that topological knot theory does not treat of physical aspects and properties of knots.]

...

'The "classical knot theory" is concerned with imaginary "closed" knots only, and hence it is not able to address the vast majority of real knots.'

...

'We have shown above that the "classical knot theory" is a theory of hypothetical knots and braids, hence a theory which has no bearing on real knots. It is therefore not surprising that it cannot give constructional directives for the creation of knots and braids. By its very nature it is a theory in a static framework (only "closed" hypothetical knots are studied). Knots and braids are dynamic objects, resulting from creative (evolutionary) processes, and hence require a theory in a dynamic framework. Therefore, it should be obvious that the required theory should concern itself with the question: *How does a* **real** *knot or braid come forth?*'

...

'We have already mentioned that knots and braids are in essence of a *geometric* nature. Hence it is of vital importance not to lose their geometrical characteristics in any modelling processes. The "classical knot theory" however, disregards the geometrical aspects, and treats knots in a purely topological manner. This leads to various results which are ambiguous when applied to physical, hence real, knots and braids.'

...

'We mentioned that in the "classical knot theory" all knots are closed structures, and hence before studying a knot it gets closed. We have already seen that a "closed" *Overhand Half Knot* and a "closed" *Overhand Knot* both result in a *Right-Hand Trefoil*. Now let's again close these two knots **after** everting* and investigate the result. In the case of a "closed" everted *Overhand*

*'everting' means 'turning inside out'; here he is referring to cylindrical braids, which have

Knot we obtain a *Right-Hand Trefoil*; but in the case of a "closed" everted *Overhand Half Knot* we obtain a closed ring with a full (left-hand) screw turn in the string (an *Un-Knot* in the "classical knot theory"). Hence we obtain two vastly different knots, even in the "classical knot theory". Of course, the "classical knot theorist" may say that the everting process should take place after closure, but not only is such an excuse obvious nonsense (since by doing so we **delete** an existing crossing and **create** a new one; hence if we allow closure before the completion of braiding procedures, we may as well start with a closed string-run in the form of a circle (their *Un-Knot*); it moreover results in the loss of the relationship between an *Overhand Half Knot* and its everted form.'

...

Schaake goes on to discuss many aspects of the construction of braids, using examples from basic knots such as the 4-crossing knots (see the 16 grid diagrams shown above, and the comments given upon them), the Overhand Half Knots, composite Trefoil Knots, Strangle Knots and Constrictor Knots. His aims are to uncover [quote] 'the Facts about some REAL KNOTS'.

9. Summary

In this Chapter we have tried to present a picture of fields of knot studies which generally are unknown to, or given only a sideways glance by, mathematical knot theorists. We believe that if these other fields are added to that which concerns topological knot theorists, then we shall have a more correct purview for a balanced and useful theory of knots.* A much wider range of problems will then present themselves to mathematicians for solution.

We began this Chapter by discussing the questions 'What is a Science? and 'Is there a Knot Science?'. Let us end with the same questions. Some might ask—perhaps with justification—Does it matter? What's in a name? Philosophers have argued about vaguer concepts, sometimes for hundreds of years (for example 'the number of angels that can stand upon a pinhead'; we maintain that our topic stands firmer than that one.) We might also remark that deciding whether a subject is, or is not, a Science, is one that exercises deeply the minds of Vice-Chancellors of Universities; because the magnitudes of their annual subject grants depend heavily upon the decision. The Sciences are much favoured over the Arts, for purposes of allocating grants.

---

an outer and an inner surface, and two lateral faces. Using grid diagrams he can define 'evert equivalency' for these braids.

*Gauss, the father of knot theory, would seem to have recognized this, when, in January 1833, he wrote that the major task for knot theorists would be to count the 'different windings' [*Umschlingungen*] of two closed or infinite curves, and he placed the problem in 'the boundary field between'[*Grenzgebiet*] Geometria Situs and Geometria Magnitudinis.

Leaving those somewhat facetious comments aside, we believe we have presented a reasonable case for the affirmative—The Study of Knots may be designated a Science! We believe that this Chapter, allied with the evidence of many of the other Chapters in this book, strongly supports our view. Moreover, we believe we have shown that the field of this Science is much wider than the restrictive one of the topological knot theorist; to use Schaake's words, it must be broad enough to deal with 'real knots', as well as 'hypothetical ones'.

It seems pertinent here to give the following quotation, from a discussion about the position which the science of number occupies with respect to the general body of human knowledge:*

> *It seems to me that what Philosophy lacks most is a principle of relativity ...*
> *A principle of relativity is just a code of limitations: it defines the boundaries wherein a discipline shall move and frankly admits that there is no way of ascertaining whether a certain body of facts is the manifestation of the* observata, *or the hallucination of the* observer.
> Tobias Danzig

In these terms, I have tried to shed light on what might be a principle of relativity for the Philosophy of Knots.

Whether you will agree with me or not on the above issues, I know I can assert, without any doubt, that whatever it is called, the study of knots will continue to fascinate and exercise the minds of men and women for as long as humankind exists. That is assured because knots will always be useful to them; because knots are beautiful and mysterious objects; and because the expanding fields of knot studies will pose ever more complex problems, challenging our descendants to solve them.

10. References

1. C. W. Ashley, *The Ashley Book of Knots* (Doubleday, New York, 1944) 620 pp.
2. C. L. Day, *The Art of Knotting and Splicing* (Dodd, Mead and Company, New York, 1st edn. 1947; Naval Institute Press, Annapolis, Maryland, 2nd edn. 1955) 225 pp.
3. D. J. Albers, 'Freeman Dyson: Mathematician, Physicist, and Writer', *The College Mathematics Journal, MAA* **25** No. 1 (January, 1944) 2–21.

*In:'The Two Realities', from *Number—The Language of Science* (Doubleday, 1954) 232–253.

4. R. Graumont and J. Hensel, *Encyclopedia of Knots and Fancy Rope Work*
(Cornell Maritime Press, 1st edn. 1939, 4th edn. 1952) 690 pp.
5. B. Grant, *Encyclopedia of Rawhide and Leather Braiding*
(Cornell Maritime Press, 1972) 528 pp.
6. P. van de Griend, *Knots and Rope Problems*
(Privately published, limited edn., Århus, 1992).
7. V. F. R. Jones, 'A New Knot Polynomial and von Neumann Algebras', *Notices of the American Mathematical Society* **33** No. 2 (1986) 219–225.
8. A. Karner, *On The Development Of Turk's Heads*
(The International Guild of Knot Tyers, 1984) 73 pp.
9. T. P. Kirkman, 'XXVI—The 364 unifilar knots of ten crossings, enumerated and described.' *Trans. of Roy. Soc. Edinburgh* **32** (1885) 483–491.
10. J. Lehmann, *Systematik und geographische Verbreitung der Geflechtsarten*
(Abh. u. Ber. d. K. Zool. u. Anthr.-Ethn. Mus. zu Dresden, Bd.XI, Nr. 3, 1907) 36 pp.
11. J. B. Listing, *Vorstudien Zür Topologie* (Göttingen Studien, 1847).
12. K. A. Perko, 'On 10-Crossing Knots', *Portugaliae Mathematica* **38** (1979) 5–9.
13. K. Reidemeister, *Knotentheorie* (First published, 1932; reprint Chelsea, N.Y., 1948; English translation, BCS Associates, Idaho, U.S.A., 1983).
14. D. Rolfsen, *Knots and Links* (Publish or Perish, Inc. 1976) 438 pp.
15. A. G. Schaake et al., [Pamphlet Series—Topics in Braiding Theory and Practice]
Pamphlet No. 1, *Introducing Grid-Diagrams in Braiding* (1991) 32 pp.
Pamphlet No. 2, *Edge Lacing—The Double Cordovan Stitch* (1991) 23 pp.
Pamphlet No. 3, *Braiding Application—Horse Halter* (1991) 24 pp.
Pamphlet No. 4, *The Regular Knot Tree and Enlargement Processes* (1991) 38 pp.
Supplement to No. 4, *Casa-Coded Regular Knots* (1995) 31 pp.
Pamphlet No. 5, *An Introduction to Flat Braids* (1991) 36 pp.
Pamphlet No. 6, *An Introduction to Evolution Processes (Part 1)* (1992) 82 pp.
Supplement to No. 6, *Headhunter-Fan Knots* (1994) 18 pp.
Pamphlet No. 7, *The Braiding of Column-Coded Regular Knots* (1992) 37 pp.
Pamphlet No. 8, *Knots—Facts and Fallacies* (1992) 45 pp.
Pamphlet No. 9, *The Braiding of Row-Coded Regular Knots*

(1993) 43 pp.
Pamphlet No. 10, *Special Braid Forms Pt. 1, End-Keepers and Mid-Keepers* (1995) 176 pp.
Special Braid Forms Pt. 2, CFC and CWH Braids (1995) 196 pp.
Pamphlet No. 11, *Braiding Application—Bridle and Reins* (1995 in preparation).
Pamphlet No. 12, *The Braiding of Wheelknots* (1994) 110 pp.
Supplement to No. 12, *Wheelknots* (1995) 49 pp.
Pamphlet No. 13, *The Braiding of Long Regular Knots with the Maximum Number of Free-run Half-cycles* (1995) 50 pp.

16. A. G. Schaake, T. Hall, and J. C. Turner, *BRAIDING—STANDARD HERRINGBONE KNOTS* [Book 3/1 of Series on Braiding] (University of Waikato, Hamilton, N.Z., 1992) 208 pp.
17. A. G. Schaake, J. C. Turner and D. A. Sedgwick, *BRAIDING—REGULAR KNOTS* [Book 1/1 of Series on Braiding] (University of Waikato, Hamilton, N.Z., 1988) 117 pp.
18. A. G. Schaake, J. C. Turner and D. A. Sedgwick, *BRAIDING—REGULAR FIADOR KNOTS* [Book 2/1 of Series on Braiding] (University of Waikato, Hamilton, N.Z., 1990) 159 pp.
19. A. G. Schaake, J. C. Turner and D. A. Sedgwick, *BRAIDING—STANDARD HERRINGBONE PINEAPPLE KNOTS* [Book 4/1 of Series on Braiding] (University of Waikato, Hamilton, N.Z., 1991) 202 pp.
20. A. G. Schaake and J. C. Turner, [Research Report Series] *RR 1/1, A New Theory of Braiding* (Department of Mathematics, University of Waikato, Hamilton, N.Z., Report 165, 1988) 42 pp.
21. A. G. Schaake and J. C. Turner, [Research Report Series] *RR 1/2, A New Theory of Braiding* (Department of Mathematics, University of Waikato, Hamilton, N.Z., Report 168, 1988) 41 pp.
22. A. G. Schaake and J. C. Turner, *New Methods for solving Quadratic Diophantine Equations: Part I - Investigations of Rational Numbers Using Rooted Trees and other Directed graphs; Part II - The Pythagorean Triples* (Department of Mathematics, University of Waikato, Hamilton, N.Z., Report 168, 1989) 99 pp.
23. A. G. Schaake and J. C. Turner, *Generalizing Euclid's Algorithm, via the Regular and Möbius Knot Tree—Order-n Arithmetics* (Department of Mathematics, University of Waikato, Hamilton, N.Z., Report 196, 1990) 61 pp.
24. A. G. Schaake and J. C. Turner, *A New Chapter for Pythagorean Triples*

[book] (pub. by Department of Mathematics, University of Waikato, Hamilton, N.Z., 1989) 155 pp.
25. J. C. Turner, 'Three Number Trees—Their Growth Rules and Related Number Properties', *Applications of Fibonacci Numbers* **3** (eds. G. E. Bergum et al., Kluwer Academic, 1990) 335–350.
26. J. C. Turner and A. G. Schaake, 'A Proof of the Law of the Common Divisor in Braids', *Knotting Matters*—Newsletter of the International Guild of Knot Tyers, Issue 35, (Spring, 1991) 6–10.
27. C. Warner, *A Fresh Approach to Knotting and Ropework* (pub. by Warner, P.O. Box 194, Picton, NSW 2571, Australia, 1992) 272 pp.

CHAPTER 13

TRAMBLES

Geoffrey Budworth

We may not have too long to wait before ... knots quit the academic backwater they have occupied since Victoria's reign, and take a place in the mainstream of scientific development.

Desmond Mandeville, 1910–1992

Introduction

I expect you have come across the pastime in which a word is changed, one letter at a time, into another totally different word. For example:

REEF
REED
SEED
SHED
SHOD
SHOT
SNOT
KNOT

In much the same way trambling is altering a real knot, one tuck at a time, to yield a series of related knots. We call these changes *tuck-transformations*.

Sometimes, in the word game, substituting a letter makes no sense; then, if a meaningful word cannot be found, progress is halted. Only proper words will do. In trambling, too, knots must be practical ones. But sometimes—just as my seventh word is vulgar—an odd or unusual knot may be unavoidable.

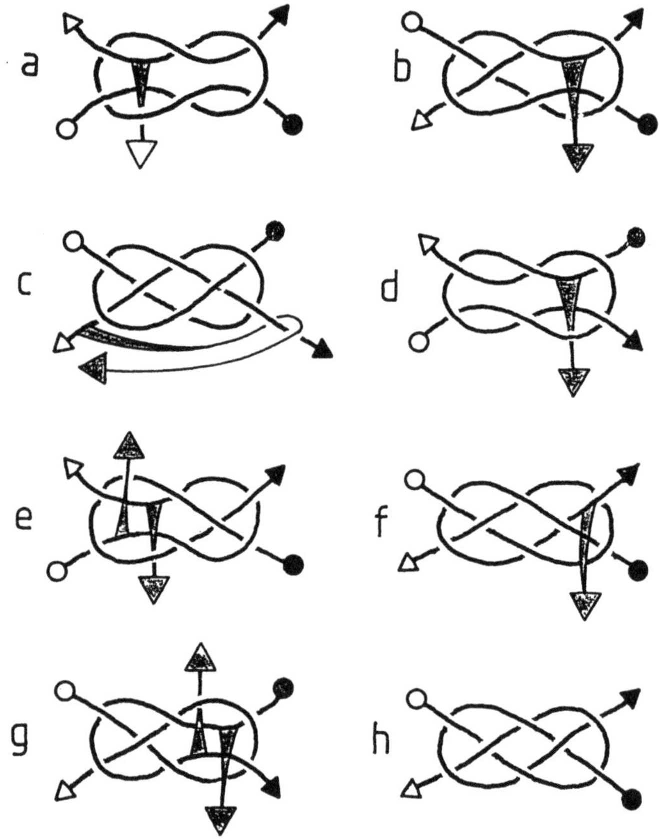

Fig. 1. The Shaw Series

The Shaw Series

The first to write about tuck-transformations seems to have been George Russell Shaw, over 70 years ago [15]. He considered 20 possible layouts of the classic Carrick and Reef patterns tied in two cords. In fact only eight proper knots or bends can be made, but he found that they were were linked by a series of tucks. Starting with the Reef Knot or the Square Knot he could proceed, without completely untying and retying, through six other bends to finish with a Double Carrick. He wrote: '... the eight Bends ... pass, each into the next, by a slight change in the lay of the strands.'

Fig. 1(a)–(h) depicts Shaw's series. He labelled the knots:

(a) Square Knot	(b) Weaver's Knot
(c) Half Granny	(d) Granny
(e) Single Carrick	(f) Double Carrick 1
(g) Half Carrick	(h) Double Carrick 2

Today many knot tiers would call both (e) and (g) half Carricks, (f) a single Carrick, and (h) a full Carrick–ends opposed, i.e. emerging on both sides of the knot.

Anyway, try it for yourself. All you need is a couple of lengths of cord. I use flexible braided stuff about 7mm diameter (that's about one inch in circumference), and it helps if each bit is a different colour. But anything easily manipulated will do.

Desmond Mandeville, O.B.E., M.A., F.R.I.C.

In the late 1960s a retired organic chemist in London, England, was concocting a knotted alphabet, each letter from A to Z represented by a knot (for vowels) or a bend (for consonants). It was mere whimsy—done for fun; a childish code to pass a secret message. The knots represented words and sentences tied in a piece of string. Desmond Mandeville called it his Alphabend [2](Fig. 2). And he contrived the knot names (ignoring vowels) to make, when read aloud in alphabetical order, a rhyming mnemonic:

A
the Barrel Knot
the Carrick
the Drawbend Double

E
the Fisherman's
the Granny
the Hubble Bubble

I
the Jinx
the Kilkenny

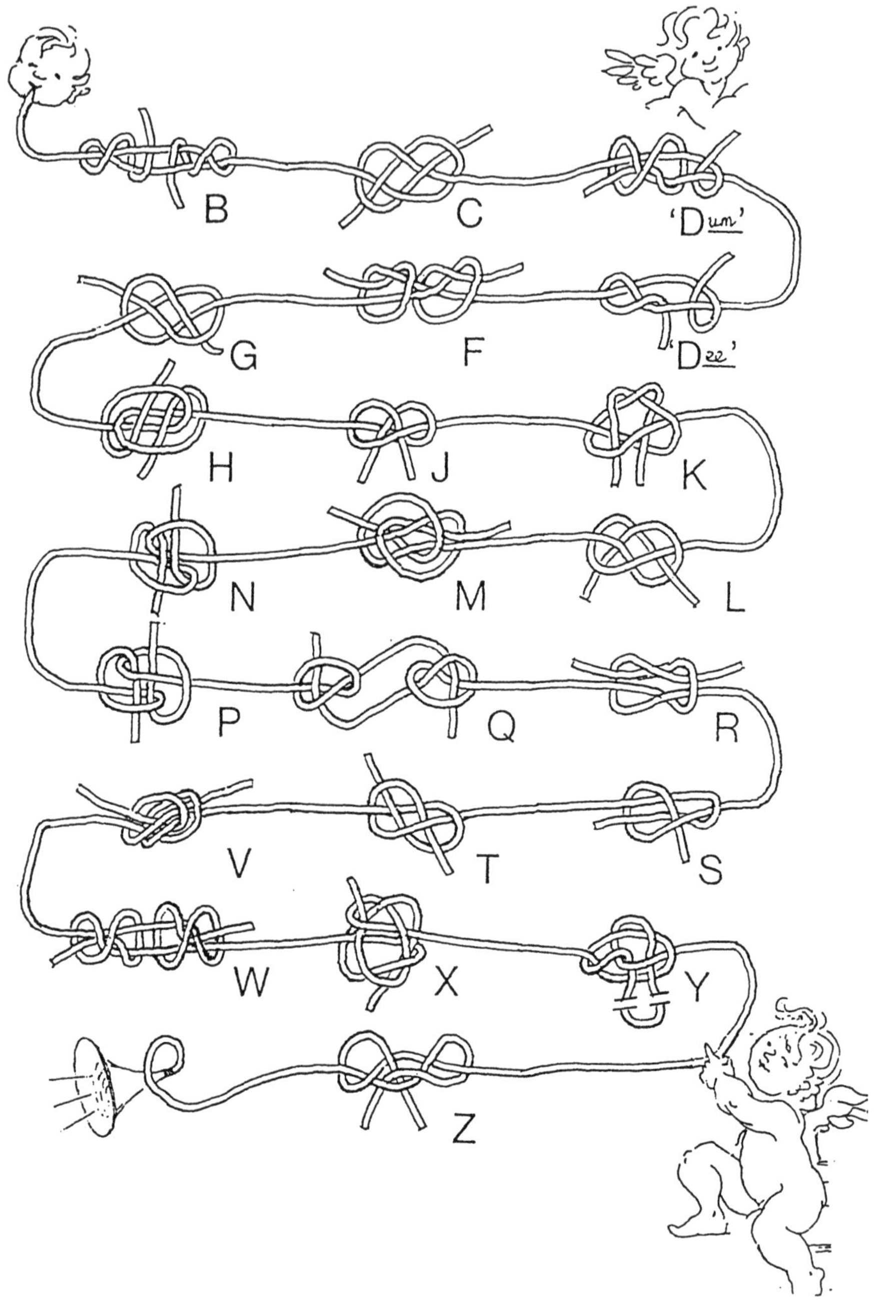

Fig. 2. Desmond Mandeville's original ALPHABEND (vowels omitted)

the Latin Lasso
the Matthew Walker
the Neat & New

O
the Poor Man's Pride
the Querry
the Reef
the Sure Sheet Bend
the Tumbling Thief

U
the oVerhand
I wonder...? and
the hot-X-bend
Yippee aYe, Yippee aYe
for the iZZard...
is the end.

If you are no knot tier, skip the following glossary. It is not vital to the tramble story. But knot tiers, puzzled by some of his cute and curious names, may find it helpful. The ones listed below in inverted commas he dreamed up—you won't find them in older knotting publications—for knots he did not know, and also as pseudonyms for existing ones that would not rhyme. Numbers preceded by the # mark are references to that monumental book—the knot tier's bible—known simply as Ashley [1].

Drawbend 'Double'	...	twin variants of the Harness Bend (#1474) which he called DeedleDum & DeedleDee;
'Hubble Bubble'	...	Hunter's Bend (#1425A, added in 1979);
'Jinx'	...	Whatnot (#1406);
'Kilkenny'	...	Original (Kilkenny Castle took the place of Carrick in Irish history);
'Latin Lasso'	...	A single Carrick Bend, named for the way it can be made by capsizing a Constrictor Knot (#1249), probably known to the Romans, over another cord's end;
'Neat & New'	...	Original;
'Poor Man's Pride'	...	Rosendahl's Zeppelin Knot;
'Querry'	...	Original;
'Sure' Sheet Bend	...	The orthodox knot with both short ends on the same side, as opposed to the variant with ends on opposite sides (which can in some

		cordage be less secure);
'Tumbling' Thief	...	His own stabilised version of an otherwise unstable knot;
'oVerhand'	...	Ring or Water Knot (#1412);
'I Wonder...? and'	...	Grapevine or double Fisherman's knot (#1415); 'I wonder ... will it hold me?' as this is a knot for rock climbers;
'Hot-X-Bend'	...	Shown to him in 1979 by Ettrick W. Thomson of Suffolk, England;
'Yippee aYe, Yippee aYe'	...	Farmer's Loop (#1054) with the ends cut to form a bend;
'iZZard'	...	Double Harness Bend (#1420) with parallel ends. Izzard is an olde English name for the letter Z.

Working on his Alphabend he now and then made a mistake in tying a knot; and often found he could correct the fault without undoing it all to start again. It's a labour-saving trick most knot tiers learn. But Desmond Mandeville—a 1930s Cambridge graduate with a double-first in Natural Sciences—grew curious about the way a tuck or two transformed one bend into another and, unaware of Shaw's earlier efforts, he decided to study the phenomenon.

In the next 25 years he learned a lot about this particular class of knots, naming up to 20 previously unrecorded ones. He also discovered and charted the steps between kindred knots.

He named these knotting routes Trambles (short for Tuck-Rambles) [3], and observed that knots a mere tuck apart could have extremely different characters. The dependable Sheet Bend—he pointed out—is cousin to the scary Thief Knot (which exists like Dr. Jekyll and Mr. Hyde in two forms, one of them stable ... the other extremely unstable).

A Basic Tramble

The sequence of Figures 3–6 illustrates a typical trip through bend territory. If you venture into it, take care not to become lost. A knot is 'either exactly right or wholly wrong' [1, #77]. One misplaced tuck or turn will lead you either to an entirely different knot—or to no knot at all. So, if at any stage you find yourself off course, stop. Retrace your steps, or start again.

It is a rough rule of trambling (not always strictly observed) that you should leave **more than half** of the original bend intact when you untuck and retuck any strand. The transformation Reef-to-Granny is not allowed, as it would involve a 50% untucking.

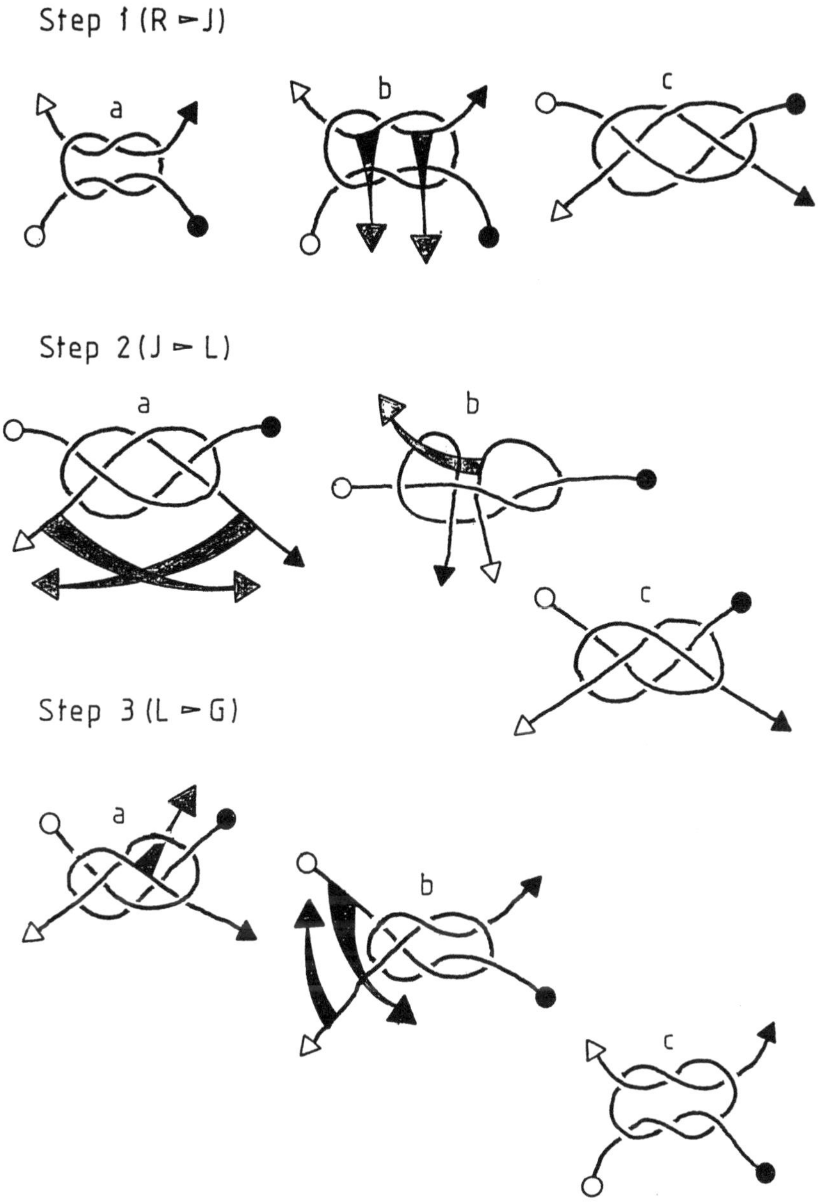

Fig. 3. The Basic Tramble (commenced—Steps 1 to 3)

Step 1 (Reef to Jinx)

(a) Start with the Reef Knot shown.
(b) Tuck both working ends (indicated by arrowheads) down through the centre of the twisted standing ends (marked by circles).
(c) The result is a Jinx (the stable Thief).

Step 2 (Jinx to Least-You-Can-Do)

(a) Uncross both working ends (or wends) so they are parallel.
(b) Pull the light wend through the dark loop.
(c) This produces the Least-You-Can-Do (which is a Single Carrick Bend).

Step 3 (Least-You-Can-Do to Granny)

(a) Withdraw the dark wend through the light loop.
(b) Uncross the light wend and standing end (or stand)
(c) You have arrived at the Granny Knot.

Step 4 (Granny to Carrick—ends adjacent)

(a) Re-cross the light stand and wend.
(b) Cross the dark stand over its own wend.
(c) Partly withdraw the light stand, as shown, to create a temporary holding loop.
(d) Tuck the light wend through the dark loop, *before* fully withdrawing the light stand.
(e) In the same way, partly withdraw the dark stand to create a holding loop.
(f) Pull the dark wend out through the light loop.
(g) Fully withdraw the dark stand.
(h) All of this yields a full Carrick Bend—with wends emerging on the same side of the knot.

Step 5 (Carrick–ends adjacent, to Carrick–ends opposed)

(a) Create a temporary loop* in the dark stand.
(b) Pull a bight, withdrawing and retucking the dark wend through the loop already formed.
(c) Eliminate the loop by removing the slack. This transfers a half-twist (or elbow) in the dark stand and wend to the light stand and wend.
(d) Treat the light stand and wend similarly.
(e) A Carrick Bend–ends opposed–appears.

Step 6 (Carrick–ends opposed, to Tumbling Thief)

(a) Pull out a holding loop* in the light stand.
(b) Withdraw the light wend.
(c) Retuck the light wend under one/over two/under one, as shown.
(d) Now fully withdraw the light stand.
(e) The result is a Tumbling Thief Knot.

Finding it hard? Trambles can be tricky! See the character in Canadian Rob Chisnall's apt cartoon—you are not alone.

Now the what goes where?!

Step 7 (Tumbling Thief to Sheet Bend)

(a) Withdraw the light wend to lie parallel beside its own stand.
(b) This is an orthodox Sheet Bend.

Step 8 (Sheet Bend to Thief Knot)

(a) Pull the dark wend out alongside its own stand.
(b) A Thief Knot is made.

Step 9 (Thief Knot to Reef Knot)

(a) Merely pull the light strand through the knot, until wend and stand swap over.
(b) Turn the result over, top for bottom.
(c) The final result is the original Reef Knot.

*These loops are my contribution to trambling. They come naturally. Desmond does not seem to have resorted to them. I think he just ignored his own 50% rule.

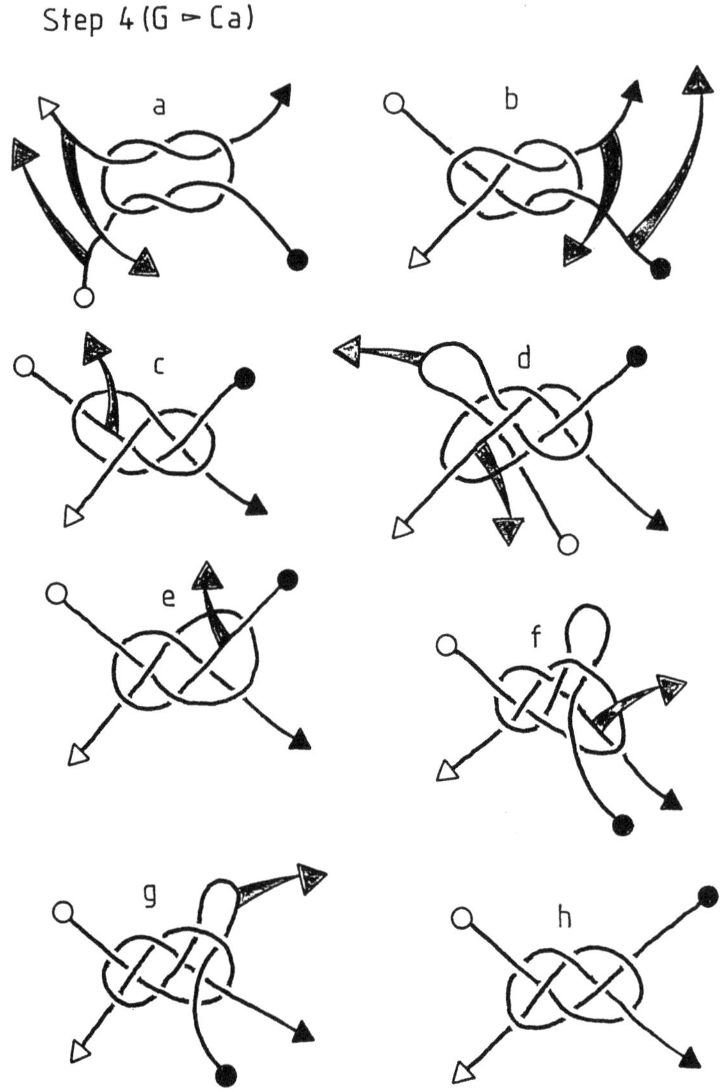

Fig. 4. The Basic Tramble (continued—Step 4)

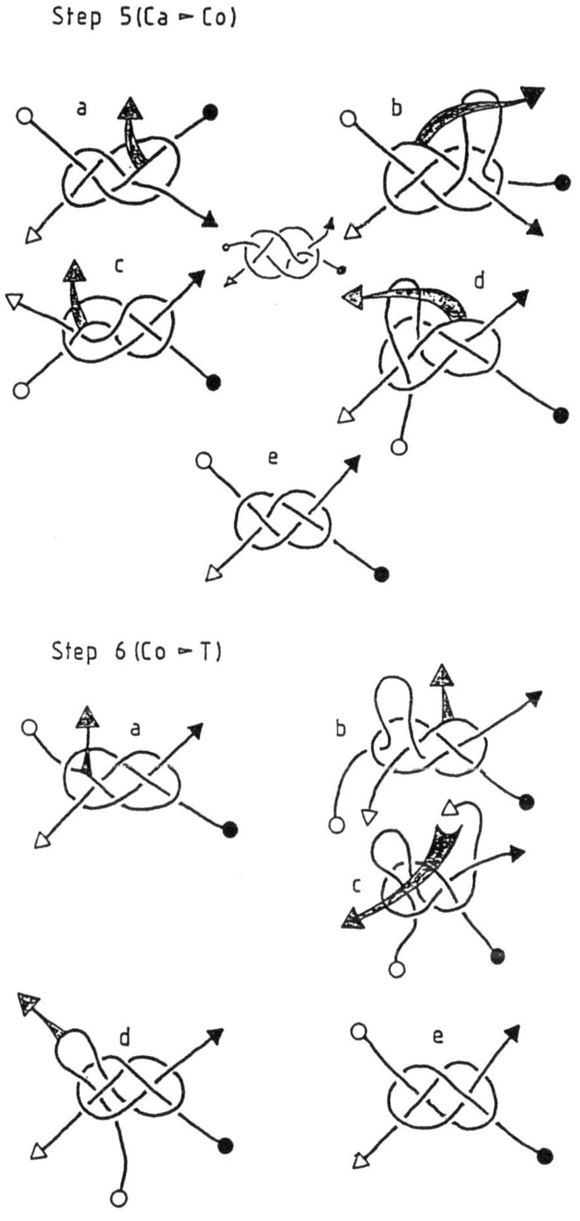

Fig. 5. The Basic Tramble (continued—Steps 5 and 6)

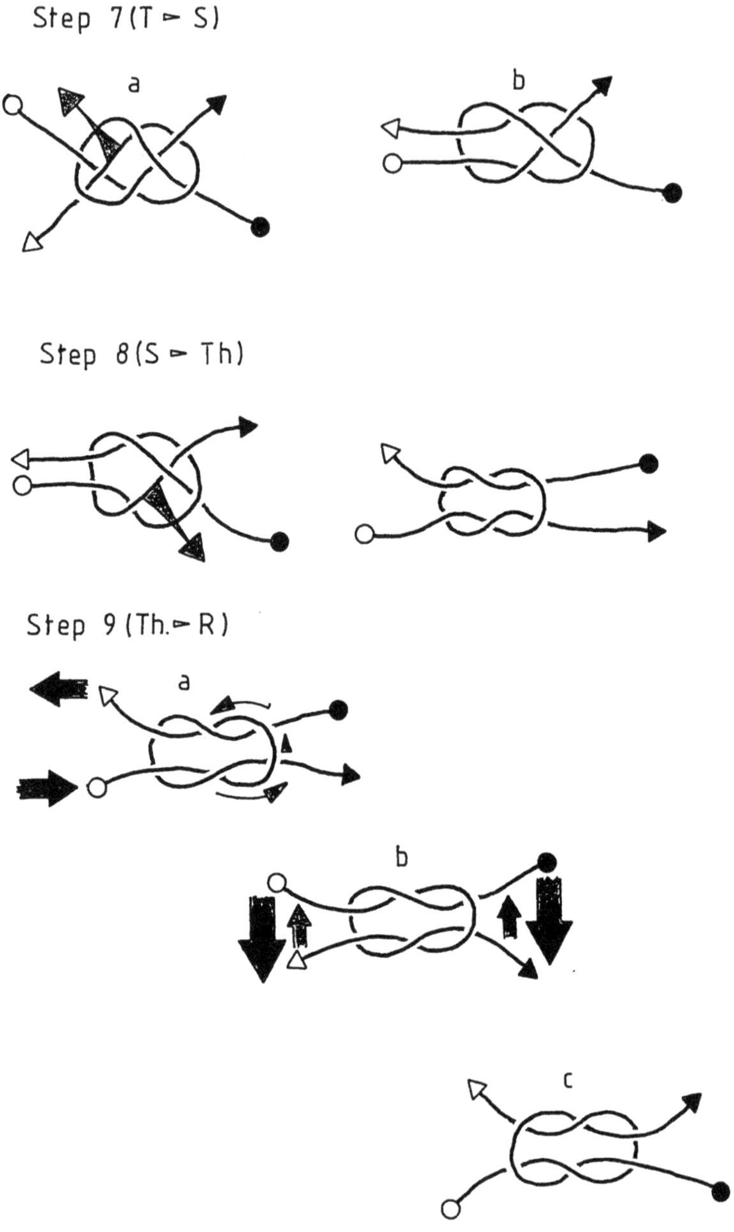

Fig. 6. The Basic Tramble (continued—Steps 7 to 9)

Welcome back. You have completed a tramble. This route can, with practice, be covered in two minutes or less. Desmond Mandeville did it between stations on London's underground train system. And it can be reversed.

Mapping the Bends

Desmond Mandeville came to believe that trambles were a serious and substantial field of study. He demonstrated many more of them [4], going off in all directions over an immense range.

There are at least a dozen distinct routes from Reef Knot to Granny Knot [5, 6]. But there is no direct route, short of untying and starting afresh, from the Reef to the common Sheet Bend. Each attempt only produces the left-handed (wends opposed) variant. Shaw found this—look back at Fig. 1(a)–(b). So the Reef is close—measured in tucks—to some knots, but remote from others; although it can be a bridge between seemingly unrelated far-off neighbours.

Bends are not simply strung together like beads; they have cross-linkages too. He charted one such network. The map (Fig. 6), based upon the Alphabend, covered a small zone around the full Carrick Bends (Co)(Ca), and revealed an almost crystalline symmetry and logic.

All the knots are reversible. That is, if both stands are shortened to become wends (and wends lengthened into stands), a similar knot results. A Reef (R) treated this way is another Reef (R'). Similarly G' and J' are bends created from G and J, and Hc, Hc' are forms of the half Carrick Bend (#1444). L and L', single Carrick Bends (#1445), are in practice identical.

But swap just ONE stand and wend for a different knot to appear. The Reef turns into a Thief (R, T'). Fl is the Flemish or Figure of Eight Bend (#1411) and Fl', M', V' are derived from Fl, M, V.

All knots have a mirror-image, and this handedness has to be taken into account in trambling. Bends on the left of the chart are left-handed, those on the right are right-handed, except for certain neutral ones (R, R'; T, T'; L, L'; and Fl, Fl').

Sl is the so-called left-handed Sheet Bend (#1432), but the name has nothing to do with handedness. It hints at an inferior performance (which irks left-handers like me). The basic tramble is shown by a broad line. It does not stop at Sl or Hc, which are intermediate steps to other bends.

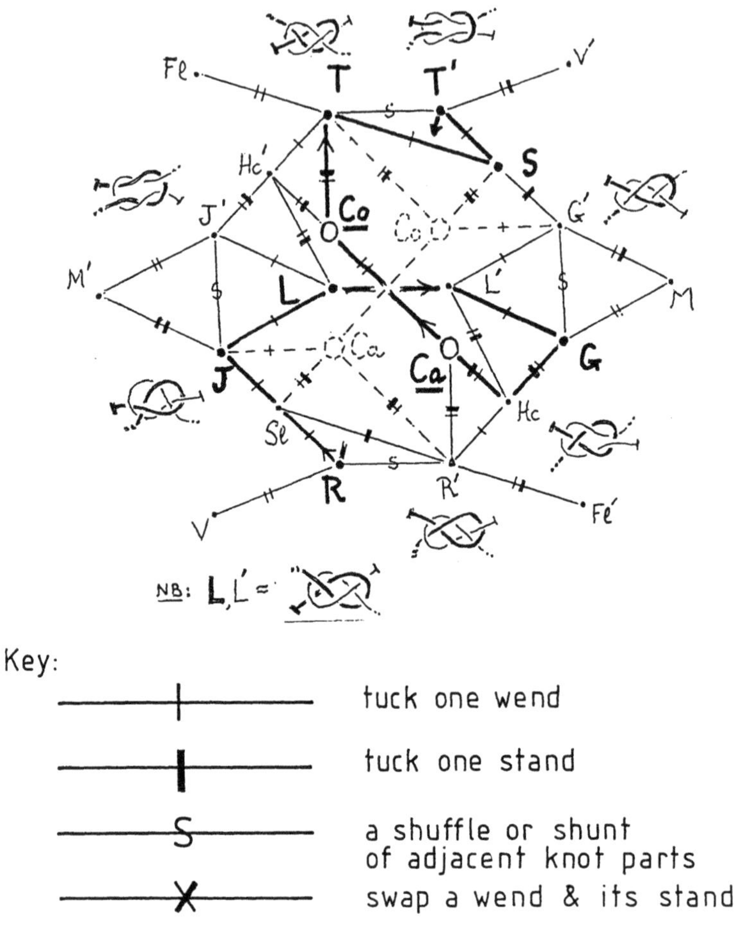

Fig. 7. A Map of the Bends

Using this map, trambles could be planned in advance. Where gaps existed unknown bends might be deduced. For example, there must be a looking-glass land containing the same bends (but with opposite handedness) and the dotted lines hint at simplified connections to this other uncharted territory.

The Tramble Talisman

But a two-dimensional map of the bends could not cope with the rich complexity of trambling. The Alphabend too (although twice amended) was soon superseded. What took their place was 25 closely related bends arranged as if inscribed on two sides of a single coin or medallion (Fig. 7). Desmond Mandeville called it the Tramble Talisman. Previously unpublished, it is another useful tool. You have already met most of the knots. Newcomers are:

front side ... Wh (Whatnot, #1407), the unstable form of the Jinx;
Kn (Knobble, #1424);
rear side ... Cf (Counter Fisherman);

The 15 front-side bends, those which share a common frontier or corner, convert directly one into the other. They are all like-handed. Opposite handedness can be an obstacle to trambling. Indeed the limited scope for trambling from bends on the front of the Talisman to the 10 other bends on the back was due to a change in handedness. Such changes could, however, be effected via Bends *C*, *G*, *L* and *R*, which is why they are shown on the front (where they belong) and also penetrating through in places to the rear.

Desmond claimed [7] to have tested out every one of the 70 possible Talisman tuck-transformations. Each worked in both directions, 140 in all.

The Tramble Drum

Perhaps a world map of bends would actually be a sphere, as several bends were later found to leak around the rim of the Talisman connecting front with rear. Desmond Mandeville began work again and in 1986 made a simple drum-shaped model (Fig. 8 shows it flattened out) of certain knot relationships he was confident he had established [8].

It was a six-sided cylinder, divided in half to give 12 faces around, plus two end plates. An A-prefix replaces the usual # Ashley reference. Only symmetrical bends were used, except perhaps L, so Sheet Bends were omitted. Most were like-handed. The sole newcomer is: Cl (Corrickle, #1451);
but note that P (Parallel Ends) was Z (the iZZard) in the Alphabend.

The full Carrick (Ca) proved to be more of a nexus than either the Reef or Granny; and so it was imagined lying at the centre of the cylinder, from where it could readily be tucked to give any one of the other 14 bends.

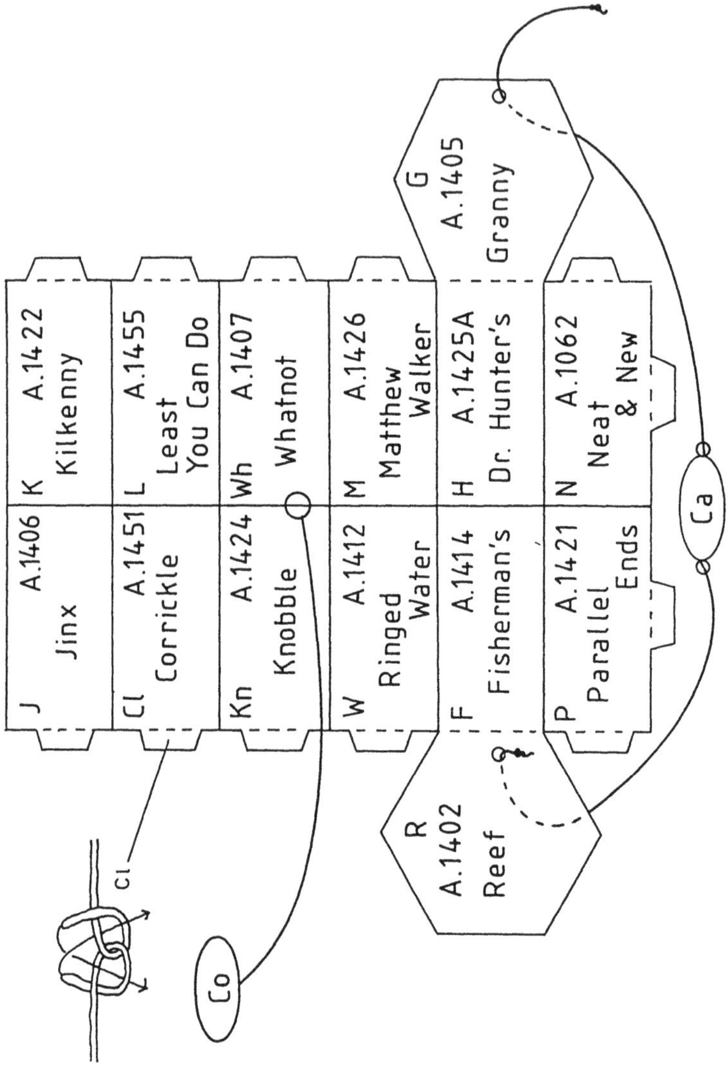

Fig. 8. The Tramble Drum ('flattened')

The other Carrick (Co) had fewer links and existed somewhere outside, adjacent to Kn and Wh. There it was the principal, if not the only, connection with skew-symmetrical knots such as the Zeppelin (originally Poor Man's Pride), Tumbling Thief, Querry and Flemish Bends. Another drum would be needed for this family. Perhaps they too would be at home on the surface of a sphere, a smaller one since they were fewer in number. If so, this part of the knotting universe would resemble Earth and Moon.

Summary

When trambling's innovator Desmond Mandeville died, at the age of 82, few knot tiers understood trambles. Many were bemused by them. He had devised shorthand symbols akin to chemical formulae for trambling techniques [11], but even his keenest followers struggled with the cryptic language and failed at fluency in it; no doubt it deterred many practical knotters, being too much like topological knot theory.

Topology is a mystery to even erudite knot tiers. They are surprised to hear that its practitioners do little or no real knot tying, and that it isn't easy—mathematically —to sort out a Reef Knot from a Granny knot. Knot theorists transform imaginary knots, without any tucking or untucking. They have no use for ends, unless those ends stretch to infinity; whereas Desmond Mandeville sought an ergonomic explanation of knot relationships [12]. You cannot tramble without ends, or with just the two ends of a single piece of cord. Trambling uses both stands and wends. Tension, flexibility and friction too—subjects not in any topology syllabus—also intrude. If knot theory is mental gymnastics, then trambling is free-for-all wrestling.

Both, nevertheless, are pure research done for its own sake. Practical uses may occur later. Meanwhile each seeks in its own way to unravel what the 19th century mathematician Tait [16] called 'beknottedness'.

Trambling shines new light on knots and makes it possible to interpolate or extrapolate where others may lurk in the shadows. Trambles open up untrodden areas for puzzle addicts to explore. They might also add to scientific knowledge. Knot classification is becoming significant in the understanding of polymers, DNA and other enzymes, and the logic of space [17]. So put aside those molecular models made from sticks and balls. Take a couple of cords and try trambling.

Desmond Mandeville left detailed notes [13], his correspondence and publications. These include a trio of testing parlour games, one of which is like a point-to-point (steeplechase) horse race [14].

Acknowledgements

The writer gratefully acknowledges Mrs. Norah Mandeville's help, making her late husband's trambling research available to him.

Knotting Matters is the quarterly journal of the International Guild of Knot Tyers, published in Suffolk, England.

Rob Chisnall's cartoon 'Now the What Goes Where' originally appeared in Knotting Matters, Issue No. 3 (April 1983), page 3.

References

1. C. W. Ashley, *The Ashley Book of Knots* (New York 1944 and London 1947).
2. D. Mandeville, 'The Alphabend', *Knotting Matters* Issue No. 4 (July, 1983) 2–5.
3. D. Mandeville, 'Trambles', *Knotting Matters* Issue No. 10 (January 1985) 5–9.
4. D. Mandeville, 'On Trambling Territory', *Knotting Matters* Issue No. 18 (January 1987) 10–15.
5. D. Mandeville, 'Granny-into-Reef - I', Knotting Matters, Issue No. 27 (April 1989) 15–18.
6. D. Mandeville, 'Granny-into-Reef - II', *Knotting Matters* Issue No. 29 (October 1989) 6–12.
7. D. Mandeville, Unpublished correspondence, Geoffrey Budworth's private collection.
8. D. Mandeville, Unpublished correspondence, Geoffrey Budworth's private collection.
9. D. Mandeville, 'Hunter's Shunts', *Knotting Matters* Issue No. 24 (July, 1988) 11–13.
10. D. Mandeville, 'Annex to Hunter's Shunts', *Knotting Matters* Issue No. 25 (October, 1988) 9–11.
11. D. Mandeville,'Trambling Technique', *Knotting Matters* Issue No. 19 (April, 1987) 13–16.
12. D. Mandeville, 'Knots Topological or Ergonomical', *Knotting Matters* Issue No. 23 (April, 1988) 8–10.
13. D. Mandeville, Unpublished notes, Geoffrey Budworth's private collection.
14. D. Mandeville, 'A Point-to-Point Tramble', *Knotting Matters* Issue No. 34 (January 1981) 10–11.
15. G. R. Shaw, *Knots – Useful & Ornamental* (New York, 1933) 14–16.
16. P. G. Tait, 'On Knots - I, II & III', *Scientific Papers* **I** (London 1877-1885) 273–347.
17. Wales, University of, *Mathematics & Knots* (Bangor, Wales, 1989).

CHAPTER 14

CROCHET WORK—
HISTORY AND COMPUTER APPLICATIONS

Sally Jo Cunningham

Crochet Technique

Crocheting is a technique for creating textiles that can vary from solid, sturdy clothing fabric to light and airy laces. The most striking way in which crochet differs from knitting, with which it is frequently confused, is that crochet uses a single hooked tool rather than a pair of straight knitting needles. Crochet stitches are based on a loop of yarn or thread pulled through existing loops by a hook (see Figs. 1 and 2). In crocheting each stitch is individually created, whereas in knitting an entire row is defined at a time.

Fig. 1. The chain stitch (Caulfeild and Saward, 1882)

The chain stitch is the simplest crochet stitch, and is based on the primitive 'finger knitting' technique. A chain is constructed by first forming a grounded loop around the hook. Additional chains are created by hooking the free end of the thread around the hook, and then drawing the free thread through the previous loop. Chain stitches have two uses in crochet work: as the components of a 'base chain', and as a means for incorporating 'holes' into the piece. Nearly all crochet pieces are started by creating a row of chain stitches that forms the 'bottom' of the piece. In round crochet work, the end of the base chain is joined to its first stitch, creating a circular foundation chain. Chain stitches are also used to achieve a feeling of lightness and airiness in the body of the crochet work, by connecting denser single, double, and treble crochet stitches with one or more chains.

Other common stitches are the single and double crochets (Fig. 2)*. These stitches are formed by: winding the loose end of the thread over the hook zero or more times; inserting the hook through the next stitch in the previous row and hooking the loose thread; drawing the loose thread back through the previous row's stitch; and then drawing the hook back through the loops on the handle, two loops at a time. The more loops that were originally wound around the handle of the crochet hook, the taller the resulting stitch will be; a treble crochet, for example, is formed beginning with two loops wrapped on the handle. While even taller stitches are possible (the double treble, triple treble, etc.), the treble is the longest stitch in common use.

(a) Single crochet stitch (b) Double crochet stitch

Fig. 2. Single and Double stitches (Caulfeild and Saward, 1882)

Textile production with crochet has a number of appealing characteristics, both for hobbyists and for professionals. These traits have contributed to the wide dissemination of crochet throughout the world, and to its enduring popularity:

- It is useful for both clothing and fine lace, and for both two-dimensional and three-dimensional work. Lacy patterns are created by using a greater degree of double, treble, and chain stitches, while sturdy cloth for sweaters is produced with a larger amount of single stitches. By

*This is the American terminology; in the British system, these are double and treble stitches, respectively.

varying the size of the crochet hook and the thickness of the thread, the same pattern can easily produce either a fine bit of lace or a heavy floor rug—and both will be equally attractive and well- constructed. And finally, the 'stitch at a time' nature of crochet makes it easy to build three-dimensional pieces (a much more daunting proposal with 'row at a time' techniques like knitting!).

- The technique is comparatively easy to learn, and crochet items can be produced relatively quickly. Perhaps because crochet uses only one needle rather than two, many people report learning to crochet more easily than learning to knit; and similarly, a crocheted item can generally be created in less time than a similarly-sized knitted piece.
- Crochet is extremely versatile—it can imitate a variety of other textile production techniques. A quick glance through a stack of crochet patterns reveals an amazing array of styles that crochet can copy: reticella lace, filet, Venetian point, bobbin laces (Cluny, Torchon, Maltese, ...), Honiton, Brussels, Battenburg, Hedebo, tatting, weaving, macramé, patchwork quilting, and more. Generally, these copies can be made in a quarter the time or less than that needed to make the 'real' lace or cloth! [10] And by using crochet, a designer can more easily mix types and styles, to give more interesting effects than is usually possible with the original methods. However, the crocheted imitations are generally much coarser in effect, since in crochet 2–3 thicknesses of thread are pulled through loops, rather than the single thickness for true needle laces or knitting.
- Mistakes can be easily corrected or unravelled. Crochet is much more forgiving than other textile production techniques; 'dropped' stitches don't ladder the fabric, as is the case in knitting, and can often be inconspicuously added back in the succeeding row.
- Crochet requires minimal, inexpensive equipment: a single hooked needle and a length of string or yarn. The crochet hook can be easily handmade: the crochet workers of impoverished 19th century Ireland, for example, created their own tools by cutting the side of 'eye' away from an old needle, and embedding it in a cork or stick handle. If thread or yarn is too expensive, then other materials can be substituted. In the U.S. South, for example, the poor used to unpick feed sacks and flour sacks for thread to crochet into fine doilies.
- Crochet requires less attention than other needlecrafts, and is easy to pick up and put down. Mary Konior reports that Victorian housewives favored crochet because it could be worked while minding children, doing housework, overseeing servants, and gossiping; at a lower level

of the social scale, crochet was popular with canal-boat workers and gypsies because it was so easy to do while travelling [10, p. 5].

History of crochet

Where did crochet come from? When was it developed? Given the relative simplicity of the technique, one would expect that it was the precursor to knitting, and that it perhaps antedated weaving as well. Surprisingly, it seems to be one of the last textile production methods to be invented! Many popular crochet pattern books cite a history for crochet dating back to early Mesopotamian civilizations, where it allegedly played a significant role in religion. This mythological origin appears to be due to a book by Heinz Edgar Kiewe, The Sacred History of Knitting [9], that contains many exciting speculations that are—unfortunately—not supported by any reputable evidence. Similarly, Annie Potter reports the existence of an Egyptian 'threadwork' specimen in the Benaki National Museum (Athens), dated A.D. 900–1100, that is similar in appearance to a common crochet pattern (the 'chevron', a rippled effect popular for afghans) [15]. Unfortunately, there is no proof that this piece was indeed produced by a crochet-like technique; and even if it had been, the method had not survived into modern times.

A more reliable, though less spectacular, estimate is that crochet was developed in Western Europe during the 16th century. It most likely developed as a form of 'tambour' work (tambour uses a similar hooked implement to produce crochet-like chain stitches anchored in a piece of cloth, giving an effect similar to embroidered chain stitches). If tambouring was worked onto a parchment background, 'chain lace' could be created by ripping the parchment away. One of the earliest printed references to chain lace is in an inventory of Queen Elizabeth I's household effects, where she was reported to own a 'gowne that was exquisitely laid aboute with small cheyne laces of gold' [10, p. 10]. The speculation is that at some point it was realized that these chains could be produced without the parchment background. The term 'crochet' is derived from the French 'croc', or hook. By the late 1700s the simple craft of 'crochet en l'air' (literally, 'crochet in the air'), or tambour work without a support, was a fashionable hobby for middle- and upper-class French ladies. The double and treble stitches appear to have developed in the late 18th century, perhaps from a slip with a tangle of thread on a hook. Their utility was quickly recognized, and by the early 1800s the basic stitch and construction techniques of crochet were well-established.

As a hobby, crochet experienced a boom of astonishing proportions during the Victorian age. From the late 1830s, countless crocheted objects poured from the hooks of middle class ladies. Lace was especially popular—both to wear, and as "doyley's"*to decorate furniture [10, p. 8]. Patterns for children's clothing in crochet were also common, though these were usually plain and practical rather than decorative: 'Simple crochet is well adapted to the wants of everyday life, as it requires little skill in execution, will resist wear and tear, and costs a comparative trifle for materials.'[2] Finally, women crocheted a plethora of more or less 'useful' household objects—everything from pillows to penwipes, bedspreads to braces (men's suspenders), 'fascinators' to imitation fur trim. It became popular to have covers, edgings, or decorations for everything inside the house—and crochet was the easiest, cheapest way to produce them [16].

The sometimes extravagant extent to which this hobby was exercised is reflected in George Eliot's somewhat jaundiced view of it:

> *'When a man is happy enough to win the affections of a sweet girl who can soothe his cares with crochet, and respond to all his most cherished ideas with beaded urn-rugs and chair-covers in German wool, he has, at least, a guarantee of domestic comfort, whatever trials may await him out of doors. What a resource it is under fatigue and irritation to have your drawing-room well supplied with small mats, which would always be ready if you ever wanted to set anything upon them! And what styptic for a bleeding heart can equal copious squares of crochet, which are useful for slipping down the moment you touch them? How our fathers managed without crochet is the wonder...'*[7]

Crochet's reputation as a genteel, respectable occupation for ladies seems to be derived in part from an association with the British royal family. Although it appears to be untrue, there was a persistent rumor that Queen Victoria and other members of the royal family were avid crocheters [16, p. 23]. Crochet was rarely truly fashionable, however, in the sense of having crocheted items worn or used by members of the upper classes. Earlier in the century lighter laces were in mode—patterns and techniques that crochet could imitate, but in a heavier effect. After the 1850s heavier machine and hand laces were at times stylish, but generally not explicitly crochet work. The exceptions were bursts of popularity for Irish crochet in the 1840s and early 1850s, and in the 1890s [17]. These movements were at least partly inspired by a desire to aid the poor in Ireland, where this lace was produced (discussed below). When the Irish economic crisis subsided, so too did the fashion for Irish Crochet. Crochet remained a hobby for the middle class, and provided a precarious occupation for the very poor. In both cases it was generally practised to allow the middle classes to enjoy cheap copies of the finer laces worn

*The term comes from the name of the famous London haberdasher of the late 1700s, and was later corrupted to the now more common "doily".

by the wealthy, and for the poor—when they didn't sell their products—to imitate the middle class. Indeed, a common derogatory nickname for crochet work was "poor man's lace"[16, p. 14].

A number of interesting variations on crochet were developed, or at least achieved widespread recognition, in the mid to late 19th century. Hairpin lace, or krotchee, was first described by Mrs Cornelia Mee in her Manual of Needlework (1856) [16, p. 15]. This technique provided a more accurate imitation of lace produced with a netting shuttle, by first winding the thread around a metal frame shaped like a large hairpin and then crocheting the loops together. Similarly, broomstick or Peacock's eye lace is created by crocheting over a large wooden dowel, which gives an exaggerated lacy, loopy effect to single crochet stitches. Ring or moulded crochet involved working single stitches over metal, wood, or celluloid ring shapes, and these rings were then joined by more crochet work. The stitch-covered rings added texture and dimensionality to heavy laces and edgings, and were also practical for use in hot plate holders [10, p. 32].

The most enduringly popular alternate crochet technique is Tunisian Crochet, also known at times as Shepherd's, Tricot, Afghan, railway, fool's, idiot's, Scotch, or German crochet. This technique is worked on a crochet hook with a very long handle; an entire row of stitches are 'picked up' from the preceding row and held on the handle, then cast off one by one until only a single stitch remains. The fabric produced is very warm and solid, and can be cut and seamed without unravelling. Obviously this type of crochet shares characteristics with knitting, and it is tempting to think of it as the 'missing link' between the two—or perhaps even a precursor to both. Unfortunately, little concrete is known about the origins of Tunisian crochet. There are undocumented speculations that it is indeed an ancient craft of Northern African origins; however, it was first reliably described in the 19th century, and it appears likely that the name 'Tunisian' was intended merely as an exotic descriptor, rather than as an indication of the technique's true origins.

During the mid-to-late 1800s the techniques of crochet work were exported worldwide, mainly through missionaries. As will be discussed below, in a number of places crochet has become entrenched as a cottage industry, and continues to be an economic mainstay primarily for poor women. In other countries, crochet remains a popular hobby, and has been adapted to these other cultures. Japan, for example, invented the graphic patterns that are in common use worldwide. In Europe, crochet patterns had originally been written out stitch by stitch, in text. Alternatively, individuals sometimes developed crochet pattern books containing small, physical samples of a variety of patterns and motifs (whose construction could then be guessed and followed by other workers); however, these books could not be used to describe very complex or large pieces of work [15, p. 78]. The Japanese couldn't read the text of written patterns, and overcame the limitations of physical pattern books by producing stylized drawings of designs: crochet items are described by arranging stitch icons to mirror the arrangement of actual stitches in a finished work.

In Turkey, crochet was quickly adopted for the construction of oyas (decorations on headscarves). In Anatolia, by tradition a young woman couldn't speak freely to her mother-in-law until she had borne a child or had been married two years. The new bride changed her oyas to express her feelings: a red rose or carnation meant happiness and love, red peppers meant that she was cross or not getting along with her husband, yellow symbolized fatigue, and so forth. Elaborate oyas have been found dating back to 2000 BCE, but are now most frequently crocheted. In fact, crochet was embraced so whole-heartedly that it is now considered a 'traditional' craft [15, p. 45].

After World War I, crochet began to decline in popularity as a hobby when lace went out of fashion. The extensive lace lappets, cuffs, and deep edgings of the Victorian period gave way to plain, simple, clothing designs, with lace decorations reduced to at most a single narrow edging. Crochet continued to be a used as a base fabric for clothing, however, particularly for infants. And the ability to easily work in three dimensions brought about a spate of patterns for crocheted hats, purses, and other personal items. Crochet remained the prime mimic, useful for providing the effects of other construction techniques, but in a fraction of the time and without the extensive training needed for weaving, bobbin lace, etc. Indeed, the 1950s may have seen the imitative power of crochet pushed to its limits, as patterns were sold that provided crochet substitutes for milk glass vases, ironwork decorations, and pottery pieces!

The period from the late 1950s through the 1970s also witnessed the emergence of crochet into the world of fine arts, through the Art Fabric movement [3], [4]. This movement questioned the historic, and largely artificial, distinction between 'major' arts such as sculpture or painting, and 'minor arts' or crafts such as weaving and knitting. The art world took a new interest in textiles, both through innovations in traditional flat textile production (tapestry weaving, screenprinting on fabric) and in using fabrics in three dimensional sculptures. Crochet's versatility in construction again allowed artists to achieve a number of effects relatively quickly through this single technique. This artistic application of crochet was heavily influenced by the rise of the feminist movement; many of the artists in the movement who incorporated crochet into their pieces were women, and were attempting to incorporate their feminine experiences and backgrounds into their art.

The early 1970s saw an explosive revival of crochet as a popular craft when the craft caught the attention of the youth movement. In the wake of this sudden tremendous interest in learning crochet, supplies of yarn, crochet thread, and hooks ran short, and stores were rationed as to the amount of materials that they could buy [8]. The work of the early 1970s clearly reflects both the counterculture orientation of crochet during this period and its influence from the Art Fabric movement: colors were bold, non-traditional materials such as jute and wire were used, and sculptural effects were produced with newly developed techniques.

Where popular crochet of the 1970s concentrated on clothing, jewellery, children's toys, and sculpture, by the early 1980s interest had revived in crochet lace. Victorian and early twentieth century patterns were unearthed and republished, and a spate of magazines arose that specialize in new lace patterns. While many of these new patterns are simply variations on old themes, others introduce exciting new styles. At both the 'fine art' and the popular levels, crochet is thriving.

Commercial importance of crochet

Crochet lacemaking has been, and continues to be, an important cottage industry in many parts of the world. In Ireland, for example, it was introduced by French nuns in the late 1700s. It remained a genteel hobby until the Irish potato famine of the 1840s, when large numbers of people supported themselves and their families by producing crochet lace articles for sale in England and abroad. At one point, 12 000 women and children in Cork alone were employed as crocheters!

The most popular form of lace produced during the famine was an imitation of Venetian Guipure, a type of bobbin (pillow) lace. This imitation guipure has since become a well-known style of lace itself, under the name 'Irish Crochet'. As the famine receded, it lost popularity and the lace industry in Ireland collapsed. The craft experienced a revival during smaller famines, however, and an extended popularity during the 1890s with the new fashion for heavy laces. It is still commercially produced in Ireland, and is a popular craft elsewhere throughout the world. The excellent works by Mrs. Bury Pallister (1902) and Elizabeth Boyle (1971) provide a detailed history of the development of Irish Crochet as an hobby and as an industry [14], [1].

Irish Crochet was particularly well-suited for large-scale production because it is constructed piecemeal: large numbers of a limited range of motifs are produced, laid out in an attractive pattern, and crocheted together with a lacy background pattern. A production line system developed whereby individual workers concentrated on manufacturing large quantities of one or two motifs, and sold their work to an agent. The agent then hired another person to piece the motifs together, and kept the lion's share of the profits from the lace. While some men and boys plied to hooks, in most cases women were contracted to produce lace for male agents. In the push for higher and higher levels of production, the quality of the lace necessarily suffered: fewer and simpler motifs were worked in coarser thread, and the motifs were often arranged haphazardly rather than in an ordered design.

Interestingly, crochet work gained a reputation as a disreputable means for a woman to earn a living—in sharp contrast to other textile-based cottage industries such as bobbin lace and pillow lace. These other laces had been praised because they kept female workers at home, safe from immoral influence of factory work. Crochet work, however, 'quickly earned a bad name—not merely on account of its artistic deficiencies, but also because of its socially demoralizing effects. Godly people held the crochet-maker in horror, and as long as travelling agents bought the work, and enabled the demoralized crochet women to thrive, there was no doubt or justification for the outcry made against the vice, which seemed to be inseparable from this branch of industry.'[17, p. 97]

Crochet may have been viewed suspiciously because it was a much more creative, less regimented craft than pillow or bobbin lace. These latter laces were constructed from ready- printed patterns, and no deviations from the pattern was permitted in the work. In contrast, Irish Crochet was worked to no fixed pattern. Instead, the motifs were combined according to the joiner's talent for producing a pleasing overall effect, and motif workers were also free to invent new motifs. Not surprising, then, the female Irish crochet workers preferred crochet work to traditional lacemaking, as being more creative and the designs were more in their own control. As one worker eloquently explained,

> *'I likes the crochet best, ma'am, because there's hope in it. I* may *get ever so much for what I makes, if I happen to hit on a new stitch, and all the time I'm at it, I don't know but I may have a lot of money coming to me, and I'm kepin spirits like, to the last moment; but that pillow-work, och, 'tis horrid, ma'am! You're made sinsible from the beginning that you're only to get the trifle of a price, no more, nor no less, and no thoughts will help you, you must go on with the thing to your* ordthers, *which is what* I won't do, *until I can't help it, please God.'*[12]

Currently, significant sources for commercial crochet work are India, Mexico, China, and the Philippines. In Narsapur, India, the lace-making industry followed much the same pattern as in Ireland. Crochet was introduced by Christian missionaries wishing to alleviate the economic distress of their congregations. Narsapur suffered famines in the 1830s, 1870s, and 1890s; during these times, the crochet work produced by women provided the sole support for many families. The lacework was sold in Europe to feed the seemingly insatiable desire of the middle class for cheap lacework, and to the Indian upper class as Indian women began to imitate European fashions. This trade continues, and is an important source of revenue for the region. Unfortunately, a major and increasing portion of the revenue for the work produced finds its way into the hands of middle-men and exporters, rather than going directly to the women who create the lace. Since the women are forced to create lace more quickly in order to maintain their already marginal standard of living, the quality of the lace has declined [13]. In an effort to change the exploitative nature of crochet work in this area and to allow the women to gain more control over their livelihoods, the Godavari Delta Women Lace Artisans Cooperative has been formed. This association can pay significantly higher prices for crochet work by buying directly from the women who produce it, and marketing the lace abroad through charity or aid organizations.

In Mexico, crochet work has only recently achieved commercial importance. The American counterculture movement of the 1960s and early 1970s created a large demand for 'ethnic' garb, particularly of the ideologically correct 'peasant' type. Female Mexican handworkers began producing traditional hand-decorated peasant blouses for the increasing tourist trade and for export. While the majority of the handwork in these blouses is embroidery, crocheting is used for edging and lace insertions. Ironically, the popularity of these blouses abroad has occurred after they were no longer worn in Mexico itself [18].

At present, the most significant exporters of crochet work are China (primarily lace tablecloths, pillows, and other household goods) and the Philippines (primarily lace clothing). Surprisingly, for such a long-lived industry, it is difficult to obtain solid figures on the economic importance of the crochet industry in these countries and worldwide. Part of the problem lies in the fact that crochet work is often a cottage industry, and cottage industries are difficult to regulate or to accumulate statistics on.

At the hobby end of the scale, a number of businesses have come into existence to support the crochet craft. A seemingly endless stream of crochet instruction manuals, pattern books, magazines, and broadsheets have been produced and purchased since the 1840s. Special crochet threads—usually in cotton—are manufactured, in a variety of thicknesses and colors. Special purpose yarns occasionally enjoy a brief fad: raffia for creating crocheted summer hats or purses, imitation fur yarn for furry trims on hats, and so forth. Novelty items abound, such as frames for hairpin crochet, rings for moulded crochet, and beads or decorative items to be crocheted into a piece. The old- fashioned doilies that once covered the Western world's furniture (and sometimes still do!) could be stiffened by a simple sugar syrup, or by a patented doily stiffening solution. And finally, the simple crochet hook has been and continues to be available in a number of forms: with varying degrees of cleft in the hook, different angles between the handle and the hook, detachable or 'Swiss-Army' style needles to allow several hooks to be used with the same handle, hooks in a variety of materials (bone, wood, plastic, metal, ...), needles with variously shaped thumb rests, and so forth. There is even a special prosthetic device available that allows one-armed crocheters to continue to enjoy their hobby! [11]

Crochet and computing

In computing, researchers generally build an understanding of complex ideas by first selecting a narrow problem or application and developing methods specific to it. With that small problem solved, the researcher looks for ways to generalize the solutions to cover larger classes of problems. At the University of Waikato (Hamilton, New Zealand), members of the Computer Science department have been using crocheting as a sample application to explore issues in the automation of creative design and image storage and retrieval.

With the construction of the CADD (Computer Aided Doily Design) system, we examined the formal rule or grammar systems have been used in the visual and musical arts to support creativity [5]. By making explicit the generally implicit rules for forming a piece, insight is gained into the fundamental structures of the art form itself. These knowledge-based systems provide a description of the art form that can be tested for validity, and that can be used to generate new instances of that art form. We demonstrated this technique by developing a program that uses a rule and grammar description of crochet lace to automatically generate new lace patterns.

Specifically, our model incorporates the following levels of description:

1) *Simple constructive constraints*: These constraints describe the possible ways in which individual crochet stitches may be connected to each other. For example, a double crochet may connect to another stitch at its base or its cap, but not by the middle of the double crochet.

2) *Geometric constraints*: As the crochet lace is designed, a graphical pattern for it is simultaneously drawn on the screen. The geometric properties of the graphical display provide guidance in achieving validity for the pattern that CADD is creating. As an example, given that CADD is attempting to place two double crochet stitches beside each other, the distance between the caps of the stitches may indicate that a joining chain is needed. This physical distance also determines the number of chain stitches necessary for the join. The geometric model also supports the achievement of symmetry in the pattern, an important concept in crochet lace.

3) *Generative motif grammar*: There are common motifs in crochet lace that can be specified by a generative grammar. The grammar provides a mechanism for incorporating high-level design information into the model, although this mechanism is not yet fully implemented.

These elements of the model give us powerful tools with which to characterize aspects of crochet construction and design. The process of building these tools has in turn provoked speculation about some of the underlying assumptions of crochet lace composition, leading to our experimentation with variations on the traditional forms.

Is CADD itself creative? Artistic creativity has been defined in many ways: as the capacity to produce compositions, products, or ideas which are essentially new or novel, and previously unknown to the producer; as an embodiment of originality, which encompasses uncommonness, remoteness, and cleverness; and as the ability to make guesses, attempt solutions and perfect solutions, and finally communicate the results. At present CADD is creative in the sense of knowing 'the rules' and being able to manipulate them to produce a physically plausible pattern. These rules are sufficiently general so as to permit a wide variety of patterns to be formed.

This generality is one of CADD's drawbacks: generative models, unfortunately, tend to produce a copious flood of possible patterns, many of them uninteresting or aesthetically unappealing. We are currently considering the style parameters and meta-design principles necessary to permit the user to drive the generation of new patterns in novel, attractive directions.

However, in its current form CADD does enable an effective creative partnering between human designer and machine. In its 'semi-automatic' mode, CADD examines the current state of a design and proposes incremental extensions to it. The user selects one of these extensions, thereby guiding CADD and winnowing out the less promising patterns. The design process is greatly enriched, however, since CADD's set of suggestions frequently contains elements that the human has not considered.

The process of building the construction rules is also useful for stirring the creative processes in the human collaborator. As we codified the parameters defining a particular type of lace—the symmetric, flat, circular doily—we were, in essence, exploring the limits of that form. Indeed, one of the early versions of CADD contained an incorrect algorithm for of the method by which the number of stitches in a round increases as the doily grows in size; interestingly, this version re-created the 'ruffled' doily popular in the 1950s. Another version of CADD contained an incomplete geometric model, and produced designs in which rows could overlap. Several of these 'mistakes' proved intriguing, and have motivated experimentation with novel forms of three- dimensional lacework. By consciously examining the parameters of a style, the human collaborator can be drawn outside a paradigm. The knowledge base provides new tools with which to characterize stylistic and construction aspects of a design, and these new techniques can inspire new ways of thinking.

Image storage and retrieval

A growing number of libraries, museums, and art galleries are maintaining image databases of their non-text holdings. The vast majority of these databases require the image to be manually given a textual description. A textual format has obvious drawbacks: a picture is indeed worth at least a thousand words, as an attempt to list all possible terms that describe even a simple image will demonstrate. This problem is exacerbated by the fact that there are often significant differences between the language used by the document classifier and the language of the client searching the database. Indeed, for many domains, such as crochet patterns, there is no standard or even widely accepted vocabulary with which to describe the images. Additional disadvantages include an inability to effectively search on spatial relationships between image elements and an inability to easily describe complex structures.

We developed a system that permits users to search through a set of crochet lace patterns for a pattern containing a specific motif [6]. The user described the motif by drawing it, and could make either exact queries ('find patterns containing exactly the motif that I've drawn'), or fuzzy queries ('find patterns that look something like this, but that might be shorter and fatter, or taller and thinner'). By combining simple exact or fuzzy queries, the user can also express desired spatial relationships between several motifs, such as 'find patterns containing a pineapple motif to the left of a column, and above a wheat ear motif'. The techniques for describing the crochet patterns, storing these descriptions on the computer, and then performing searches can be applied to other types of complex pictures as well: for example, electrical diagrams, architectural designs, and CAD (Computer Aided Design) diagrams.

New directions?

Surprisingly, the versatile, popular, and long-lived craft of crochet has received little scholarly attention. Few books or articles seriously explore its history, and unlike knitting or knotting its mathematical characteristics have not been investigated. These omissions may be partly ascribed to its status as a craft, rather than as a fine art; partly to its association with the lower levels of society (the poor, and women); and partly to the sheer difficulty of locating primary historic materials. Even acquiring a good set of reliably dated patterns is difficult: many were printed informally or scattered through popular magazines, and pattern pirating has always been common. The same patterns crop up through the decades, sometimes with new authors' names and publishers, often without referencing their earlier publication venue. The crocheted pieces themselves are anonymous and undated. There is still a large amount of work to be done in authenticating the history of this craft.

At a more theoretic level, we have shown that crochet is a fertile field for the application of existing computational techniques, and for inspiring the development of new methods. Again, there is room for much more exploration and learning!

References

1. E. Boyle, *The Irish Flowerers* (Ulster Folk Museum and the Institute of Irish Studies, 1971).
2. S. F. A. Caulfeild and B. C. Saward, *Encyclopedia of Victorian Needlework* (First published as *The Dictionary of Needlework: An Encyclopedia of Artistic, Plain and Fancy Needlework* (AW Cowan, 1882). Facsimile reprint (Hamlyn, 1972). Quotation from p. 102 of the facsimile.
3. M. Constantine and J. L. Larsen, *The Art Fabric: Mainstream* (Van Nostrand Reinhold, New York, 1981).
4. M. Constantine and J. L. Larsen, *Beyond Craft: The Art Fabric* (New York, Van Nostrand Reinhold, 1973).
5. S. J. Cunningham and P. Denize, 'A Knowledge-Based Approach to Crochet Doily Pattern Generation: Creative or Knot?', *Proceedings of the Second International Round-Table Conference on Computational Models of Creative Design* (Heron Island, Australia, December 1992).
6. S. J. Cunningham and P. Middlemis, 'Grammar-based Queries for Image Database Systems', *Proceedings of the Conference on Robotics, Vision and Parallel Processing for Industrial Automation (ROVPIA '94)* (Ipoh, Perak, Malaysia, May 1994).
7. G. Elliot, 'Jane's Repentance', in *Scenes of Clerical Life* (Clarendon Press ; Oxford University Press, 1985) 209. First published in Blackwood's Edinburgh Magazine, July-November 1857.
8. A. Feldman, *Crochet and Creative Design* (Harer and Row, 1973).
9. H. E. Kiewe, *The Sacred History of Knitting: Recent Discoveries. 2nd edn.* (Oxford: Art Needlework Industries Ltd, 1971).
10. M. Konior, *Heritage Crochet: An Analysis* (Kangaroo Press, 1987) 18.
11. D. S. Matsushima, 'Crochet aid for the amputee', *American Journal of Occupational Therapy* **40** no. 7 (July, 1986) 495–496.
12. Mrs Meredith, *The Lacemakers* (London, 1865).
13. M. Mies, *The Lace Makers of Narsapur: Indian Housewives Produce for the World Market* (Zed Press, 1982).

14. Mrs B. Pallister, *History of Lace* (First published 1875: revised edition, re-written and enlarged by Jourdan and Alice Dryden, 1902).
15. A. L. Potter, *A Living Mystery: The International Art and History Of Crochet* (AJ Publishing International, 1990) 19-21.
16. P. Turner, *Crochet* (Shire Album 126, 1984) 23.
17. P. Wardle, *Victorian Lace* (Herbert Jenkins, 1968) 29, 41.
18. R. Waterbury, 'Embroidery for Tourists: A Contemporary Putting-out System in Oaxaca, Mexico', in *Cloth and Human Experience* A. B. Weiner and J. Schneider, eds. (Smithsonian Institution Press, 1989).

PART V. DECORATIVE KNOTS AND OTHER ASPECTS

This Part contains four chapters on mankind's uses of knots for decorative purposes, and as symbols for use in heraldry or love.

It begins with histories of two decorative knotting crafts which have been practised, more or less intensively, at various times and by various races, during the past five thousand years. The first chapter traces the history of macramé, or square-knotting as it is still known by seamen. The second deals with the story of lace—its manufacture and uses—in particular dealing with the history of commercially successful lace.

The third chapter briefly outlines the system of European Heraldry, and then describes the various ways that knots can be used in heraldry. Details are given of many knots that have been depicted on elements of armorial bearings.

The final chapter deals with the widely-known and historic concept of the *True Love Knot*. The author gives many examples of occurrence of the concept in literary works, and diagrams of several actual knot forms it has taken over the past five hundred years or so.

CHAPTER 15

THE HISTORY OF MACRAMÉ

Roger Carter

Early History

Knots have played a vital part in the culture of man since the dawn of history. Very early in the development of our species, men and women would learn to copy natural growths such as entwining creepers, in order to form ropes and to carry out simple weaving processes. They would employ their primitive cordage to tie animal skins to themselves, to help build shelters, and to construct traps and nets.

Over the centuries they would gradually develop aesthetic senses, and discover possibilities to exercise them through the construction of woven patterns, using fibre strings of different colours and strengths. It is easy to believe that knotted or braided belts, bracelets and anklets have been used for personal adornment, as well as for practical purposes, by virtually all humans and for almost the whole period of mankind's existence.

Because of natural decay, no examples of knotting are left to us which are much older than 5000 years. We cannot know the exact period when ancient man first began to tie square knots around cores of strands, to produce some kind of decorative knotting. The earliest example we have is a piece of open woven netting from ancient Egypt, which has been dated to circa 3 500 B.C. A much later portrayal of the art which came to be known as macramé can be seen in the British Museum, on a bas relief from Assyria, which is dated 850 B.C. It shows an Assyrian warrior wearing a macramé-fringed tunic; there is also fringing on his horse's harness. The art in its simplest form was first seen when Arab weavers finished off the edges of cloth removed from the loom by knotting the warp and weft threads into a fringe. From such meagre evidence it has been surmised that the Middle East was the cradle of *square-knotting*, and that from there it was introduced into Europe.

No-one knows with any certainty where the name *macramé* came from, but it is generally agreed that it is not a French word, as it appears to be. The majority opinion is that it derives from the Arabic word *migramah*; this seems reasonable, in view of the word's long use in Arab weaving. *Migrama*, which literally means 'protection', came to mean the fringed shawl or headcovering which is common to the Arab. The derivation is reinforced by the fact that in the Turkish language, which is allied to Arabic, there is the word *makrama* which refers to a fringed towel or knapkin, 'fringe' being the key connection [2]. There is another naming theory, which will be referred to later.

It appears that square-knotting first arrived in Northern Europe with the Moorish invasion of Spain, in A.D. 711. From Spain it spread to France, where it enjoyed a brief period of popularity in the 14th Century, there being known as *filet-de-Carnassière*. The nuns of both Spain and France, expert in all forms of needlework and lace-making, were quick to see the possibilities of this decorative knotting, and they developed it from its original fringe-work in course wool into a beautiful, fine, knotted lace. It became part of the general Church scene; and there exist today several contemporary paintings which may be viewed in Museums and Cathedrals, in the Northern Hemisphere particularly, which show religious themes with macramé featuring in clothing and altar cloths.

The Crusades and Renaissance Periods

Some centuries later, at the other end of the Mediterranean, the craft was discovered again by Europeans, this time 'on its home ground'— the Middle East. It seems that they discovered it there during the Crusades, in the period A.D. 1095–1228 In those days, wars were more leisurely, long-drawn out affairs than they are today; and the Crusaders took their wives and servants along with them when they set forth. Whether it was these ladies, or the sailors, or both, who discovered and copied the practise of macramé, and brought it back onto the ships, we do not know: but from then on until the present day it has been a traditional sailor's pastime.

It was also during that period that returning crusader's ships carried the macramé art to Italy, notably Genoa, where once again it was taken up by the nuns. Initially it was employed as it had been by the Arab weavers; but again the nuns developed it with other forms of lace. And because of its beauty, and its comparative ease of manufacture, it was soon being made separately as decorative panels in such materials as silk and unbleached linen. By the Renaissance period it was known as *punto á gruppo* (knotted lace), and was used to trim garments and household linen. As in Spain, and no doubt with the help of visiting sisters, nuns made fine Church decorations and vestments; and again, because macramé was so typical of the Church scene, it featured in many paintings of the day. Two notable pictures from the period which include knotting work are 'The Last Supper' by Sebastian Ricci (1659–1734), and an earlier one by Paul Veronese (1528–1588), entitled 'The Repast of Simon the Canaanite', which shows the supper being taken at a table covered by a macramé-fringed tablecloth.

As has been the case all through its history, the craft has known no class distinctions in its practitioners. Nuns, often expert needlewomen, would make fine articles for the Church; women from all walks of life would trim their clothing with knotted designs. In Rome, peasants made head-scarves with knotted borders, much in the manner of the Arab's course-wool style.

From the 17th to the 19th Century

The craft travelled North from Italy, and it may have been adopted from them by the Flemish lace-makers. The next time we hear of the art is when it is introduced into England. Tradition says that macramé was taken into England by Queen Mary, wife of William of Orange who took the Throne in 1689. Apparently she was addicted to lace, and to the making of it. Court records show that her lace bill for the year 1694 was 1918 pounds sterling. Being unused to an 'industrious' Queen, her homely habits were ridiculed by the Court wits; for example, Sir Charles Sedley, a notorious Court satirist and poet, wrote a derisive verse, a portion of which reads [1]: '... but here's a Queen when she rides abroad is always knotting threads.' Her biographer, Gilbert Burnett, wrote of her hobby: 'It is a strange thing to see a Queen at work so many hours a day.' The King, too, had a passion for lace, though there's no suggestion that he ever made it. He was very extravagent in his use of knotted materials, as records of his annual expenditure upon them show. One year his bill for lace, with which to trim 'assorted cravats, handkerchieves, combing and barbing clothes', totalled 2459 pounds sterling—a huge sum indeed.

From this introduction into England, macramé's popularity waxed and waned in the ensuing years; but it never died out completely. There are many examples still in existence from its use in the seventeenth century. For example, in the Victoria and Albert Museum in London, there is a beautiful sampler with 75 tiny macramé panels; and a pair of kid gloves, with macramé panels let into the gauntlets, which are dated 1749.

Macramé knotting again became a royal pastime in the late Georgian period. Queen Charlotte, wife of George III (1760–1820), was reputed to be an enthusiastic knotter; and since royalty are always trend-setters, knotting became a fashionable pastime in the candle-lit drawing rooms of Georgian England. When the light was too poor for the fine needlework that the ladies usually undertook, macramé was the work that they would turn to. The redoubtable Dr. Samuel Johnson was not impressed by the ladies and their fripperies; he said of them: 'Knots ought to be reckoned in the scale of insignificance as next to mere idleness.'

Fig. 1. Examples of 18th and 19th century macramé, housed in the Victoria and Albert Museum, London

Up to this point we have only looked at knotting ashore; but its universal popularity and long survival really owe much to the sailor. It is not known precisely when the craft went to sea. We have already hinted that it might have been taken up by sailors on the Crusader's ships. Another suggestion is that it was picked up in Turkey by sailors from ships of the Levant Company, which started trade to the Golden Horn in 1579 [2]. It was, and still is, known at sea as *square knotting*, because of its basic knot, the *square* or *reef knot.* Sailors developed it and combined it with their other knotting styles; and they transported it all over the world. The Spanish took it to the New World, where it became known as *Mexican lace.* French sailors took it to North America and introduced it to the Indians, who were already expert knotters and cordage makers. Examples of this work, including an Inca hat and an Eskimo fringed fur rug, may be seen in the collection of New York's Museum of American Folk Art.

The craft then continued to enjoy a very quiet and restricted popularity, being kept alive by sailor's work going ashore, either for sale to supplement their miserable pay, or as gifts for wives and girl friends.

In the 19th Century, Mrs Bury Palliser [3] published a comprehensive book on the history of lace. Of macramé she wrote:

> *There exists a beautiful and ingenious work taught in the schools and convents along the Riviera. It is carried to a great perfection at Chiavari, and also at the Albergo de Povari at Genoa. You see it at every stage; it is almost the first employment of the fingers which the poor children of either sex learn.*

The designs at this stage were, on the whole, simple ones; apparently the intricate Renaissance styles had been forgotten. Then in 1843 a richly ornamented piece of old *punto à gruppo* was brought to the school in Genoa by the Baroness d'Asti, and a young pupil, Maria Picchetti, had the patience to unpick the work and discover how the patterns were formed. From then on the children's work there grew more ambitious. The older children invented new and beautiful patterns, and some of their work was used for church decoration. Many items were sold locally; others were exported to South America and California. Interest in the craft grew, and it spread rapidly. A Genoese Lady's trousseau was not considered complete unless she had some undergarments trimmed with macramé lace.

It was at this time that the craft received its present name. One source suggests that it was named after a village on the river Macri, where it was very popular. Whichever theory is correct, it was certainly named in Italy, and called *Macramé*, or *Macrami*, although the accented 'e' suggests that it was coined in France. At an early period the word was used as a noun denoting the homespun huckaback towels with plain fringed edges which were made in Genoa. Gradually the name came to be applied to the distinctive fringes, and then to the knots themselves.

Once again, the popularity of the macramé craft spread throughout Europe in the North; and there were many elaborate specimens of it submitted to the Paris Exhibition of 1867. Then it crossed the English Channel for a second time. It could not have arrived at a more fortuitous period, the mid- to late-Victorian era. At this time everything was over-decorated; no bracket, mantelshelf or piece of furniture was safe; and the social system was more than appropriate for it, too. In middle-class Victorian society, the woman's place was in the home. An independent career was unthinkable, for not only was the husband's comfort considered to be his wife's career, but also in this late industrial revolution time a man had to show off his successes; to have a wife working would indicate to his peers that she needed to work, with a consequent loss of status for him. The period even came up with bigger furniture, so that he could buy grander items for his money, and thus demonstrate his opulence to visitors to his home.

Modern Times

So the wife ran the home, which task consisted largely of organising the servants; even quite modest households had their complements of these. She also played the pianoforte or other musical instruments, read 'improving' literature, and did needlework. In the early- and mid-Victorian eras, *Berlinwork* was overwhelmingly the most popular form of needlework practised. It required little skill other than the ability to hold a needle and follow a pattern, stitch by stitch. By the 1880s Berlinwork was replaced in popularity by art needlework. In 1872, the Royal School of Art Needlework was founded, and in a very few years a real talent was needed to produce other than mediocre embroidery; so macramé arrived on the scene to save our lady's reputation.

Fig. 2. A Picture Frame and a Wall Pocket—Examples from Louisa Walker's Book

It didn't require the fine needlework, the good light, and the good eyesight that was necessary for the first-class embroidery; and, after all, it had been a queenly pastime, so it was readopted with a vengeance. It rose in popularity so swiftly that many people thought it was a new invention; so while servants did all the humdrum sewing, darning, dressmaking and similar duties, Madam, with an abundance of spare time, made fringes, frills and flounces. With the work she produced, she had the servants adorn mantleshelves, curtain rails, gaslamp brackets, and four-poster beds; her macramé designs were even used to cover up the legs on those obscenely naked pianofortes. She made night-dress cases, handkerchief sachets, pincushions, workcases and babies' bonnets; macramé insert-panels were made for table runners and antimacassars; and the macramé craft was used to produce quite large pieces such as counterpanes. Some of these last were quite beautiful. Examples are to be found in many museums throughout the world, particularly in the United Kingdom, and in those countries that were Her Colonies and Dominions. This was the time, at the height of the popularity of macramé, that the major emmigrations took place; and emigrants took the craft with them to their new domiciles.

A whole industry sprang up in the United Kingdom, to produce special yarns, and elaborate equipment such as frames, bobbins and clamps; and a wide selection of books, magazines and pamphlets were devoted to the art. One of the books [4] was written by an English schoolteacher named Louisa Walker, and published in 1896. It was intended as a teaching aid, with degrees of difficulty to suit the class ages of the children. The book was full of patterns (see Fig. 2 for examples) for the bric-a-brac of the period (today we would label it 'kitsch') which could be associated with the knotting art; one such pattern was for producing a macramé *fly rest*! Perhaps this book contributed to macramé's fall from grace in the Edwardian period, when Britain was throwing off the genteel yoke of Victoriana.

During this period, the art was also very popular in France, Germany and the United States. And by this time, square knotting had become an integral part of the sailor's repertoire of handicrafts. Earlier, during the Napoleonic wars, British ships sailed to and fro for months on end, patrolling the same stretches of water on blockade duties. Their crews had none of today's entertainments on board to relieve their boredom; radios, T.V.s and video recorders were a long time in the future. Sailors were among the simplest and poorest of men, and they were usually illiterate, so they could not entertain themselves by reading or writing. There may have been a few simple musical instruments aboard, but for the vast majority, living in cramped quarters, the only things they had to occupy their minds and hands were the things about them. While a few, with talents in that direction, carved snuff boxes from their so-called 'beef ration', and buttons from aged pieces of cheese, the majority, battened below in bad weather, confined to their cramped, damp and smelly quarters, turned for entertainment to the only skills they had, namely working with ropes and canvas. With only waste ship's materials available to them, such as unwanted pieces of sails and cordage, they adapted the severely practical techniques they had painfully acquired for maintaining the ship's gear, and constructed decorations for their small personal items such as knives, needle-cases, kitbags and the like.

In much the same way, the men of the Merchant Navy, who had long periods of inactivity when becalmed, or when at anchor they awaited a favourable wind or a berth in a harbour, turned to the same kinds of pastime. Up to the 19th century a seaman was very interchangeable; he may have found himself aboard a merchant vessel one day, and in a navy vessel the next. When the press gangs were abolished, and long term engagements were introduced in the Royal Navy, then the two navies began to diverge. The Merchant Service changed; speed became all, and ships were driven hard by crews that were kept to minimal levels by penny-pinching Owners. In those conditions, sailors concentrated their spare time occupations on the intricate, fancy knots which could be made at one sitting, or over a short period. While in the Royal Navy, and to a lesser extent in the superior passenger lines, conditions did not change so much except that discipline became less harsh. However, in both services a certain pride in profession arose, now that sailors were there from choice. In this atmosphere, the Victorian Navy adopted the fashionable macramé crafts from ashore, and set about developing them with a will. Both the British and American Navies specialised in square knotting, and sometimes called it *Mc Namara's Lace*, a humorous corruption of Macramé Lace. Much of the work done aboard ship took the form of fringes for sea-chest covers, tablecloths, shelf covers; binnacle and skylight covers; and even capstan covers, wheel covers and bell covers. They made spray screens for their launches; and also 'eye screens' for ladders, because Navy ships were often visited by ladies, and this was a time when ladies were not acknowledged to have legs (or at least, legs were considered to be in their complement of 'private parts'). It was probably during this period, when sailors' fancy knotting was at its pinnacle of development, that square knotting was recognised ashore as being a sailor's art. And Jack Tar, ever resourceful, made for sale ashore handbags, belts, watchguards, dog leads, lampshades, bell-pulls, wall tidies, and a dozen other articles which by today's standards were rather tasteless; but they were often quite intricate, and cleverly done, and so were in good demand. From this time on, macramé has remained a sailor's art, in that it has been practised continually at sea ever since, whilst ashore its popularity has waxed and waned.

After the Victorian and Edwardian periods were over, the First World War changed tastes; soon thereafter glass and chrome replaced lace frills in the public fancy. Throughout Northern Europe, and also in the United States, macramé faded from the scene; but it persisted in Italy; and it remained a popular sailor's art.

In Italy a particular kind of macramé evolved under the auspices of a Madame Cavandoli, who was head of the Casa Del Sol, an open air school in

Turin. The work, now known as *Cavandoli work*, is knotted in close horizontal and vertical cording, in two colours, one for the background and one for the design. The finish is a closely woven fabric, similar in appearance to tapestry. This Cavandoli work is now a part of macramé, and it is frequently combined with square knotting [3].

Fig. 3. From the Introduction to the Chapter on Square Knotting in *The Ashley Book of Knots*

Macramé then assumed what we would now call a low profile until the Second World War, when it enjoyed a brief revival on both sides of the Atlantic. A book on lace published in 1940 stated that: 'Macramé threads are available in abundance, and colours are plentiful.', which suggests that the hobby was quite popular at that time. But there seem to be few examples of the period in existence; and the practise of the art had faded out by the closing years of the war. Perhaps history had repeated itself in that era. The craft's suitability for long hours in dimly-lit drawing rooms of Georgian England made it equally suitable for dispelling boredom during times spent in dimly-lit air raid shelters. Curiously, one macramé item that was frequently seen during the war, and is still with us, is the shoulder lanyard on a soldier's uniform.

For the following 30 years it seems that the art was practised almost exclusively at sea. This author, as a boy seaman, then supplemented his pittance by making and selling knife belts, lanyards, and dog leads, and covering telescopes for young Midshipmen.

In the late '60s and early '70s, a 'counter culture' boom began. People rebelled against mass production and all that it stood for; all the old handcrafts were revived, and 'do-it-yourself' practices emerged. Spinning, weaving, copperwork and leatherwork, batik and other crafts, all became popular; and macramé again came into its own. This time, it seems that the resurgence began in the United States; and it appealed to all types, again without class distinctions. It became enormously popular within a very short time. By 1971 there had been at least fifteen books, and six periodicals, published in the States which were entirely devoted to the macramé arts: one paperback on the craft sold over 500 000 copies. Again an industry sprang up to supply the necessary materials for macramé practitioners, and for all the other related crafts which were enjoying renewed popularity at that time.

This recent revival has demonstrated, perhaps better than ever before, just how versatile macramé is. Basically, it consists of combinations of only a few easy-to-tie knots, and requires no other equipment than a length of cord and a hook or doorknob. At this simplest level, it is taught in primary schools; the children make samplers and uncomplicated belts and the like. At a more advanced level, teenagers and adults can develop their skills by learning rather more complicated designs, and by making useful, attractive articles such as handbags, jackets, shawls and jewelry. These articles are frequently of a saleable quality, and a great many of the goods purchasable from craft or boutique shops are the work of this class of knotter. They are made by men and women, of all ages, in their spare time as a paying hobby. Well made articles can fetch good prices; for example, a shop in New York's Manhattan, selling a range of these goods and appropriately named *Macramania*, was recently recorded as selling scanty macramé bikinis for $32.50; and in most western countries shops abound which sell hanging baskets, hung by intricate macramé work in manmade fibres and offered at a wide variety of prices.

The highest levels are reached by fulltime professionals, who have developed macramé from a craft into an art form, requiring great manual skills and aesthetic and design sensitivities. They are often now commissioned to make intricate art-objects, such as three-dimensional wall hangings, or other large forms involving macramé work, to decorate and enhance walls and open spaces in office towers, theatres and other public buildings. For instance, in the late 1970s a notable American practitioner named Libby Platus was commissioned to make a room divider for a Californian Country Club, at a fee of three thousand dollars [5].

Now that the art has its professional practitioners, it also has its purists, and its critics; and, as in other art forms such as music and painting, it has attracted its 'art snobs'. The Curator of Textiles at Chicago's prestigious Art Institute has said of it: 'Macramé is a beautiful and old art form, but it is being cheapened by 'How-To' books.' But it is still just as it has been for over a thousand years—all things to all fingers; and whether as a craft or an art form, macramé will surely continue to exercise a fascination for mankind as long as there are fingers to tie square knots.

References

1. C. W. Ashley, *The Ashley Book of Knots* (Double Day Doran, New York, 1944).
2. S. Banks, *The Handicrafts of the Sailor* (David & Charles, London, 1974).
3. J. Short, *Macramé—The Craft of Knotting* (Blandford Press, London, 1973).
4. L. Walker, *Varied Occupations in String Work* (MacMillan & Co., London, 1896).
5. *Knotty but Nice* (Unsigned article in *Time Magazine*, 17 May 1971).

CHAPTER 16

A HISTORY OF LACE

Europa Chang

Preliminary Remarks

Lace, by its very nature, can be made sucessfully only with thread spun from long fibres. Moreover, lacemaking is a labour-intensive craft: from the spinning of the fine thread to the finishing of loose ends each stage involves its special skill. Thus, our knowledge of the history of lace comes mainly from accounts, sumptuary laws, early pattern-books, drawings and such pieces as have survived the destructive effects of moth, laundering, sunlight and general use. This chapter on lace, therefore, does not pretend to be a comprehensive account, but tells only of such lace as was made in quantities large enough to have left some record of its production. In effect, this limits the topic to a history of commercially successful lace.

In the Egypt of the Pharoahs, paintings on the walls of tombs show important folk wearing dresses of sheer gauzes. Excavations have revealed that these sheer fabrics were made by the technique known as Sprang with bands of tabby (basket) weave in between to hold the spacing of the lattices. Ancient Coptic textiles have also been discovered showing similar Leno weaves (both having been labelled 'Babylonian' weave!) Although, strictly speaking, there are reasons for not counting these as lace, such weaves still demonstrate a stage in the evolution of lacemaking. Incidentally, these also give some indication of the degree of sophistication of the ancient civilisations.

Lace as we understand it today may be loosely grouped into the following categories:

(I) Needle-point (also known as needle lace, and not to be confused with canvas work—which is a form of counted thread embroidery);

(IIa) Continuous Bobbin Lace, where the design (toile) and the ground are made in a single process;
(IIb) Bobbin Lace, where the motifs are made individually and then joined together by the background (reseau);
(IIc) Bobbin-made Braid Lace;
(III) Mixed Lace, in which needle and bobbin lace have both been used; machine woven braids forming the pattern and linked by needle-made bars are included in this set;
(IV) Knotted Laces, such as macramé, tatting and netting;
(V) 'Embroidered' Laces, where a sheer fabric, such as muslin or net is decorated by surface stitchery, or where a more opaque fabric has threads withdrawn to make an open lattice for decoration;
(VI) 'Chain' Laces, such as are made by knitting or crochet;
(VII) Chemical and Machine-made laces.

There are subsets of each category, and further subdivisions formed by the intersections of sets. In addition, there are many characteristics peculiar to 'social class', the designs of the era, regional costumes and the locality where the lace was made. Many books have been written studying such details, and a short reading list is appended; only those particulars necessary to an understanding of the craft and its development will be mentioned. A brief description of the techniques and their comparative merits is all that will be attempted here.

All needle-made laces (I) begin from a design drawn upon parchment (nowadays architects' linen is used as a substitute) pattern. A double thickness of foundation thread (cordonnet) is couched on to the pattern in such a way that each section of the motif is enclosed while the cordonnet runs continuously round the motif. The resulting cloisons are filled with variations of buttonhole stitches (half-hitches) and spacings to give the required effect (Fig. 1).

When each cell has been filled, the cordonnet is covered with tightly packed half-hitches to give a firm outline to the motif. Some forms of the laces also have padding over the cordonnet, giving a relief effect (rose point), while others achieve variation by picots or applied sections. There are fewer than 100 stitch variations and characteristic groupings so that the quality of design gains in importance. Very little equipment is required, making this an inexpensive craft to set up, but the stitching demands great concentration and meticulous cleanliness if the completed piece is to look fresh; consequently this type of lace was the most expensive to produce.

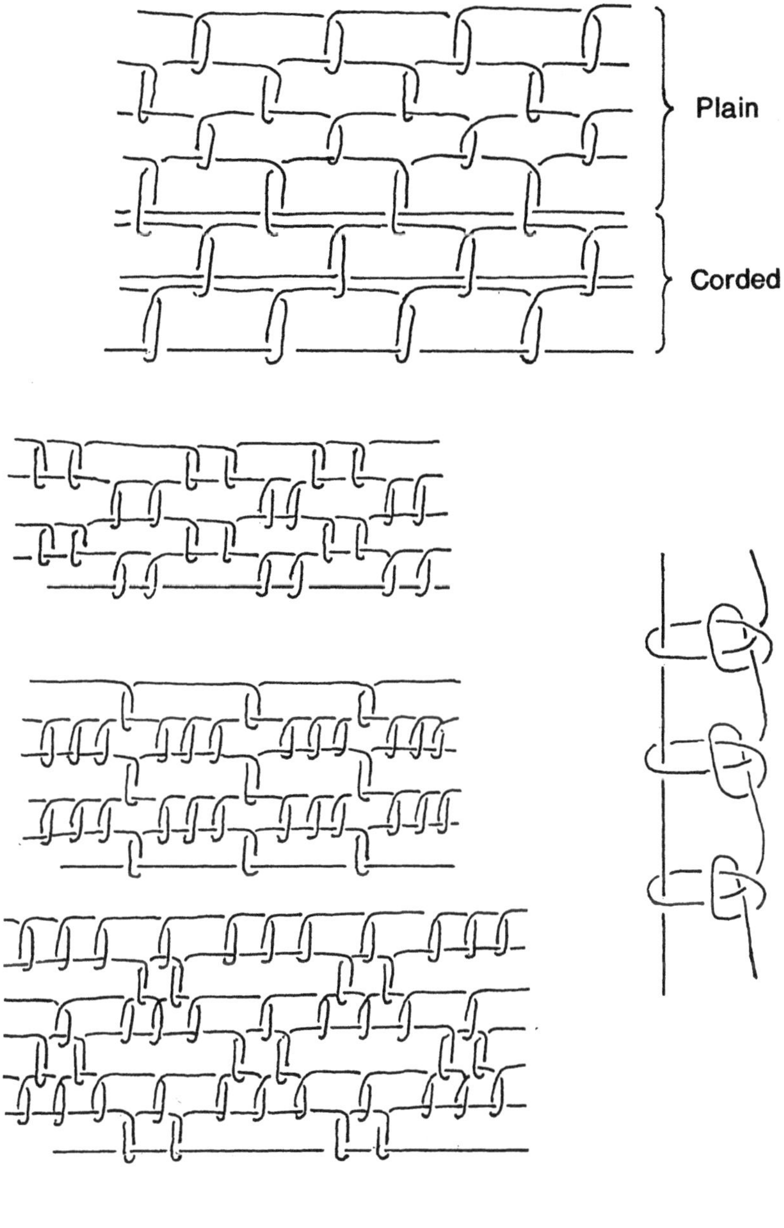

Fig. 1. Basic Half-Hitch and Variations

Bobbin lace (II) may be considered as a form of 'off-loom weaving'. Bobbins are wound with thread and hung from pins whose positions are pre-set by a marked card (pricking) encoding the design. Essentially, there are only two movements in this craft: twist—when the thread on the right is lifted over its left-hand neighbour, and cross—when the thread on the left is lifted over its right-hand neighbour.All the several hundreds of bobbin lace 'stitches' are built up from these two movements (much as all knitted patterns are based upon permutations of knit and purl stitches) and many variations of a design may be made from a single pricking. In some patterns a heavier thread (gimp) is carried through to give an outline or raise the lace, while in others raised bobbin stitches perform this function. Yet more interest may be given by use of picots, coloured threads and changes of scale. Bobbin lace has two extra sections: footside—the straight edge where the lace is stitched on to its mounting fabric, and headside—the free edge, often scalloped and trimmed with picots.

When the design and ground are made together (IIa) the work is sometimes known as 'trolley lace' (from 'trolle kant'.) Here, a typical thread path will travel from footside to ground, possibly over a gimp into the design area, through the motif into the headside, and back again. Thus the whole width of the lace is worked together, and large numbers of bobbins are required. [A rule of thumb for Torchon (the simplest of the trolley laces) is: '3 pairs for the foot + 7 pairs for the head + 2 pairs for each pinhole on the longest diagonal and about ten pairs over for luck!'] For Valenciennes, a lace much prized in the C18th, 4 pairs per pinhole would be required for the ground, and a piece of 10 cm width could involve as many as 1200 ends. Moreover, only one person could work on a piece at any one time, so that progress was slow. The main expense in setting up this style of lace is the cost of all the pairs of bobbins and pins, particularly in the days when pins were individually crafted. Sometimes, where a great width of trolley lace was required, strips were made and the sections joined.

Where toile and reseau are made separately (IIb), the design motifs may be worked with fewer bobbins, and at each stage the lacemaker needs only to concentrate on one part of the finished piece. Among other advantages, this speeds the unpicking of a mistake; thus progress is much faster and the skill more accessible to the lacemaker. It is also possible to achieve extra variation by using different scales or threads for the motifs and ground. The completed piece had the additional advantage that in case of snagging, only a part needed to be repaired—where in a trolley lace the continuous thread run could pull a whole trim out of shape. This is likely to have been an important consideration in the purchase of a piece of lace.

'Peasant laces', as some forms of braid lace (IIc) are called, are worked continuously with the design (often symmetric) being formed by a meandering braid woven with the bobbins, and linked to a previously worked part of the same braid to hold the piece in shape. Lace in this style may be made with as few as six pairs of bobbins, and grows fast. In the more sophisticated types of braid lace, pairs of bobbins are added into or taken out of the braid to give variation to the width of the 'drawing line'.

Mixed lace (III) needs little explanation. The lace producers found the most cost-effective combination to be bobbin-made motifs joined by needle-lace links, for the individual motifs could be made by several workers and linked into the design by another person. This team-production resulted in a far greater output, for each member specialized in a limited number of processes. Some styles also used a braid tacked to a parchment design linked and filled with needle-lace. This technique became popular only after machine-woven tapes were readily available.

Knotted laces (IV) are something of a misnomer for all needlepoint is made with knots. However, this set of techniques covers those laces which are made by knotting the thread without using a needle.

Macramé derives from the Arabic word 'migramah', meaning fringe. As we practice the skill today, there is little relation to lace; but worked in fine threads, the pieces are both strong and delicate. The topic is being discussed elsewhere in this book; but it should be borne in mind that when fine threads of any length are used, reels of some sort are required to simplify their handling.

Tatting derives its English name from the Anglo-Saxon for tufts (taeta), its Italian name (occhi) from its characteristic form, and its French name (frivolité) from the appearance of both the completed lace and those who made it. The lace is formed by cow-hitches (larks' heads) on a running loop which is then pulled up to close the ring. The special tool designed to hold the thread is boat-shaped and called a shuttle—which gives the craft its German name of schiffchenspitze. Today two shuttles and a fine crochet hook (for linking) are the standard requirement, but originally a single shuttle and a purling-pin would have been used, with a needle to link sections.

Lace netting is the craft of net-making with a very fine thread. A fine netting shuttle (made of ivory or steel) and gauges to give a regulated mesh size are the tools needed for this craft, and the one knot used is a sheet bend. Many variations may be obtained by spacing, scaling, and embroidering upon nets of square mesh. This last variation is known as filet-work. One style of lace has a sunburst design formed by laying a continuous thread across equidistantly spaced diameters of a circle, and joining them into a geometric pattern by knotting and weaving across the thread diameters to make a web. These 'sols' laces are attributed to Brazil and Teneriffe, where specialists make the little sun motifs while others join them into larger pieces.

The embroidered laces (V) are a subset of what needlewomen describe as 'white-work', and range from fairly heavy designs on coarse linen to the very delicate tamboured laces.

The fabric upon which the stitching is worked may be cut (cutwork), have threads withdrawn (drawn work), have ground threads distorted from true by tightly pulled stitches (pulled work), or be padded before stitching to give a raised effect. All manner of embroidery techniques are used and the best examples of this work are rich and elegant without any ostentation. The tamboured laces are made with a hook very similar to a crochet hook, using chain-stitch in a controlled sequence of movements.

Knitting (VI) is so much of a household skill that it requires no explanation. There are two outstanding areas of knitting design: the Austrian style, which gives a stocking-stitch basis and uses a cotton yarn; and that of the Shetland island of Unst, which uses a very fine wool and a garter-stitch basis. Squares made in the first style often grow from the centre, and five double-ended needles are used for the knitting. The Shetland square shawls are always made on two needles with all finishing edges being grafted so as to produce a very soft, pliable fabric. These shawls are often heirloom pieces, with certain patterns being named after the knitters who designed them.

Crochet is discussed in detail in another chapter. Also, the chemical and machine laces (VII) mentioned are those which have led to the manufacture of the 'shop' laces. These will only be given passing reference here.

The Deep Past

Although bobbins shaped like those we use today have been found in C19th excavations of Etruscan sites and in C20th 'digs' in Cyprus, it is suggested here that these were used for weaving inlay rather than for lace. The clothing of ancient times was shaped very simply from rectangles seamed along the edges and bound with ribbons to fit the wearer. It stands to reason that the cloth, being the result of a season's shearing, spinning and weaving, was treated with respect; no housewife would wish to waste even one square inch of her hard work at the loom, and the loose rectangles gave room for growth, while the selvedges minimised the amount of sewing (and thread) required. The same need to conserve time and yarn makes it reasonable to suggest that all girdles (in the sense of bindings) would have been woven off the loom rather than on the tall, wide, standing looms illustrated in paintings of the time. The friction at the edges of a braid that is frequently knotted could have been a good reason for developing the trick of an extra twist at the selvedges; and so, thereafter it would have been a simple step to weaving with two worker bobbins instead of one. Furthermore, pins were not manufactured to be sold in numbers, but as individual (and expensive) items. Whereas a ribbon may be woven with a minimal number of pins to hold the width, or even with no pins at all, lace made with bobbins requires pins by the hundred. [Although the ancients knew how to draw fine wire from soft metals, they lacked the technology to harden that wire sufficiently for pins.] Isaiah XIX 9 is sometimes quoted as evidence for the existence of filet-work, but it should be remembered that the King James translation dates from 1611; the New English Bible gives this verse as: 'The flax-dressers shall hang their heads, the women carding and the weavers shall grow pale,'—we have no record there that netting was anything other than a way of making fishing-nets.

The Early Records

Let us now consider the 'known' world of the C12th: although civilisations were well established in many areas, they were as yet undiscovered. The great overland trade routes of the Silk and Turquoise Roads provided an exchange of goods and culture with the east, while round Europe coastal sailing vessels also helped to carry merchandise. [The Toltecs in America, the Edo period of Japan, and the great kingdoms of India and south east Asia were unheard of; but distant Cathay (China) was the stuff of legend, and nearly as real as Avalon.] Kingdoms were settling into some stability after the invasions of the previous centuries, with established links between Europe and the Near East. The trading centres were Venice, Genoa and Pisa, there was an 'institute of science and technology' in the Moorish kingdom of Granada in southern Spain, Constantinople retaining its Byzantine habits was still a centre of theology—though now firmly seated in the wealth of its own silk production—and the troubadours of Aquitaine carried the news as they travelled from place to place with their songs and stories. Disruptive peasants were outlawed, over-powerful barons sent to the Crusades, and ambitious kings started little wars on their own account; but, by and large, people enjoyed the improved material conditions that came with peace. Silk stockings and laces were knitted in Granada and sold to the wealthy of Europe, as wardrobe accounts show.

Our knowledge of these times comes mainly from songs and paintings. From 'The Ballad of El Cid' (c.1140) we learn that the nobility wore silk shirts and brocade tunics decorated with worked gold, and fine cloth cloaks with worked gold trimmings, while cloth of gold was given as part of a dowry. Sermons preached against tight-lacing tell us that the fashionable female silhouette in northern Europe included a waistline, ['lace' at this time was used in the sense of ribbon fastening,]—suggesting a more relaxed attitude towards the practical aspects of survival. These costume details are recorded in marginal drawings of manuscripts, too. Farther east, icons of the Virgin Mary show her wearing a silk cloak with gold fringing: the style of the garment is very similar to those drawn in N.European MSS., but the icons give far greater detail, being on a larger scale.

The fringes of the Virgin's cloak (Fig. 2) are of interest, for the designs change gradually over the centuries whereas the drapery of the cloak itself remains the same. The paintings show the fringes to be of openwork, for parts of the edgings fold back on themselves and the design is shown through the overlap; the use of gold paint throughout suggests they are of gold thread or wire; and the changes in patterning suggest each icon-painter used the designs most familiar to his eye, an idea confirmed by the decorative styles through the centuries. The mid-C12th icon of the Annunciation in the Ohrid icon gallery, Yugoslavia shows a geometric fringe, which by the C15th has become a stylised pattern of lilies. While it is conceivable that the wealthiest members of society dressed their womenfolk in pallae with fringes of individually forged gold pendants, the practical problems of weight, detaching, theft and even laundry make macramé fringes a more likely solution. That the designs can be knotted using macramé techniques is self-evident.

Another snippet of evidence confirming the manufacture and use of openwork macramé patterns is the word itself, which derives form the Arabic word for fringe. Marco Polo (mid C13th) writes of saddles made in 'Kierman—a kingdom on the eastern confines of Persia'; while from the late C12th on, minstrels' lays make passing reference to noble destriers with fringed saddle and trappings, generally as an indication of status. Macramé, then, may be the first of the lace skills to be recorded. In Italian, macramé is named 'punto groppo', though this name cannot be traced back to the times before the printing press.

In 'The Romance of the Rose' (c.1240) Guillaume de Lorris describes Idleness as wearing sleeves sewn elegantly, and in the continuation by Jean de Meun the (ideal) woman is described as wearing a cloak decorated with silver-gilt thread-work and seed pearls. In the mid C14th, we have the tale of Sir Gawaine and the Green Giant, in which the lady's token is referred to interchangeably as a girdle and a lace. More precise still, in *The Canterbury Tales* Chaucer describes the Knight's shirt as being embroidered like a meadow filled with flowers, and of the Wife of Bath he says 'Her coverchiefs (veils) full fine were of ground; I dare swear they were worth ten pounds'(each). The word 'ground', here used to describe a particular fabric, would suggest some kind of lace reseau. (Chaucer worked as an official in Customs and Excise to earn his living, and wrote for his own amusement.)

By the beginning of the C14th, technical skills were sufficiently advanced for needles and pins to be available (though very costly—hence the expression 'pin money') for purchase in quantity. According to an inventory of 1347 the Princess Joanna, daughter of Edward III, was given 12000 pins as part of her trousseau. While she and her ladies must have needed a considerable number for holding their wimples in place, it is tempting to think that some of the pins were used for the making of braids. In the Harley Collection of MS. in the British Museum is one dated c.1471 which gives instructions for the making of laces including 'an open lace, lace for hattys—' and begins with a decorated capital showing a woman making one of these forms of lace. This may be our first record of the craft; the lace shown is a simple set of loops made by leaving longer sections of yarn at the edges of the weaving. It is around this time that lace takes its modern meaning of an openwork fabric. At this time, 'passementerie' was a word used for any braid woven, knitted, plaited, close-textured, or open.

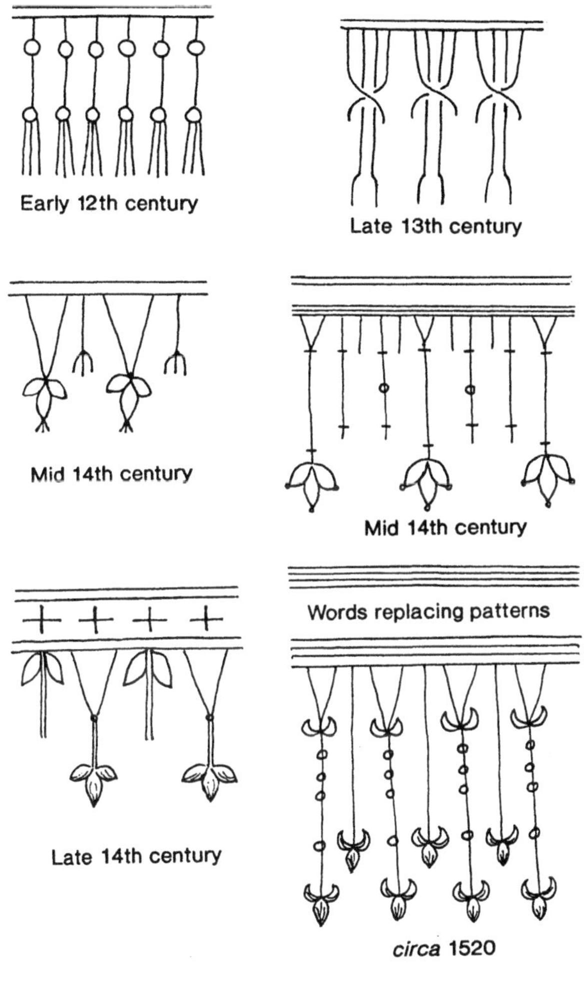

Fig. 2. Fringes from the Virgin's Cloak

From the end of the C13th mention is made of networks: inventories of old St Paul's and Exeter cathedrals refer to cushions of opus filatorium and opus araneum—possibly the distinction was between darned square meshes and spider-web patterns formed by working in circles from the centre, and making the increases at set places, or with differing sizes of mesh.

By the end of the C15th wood- and copperplate-engravings were sold with patterns suitable for use as surface decoration on fabrics. Artists who were good enough to merit patrons designed exclusively for them, while others would sell through workshops, or under the aegis of a famous name (such as the Academy of Leonardo da Vinci.) Among others, Dürer designed many pieces. It was also at this time that all the Moors were driven from Granada, (and, her vow fulfilled at long last, Queen Isabella could change her body-linen), leaving behind their traditions of craftsmanship. One legacy of the Moors was the silk stocking knitting industry of Spain, which continued to flourish and to bring wealth to the country.

The traditions of Le Puy in the Auvergne region of France state that their first lace-making began around this time, the designs being of the style now known as Cluny. The Lace College at Le Puy still maintains the teaching tradition, and boasts of more than four centuries' continuity, though students today learn more for pleasure than for a trade.

The Sixteenth Century

At the end of the C15th printing with moveable type was established; this resulted in more people learning to read, as well as in a more 'scientific', less superstitious, attitude towards learning. The Spanish Inquisition under Torquemada upheld the authority of the Church with grim determination, but in Germany one Martin Luther was beginning to challenge the teachings of the clergy. As people moved away from unquestioning obedience to their spiritual leaders they also began to measure themselves against human achievements, and material wealth became more important as a result. A show of generosity for the good of the soul and the benefit of the Church became less fashionable than a display of wealth for self-aggrandisement. In Italy, the centre of the Christian world and of trade routes east and west, this was particularly evident and even today words such as 'bankrupt' and 'florin' show the extent of the Italian influence at that time.

One way of displaying wealth was in the ornamentation of clothing, for it showed that the wearer could afford to spare the thread used in decoration, and that the women of the household had the leisure to make (or that the family had the money to purchase) the trimmings. The fashionable fabrics at this time were English wool (Staple, the wool-marketing consortium, had been established for over a century), Flemish linen, cotton from the Turkish

Empire, Spanish (Moorish) and Italian silks and silk fabrics and threads for needlework, and Italian lace. Since fabrics indicated class distinctions and were strictly regulated by sumptuary laws, the embellishment of shirts and shifts became a particularly important detail; a rich merchant could show off by wearing a more elaborate lace than an impoverished scion of the nobility.

The lace and designs of this era are generally designated by their Italian names: lacis (filet), punto tagliato (cutwork), etc. Although filet-work is now considered to be only a crude form of lace, it must have been innovative and fashionable in the mid C16th, for in Bronzino's portrait of Agnolo di Cosimo Turi, the part of her shift showing above her bodice is made entirely of lace with a diamond mesh. From the names in use at the time, it is thought to be a darned lacis, and the design is consistent with this style of work. (While the design is also consistent with the style of Torchon lace, there would have been difficulties in the construction of the shift which make the possibility less likely.)

More frequently recorded are the techniques of cutwork and reticella. In the former, a design is drawn upon linen in such a way that the elements of the pattern are solid, with spaces defining the background.The edges of the pattern are bound by buttonhole stitches packed close (and sometimes padded to give relief), threads are laid connecting the spaces of the background and buttonholed (half-hitched) over, and then the linen in the background is cut away. This proved so popular a style of lace that in England Queen Mary Tudor issued a decree forbidding the wearing of 'the lace commonly called cutwerkes' to any person below the rank of baron. Reticella followed logically from cutwork: as more and more linen was cut away, it became necessary to build up a lattice to hold the weave. Sometimes threads were withdrawn, the remaining strands buttonholed over, and the squares then filled with a spider-web of stitching; then it was seen to be simpler to lay a lattice of threads directly on to a drawn parchment and working over the grid so formed. (The squares of the grid gives this form of lace its name and its character.)

By the last quarter of the century, the lacemakers had realised that the work could spring free from its reticulated lattice, and heraldic devices, monograms, flora and fauna were all incorporated into the designs. Edges developed little scallops, then pyramids of scallops, to give deep borders. Lace-making was established as a highly specialized set of skills. Domestic accounts of 1586 show that the cost of 6 yards of fine lawn together with cutwork and gold lace to make a ruff cost 60 shillings. [At this time, Shakespeare's hired actors drew a basic weekly wage of 5 to 8 shillings.]

We know that bobbin lace was also in production by this time, for in 'Twelfth Night' this is mentioned as a group activity: '—the free maids who sit i' the sun weaving their thread with bones.' In Germany, Barbara Uttmann taught lacemaking and so founded a flourishing industry which gave employment and brought prosperity to Saxony. The Spanish Inquisition had forced vast numbers of Dutch Protestants to flee, and in 1563 many of them settled in Kent and spread westwards along the south coast, where the weather was more clement. They brought their textile skills and taught the women of the villages which gave them hospitality, so that individual techniques, not all of which have survived, were associated with particular villages. In 1568 and 1572 two other groups, Protestants from Mechlin and French Huguenots were welcomed into the Protestant England of Queen Elizabeth I, and settled in the counties of Buckinghamshire and Bedfordshire. To this day, there are bobbins called 'Huguenots'. The parchments on which the patterns for bobbin lace were pricked could be used again and again, and when the holes had worn too ragged for the design to be accurate a fresh parchment was laid underneath and a new pricking made. At a time when writing was still a comparatively rare exercise, this was the easiest (if not the only) way to take a copy, and the prickings were jealously guarded.

Pattern-books were published around this time too, showing designs of great complexity; it is interesting to note that all the designs of that era are workable, showing that all pattern-drawers had a basic training in the craft before being allowed to design any pieces. Those that have survived include designs by Nicolo Zoppino, Giovanni Vavasore, and Matteo Pagan; their designs specialising in strapwork interlacing, filet-lacis and punto in aria, respectively. In Germany, too, filet designs by Sibmacher were published at Nuremberg, while in France, patterns for cutwork and reticella were designed by Foillet. Although all these pattern-books provide inspiration even today, probably the best known remains the collection that Vinciolo of Venice published in 1587, with a set of patterns for reticella and filet-lacis. He was one of the foremost designers of the late C16th, and upon the marriage of Catherine de Medici to Henri II went to the French court to be given the monopoly on the lace for ruffs. There is some confusion between Queen Catherine of Aragon and the patron saint of lacemakers (St Catherine of Sienna) in the later English histories of this era; for Ampthill in Bedfordshire, where Catherine was sent to live after her divorce from Henry VIII, always claims to have learnt lacemaking from the Queen in 1531, while the rest of England says the craft was first taught by Flemish immigrants in the 1560s.

Perhaps the best indication of the truth lies in the fact that the Spanish laces were needle-point, made with gold or silver wire and fine thread, while the laces of Bedfordshire are made with bobbins and the early designs have a marked family resemblance to those of Cluny in France. The first record of lace being taught as a trade is in 1596, when a woman was employed to teach bobbin-lacemaking to the orphans of the parish of Eaton Socon. She was paid a weekly rate of 2d per child by the workhouse authorities, and the monies from any lace that was sold went to offset the orphanage expenses.

The fashion for lace brought many ancillary employments, and some changes in the established methods. The fine smooth linen thread required resulted in a greater awareness of the quality of spinning. Hand-spun threads tend to show a tiny irregularity at roughly 50 cm intervals, this being the approximate drop of the spindle; these nodes may be minimised by twirling the spindle in a bowl with one hand while using the other to pull out the raw fibre. Flax spun wet gives a finer thread than when dry-spun, so that the spinsters often sat in cold, damp rooms to achieve the required standards. The great ruffs themselves needed special laundering, and merchants opened specialist shops selling trimmings, mainly from Milan—hence the word 'milliner'. The domestic accounts of Hengrave Hall, near Bury St Edmunds (in Suffolk, England),show that when Queen Bess stayed there in the course of a royal progress, a woman from the nearby village of Flempton had to be brought in specially to starch the Queen's ruff, and for this task she was paid 5 farthings (quarter-pence) and her dinner. Saffron starch, used to give the lace its distinctive creamy colour needed special preparation and the growing of the saffron crocus gave a village its name (Saffron Walden) and livelihood; while laundresses could earn the equivalent of a month's wages for teaching the secrets of starching lace. For those who could not obtain saffron, the roots of the wild arum were used—hence its country name of 'lords and ladies'.

Many miniatures and portraits of this era record the clothing in such detail that it is possible even to recognise the pattern-book from which the laces were made! We also have a few domestic samplers of the era still in existence; the earliest one we know of being worked before 1544. The embroideress, Isabel Fleming, bequeathed her 'beste sampler' to her daughter Anne, and the will was proved on 8-vii-1544. This piece, and others of the era, shows the standard of needlework achieved by women at home, and the sample-bands include cut-work and drawn-thread work, and punto in aria, as well as stitched patterns and elaborate ways of making a decorative seam.

One technique which appears to be peculiar to England is also found on samplers of this era: Hollie point. This produces a heavy lace in which the design is made by a series of holes. Until the Dissolution of the Monasteries its main use was in church furnishings, and the designs were almost invariably biblical—Adam and Eve, the Tree of Knowledge, the Dove, and sometimes the lily in a pot (a mediaeval symbol of the Annunciation.) Later it was used as a technique for children's bonnets, etc, when the designs were simple floral borders, sometimes with lettering in between the flowers.

The Early Seventeenth Century

Exploration, the circumnavigation of the world and the challenging of the Church, had changed the balance of power and wealth in Europe. The C16th had ended with the Edict of Nantes, which gave Protestants freedom of worship, and the C17th began with the granting of charters to the East India Companies. Spain, having expended her gold from the Americas, and lost her Armada fleet, no longer had her former power. France, spiritually united by Henri IV's edict, had attracted back many of her Huguenot craftsmen and was regaining some of her prosperity. In Britain, the kingdoms of England and Scotland were united under James I and VI; the stability so gained enabled enterprising merchants to develop the London textile trade till it became the fashion market of Europe.

Venetian lace remained the fashionable fabric for those who could afford the price. As the stiffly starched ruffs gave way to falling collars, the designs became less geometric and more formally floral, sometimes with heraldic motifs figuring in the centre. The outlines of the motifs grew more heavily padded, becoming part of the design, and Venetian gros point developed; this extra weight of padding gave body to the lace, and so style to the garment. To speed production and to economise on the cost of materials horse-hair was sometimes used in place of thread in the padding, but this would shrink with repeated washing, destroying the lace in the process. Such great sums of money were expended on the purchase of lace that sumptuary edicts were published in France in 1594, 1600, 1601 and 1613 limiting the amounts of lace used at the French court.

Meanwhile, the French 'blonde' laces (which took their name from the pale gold colour of the silk with which it was generally made, though sometimes black silk was used) were being imported by Spain, for the softer character of this style of bobbin lace and the draping qualities of the fibre made it perfect for use as mantillas; and the designs of the French laces were more sophisticated than those of Spain. Collars at the Spanish court followed the French styles and used the Italian laces; so much money was being drained out of Spain that in 1623 Philip IV issued a decree banishing French styles from the Spanish court. The huge lace collars, slashed coats and elaborate embroidery were banned, as were gold and silver lace; thereafter courtiers were expected to wear plainly tailored suits of rich fabrics, while their collars were of plain white linen stiffened with shellac.

In Britain, the French influence had been increased by the presence of Queen Henrietta Maria. While court fashions retained the sober elegance favoured by Charles I, the younger courtiers were using lace in more and more exaggerated forms—as trims to boot-tops, deep collars falling to mid-chest level, etc. By the time of the English Civil War the collars were so large as to prevent gentlemen from drawing their swords easily, and Cavaliers were tying their lace into bunches under the chin with a ribbon bow: this gave rise to the lace cravat/jabot that is still worn by the Lords Mayor of English cities on ceremonial occasions.

At this time, too, the embroidery style known as Stumpwork began to be taught in the home, and haberdashers sold calico pieces with the (usually Biblical) scene ready stamped on the fabric. The figures would be worked in raised and detached techniques, often with clothing made in needle-lace, and the completed pieces would be professionally mounted as panels on a mirror frame or a box—sometimes shaped as Noah's Ark—or casket to form part of the furnishings of the lady's room. After the execution of Charles I, favoured themes amongst those who secretly supported the Royalist camp were those featuring Solomon, the Queen of Sheba, Moses in the bulrushes (where Pharoah's daughter was taken to represent Henrietta Maria), with the Biblical royalty dressed in the style of the Stuart monarchy. There is even one example in existence which claims to have the hair of 'martyred King Charles' stitched into the head of the figure of Solomon! This style of work appears to be exclusive to England, and crosses into the realms of lace for amusement.

One other point should be made: when the Pilgrim Fathers sailed to America in the Mayflower, their roll did not include a single lacemaker. While they were all emigrating for ethical reasons, it is conceivable that some of the passengers would not have taken this step if they could have earned sufficient wages to continue living in England; that the lacemakers did earn a good wage, even on piece-work, is confirmed by this. Furthermore, there would not have been much demand for lace in a pioneer community, so the lace crafts had a very late start in the New World.

The Latter Half of the Seventeenth Century

In 1643 Louis XIII died leaving a child king on the French throne under the regency of the Queen Mother Anne of Austria. Cardinal Mazarin, the chief minister, made some attempts to curb extravagances of dress, and it is recorded that in 1648 the Swiss Guard had their new uniforms made with trims of 'counterfeit lace' instead of silver-edged ribbons. However, he had more pressing matters to distract his attention: the English prince Charles Stuart (later to be Charles II) and his court were exiled in France, there was civil war (the wars of the Fronde), and there was war with Spain. It was not until 1661, when Jean-Baptiste Colbert replaced Fouquet as Minister of Finance that any action was taken to deal with the flow of money to Italy.

Colbert suggested to Louis that it would be profitable to tax 'all that wear clothes beyond their quality'. On 30-vi-1661 Louis XIV issued a Declaration du Roi in which he stated the wish that his French subjects should wear French trimmings and lace of no more than two fingers' breadth and no more than 40 sols' value, when decoration was applied to clothing. [At that time, 20 sols = 1 livre; fine hollands (linen) for shirts cost less than 14 sols per 45 inches length.] Lace trims for men were to be limited to the collar edging, the hem of the cloak, down the sides of breeches, along the seams and round the head of each sleeve, down the centre back seam, round the basque seam down the button-front, and round each buttonhole. The corresponding restrictions for women were the hem of the petticoat, down the fronts of gown and skirt, round the basque, and on bodices. [It is worth noting how this shows that lace was applied to cover all seams, so suggesting that tailors may have used long stitches for seaming.] To encourage his subjects to carry out his wishes, Louis XIV also imposed the heavy fine of 1500 livres (= 30,000 sols) upon any merchant caught selling foreign laces. To stimulate interest in the French textile industry, Colbert also imported skilled craftsmen from all over Europe, giving them special privileges and tax exemptions to persuade them to settle in France. Although the experiment failed in some areas—for instance, French serges lacked the texture of those made by the finer wool from sheep bred in the English Cotswolds, so that the waiving of loom-taxes for the immigrant Dutch and Flemish weavers was simply money wasted—the lace-making, silk-hosiery and tapestry industries were well established by these tactics.

At this time too, court dress was made into something approaching uniform. Only those who had warrants signed by the King could wear these Justaucorps à brevet, and warrants were only issued to those who could pay for the signature! The wearers of such coats-under-warrant could join in any royal party without invitation, their status being assured by their coats; the double advantages to the throne were the swelling of the royal coffers and the corresponding assurance that the princes of blood were now too poor to raise private armies. The lady's court dress (grand habit) was the style of 1670, for the manteau with its uncorseted comfort was thought to be unsuitable. The rigidly boned court dress left the shoulders bare; however, Louis XIV's sister-in-law went to mass one morning wearing a lace scarf over her shoulders, giving rise to the fashion for fichus. (When Louis complained of her casual appearance, Madame retorted that God would surely be less disturbed by her wrapped figure than by her chattering teeth!) One further touch was made by la Duchesse de Fontanges, whose hair caught in a branch while riding; to hide the disarray, she arranged a lace scarf over the top of her head, so giving rise to the fontange cap. The fashion for Court dress spread all over Europe, and Peter the Great brought the idea to Russia in 1698.

In 1667 the request that the French nation should wear French lace was repeated, this time with expressions against the use of foreign laces; and in March 1668, an edict was published forbidding the importation of Venetian point and Genoese laces. A copy of Venetian point plat (in which the outlines are left unpadded, so producing a flat piece without any relief), was made at Alençon; this soon developed its own characteristics, and was known as Point de France. In general, the designs were more fluid, with a crisp ground, stiff outlines and delicately patterned fillings. Between 12 and 18 workers were required to make any one piece, each person taking a different task: tracing the design on the parchment, laying the foundation cordonnet, making the different fillings, outlining the fillings, joining the motifs into strips of ground (each strip less than 2.5 cm width) and fine-linking the strips into a whole. Apprentices would learn each process, then work on that stage of each piece of lace until they were proficient, before learning the next stage. Supervisors, who had acquired skill in every stage, examined each stage of every piece and maintained high standards. Workers were required to conform to stringent standards of cleanliness, and soap and water for washing the hands were provided on the workshop premises.

The Low Countries (Flanders, Belgium and the Netherlands) were producing fine bobbin lace, for the damp atmosphere was particularly suited to the needs of linen thread. Valenciennes (in Flanders, but ceded to France) and Brussels laces were worked in damp and darkened rooms so as to prevent the thread from snapping or turning colour—it was said that a piece begun in the town and taken out to be completed by the same worker using the same thread would show a change of colour simply from the change in atmosphere. It was also said that Valenciennes lace (which was tougher than other laces thanks to its reseau of four-thread plaits) gained its life by draining vitality from its workers, whose long hours in cold damp rooms must have shortened their lifespans. A pair of Valenciennes lace ruffles for a gentleman's coat could cost as much as 4000 livres (80,000 sols) and would have been considered as an investment. The near-invisible thread that was required for the very best quality bobbin laces could only be achieved if the flax used in the spinning never dried out from its first moment of retting, when the casing was destroyed leaving only the linen fibres. Consequently, the spinsters producing this thread were standing in water for long hours each day; but where the lacemakers were allowed to sit above the byre, warming themselves in the breath and body heat of the cattle below, the spinsters could not have even this concession to physical needs, and pneumonia must have wreaked havoc amongst them each winter.

About this time, the import duties on English and Dutch goods were raised from 25% to 50% of the trade value: by way of retaliation, William of Orange declared a complete boycott of all French goods. This resulted in much smuggling of lace through the frontiers of the Low Countries using a variety of methods, the two most common of which were the coffin and the dog. In the former, a coffin was packed with lace and escorted across the border accompanied by 'bereaved relatives' some of whom may even have worn new and expensive laces. Once the authorities became aware of the deception, the smugglers attempted a new trick: a puppy was fattened and pampered on one side of the border, then taken across to its new owner who would starve the beast, wrap a consignment of lace round its body, and sew a skin over the lot to give the impression the dog was still well-fed; the animal was then released, and would return to its first owner who would welcome it with a hearty meal and free it from its parcel of lace. Smuggling to England was a more elaborate matter, but Romney Marsh in Hampshire was notorious amongst the Excisemen. Despite the note dated 26-x-1666 in which Lord Arlington informed the Lord Mayor of London that Their Majesties (Charles II and Catherine of Braganza) had decided to wear no more foreign stuffs excepting linen and calico, and wished the London merchants to know so that they would not waste their efforts by sending abroad for laces and silks, French fashions continued to be popular. By 1674, the imbalance of trade showed a deficit of 1,000,000 on the British side. The French Treasury subsidised Charles' II Privy Purse to continue this situation, and, despite Lord Arlington's efforts, Charles II continued to wear French lace.

In 1685 Louis XIV was seized with intimations of mortality and revoked the Edict of Nantes. Once again non-Catholic lacemakers fled France, this time settling in their host countries. Switzerland was one country which benefitted from this. Meantime, Louis' second (morganatic) wife, was quietly modifying the court style to something more in keeping with the rest of Europe, where court dress was made in the style of the manteau/sacque (which allowed the wearer to sneeze in comfort.) Shoulders were covered over, sleeves descended to the elbow with 'pagoda frills', hair was dressed to a lesser height to allow ladies to sit in sedan-chairs, and petticoats were limited to nine rows of trimmings, usually 3x(braid, lace, frill.) The skirts of the manteau were pulled back to display the lace on the petticoat, and the train developed from this. By 1693, the train had developed into a cascade descending from a bustle, and an engraved fashion-plate in the Victoria & Albert Museum dated 1697 shows a manteau of velvet edged with metal lace while the petticoat is trimmed with net frills. Silver lace was much in fashion in the last decade of the century, so much so that two duchesses kidnapped tailors to be certain of having the correct trims added to their grands habits at the wedding of Marie Adelaide of Savoy! In Madame d'Aulnoy's collection of tales 'Le Cabinet des Fees', lace is mentioned as an article of value; treasure casks contain diamonds, pearls and lace, the heroine invariably has rich lace decorating her ballgown, or her lace is stolen by unkind friends. The true significance of this repeated emphasis is the indication it gives us of how greatly lace was prized in this era. The fontange head-dress with its towering cap-frills and streamers provided another way of wearing lace.

Among other trade goods brought to Europe by the East India Companies were the knotted embroideries of north China. These are characterised by densely-packed 'french knots' which follows the contours of the design, shading the motifs by stripes of colour, and giving a mossy effect in contrast to the better-known sheen of satin stitch. The hard-wearing but costly pieces were imitated in Europe, but the time expended in copying the knot-stitches and variations soon led to a substitution resulting in a similar effect. Strong threads were closely knotted in overhand, figure-of-eight, premium cow-hitches, bullion knots, etc. and the resulting cord couched down in imitation of the oriental pieces.

The craft was given importance in Britain when William and Mary arrived from Orange in 1688; a poem ascribed to Sir Charles Sedley, and entitled 'The Royal Knotter' describes how the Queen travelled in her coach, knotting busily all the while. Tatting, at that time no more than a series of rings of cow-hitches, came into being at about this time. In Britain, Queen Anne tatted with enthusiasm, and to this day the flower of the cow-parsley, with its resemblance to tatted work of the period, is known as Queen Anne's Lace.

Knotting- (and the smaller tatting-) shuttles of this time had open ends, so that the side elevation resembled H; the thread was wound round the central pillar and its flow controlled by the fingers. The surfaces of the shuttles provided another area for jewellers to decorate; that they did so is shown by inventories listing shuttles of gold inlaid with branched coral or gemstones in a pattern of flowers or singing birds. Portraits of ladies sometimes depicted them with their shuttles and workbags, showing that they had the leisure to sit knotting or tatting, and incidentally making the point that these two forms of work were for the amusement of the wealthy rather than for commercial gain. Possession of a costly shuttle became a status symbol, (at the French court these were flaunted by royal favourites), and it is easy to imagine some owners using their shuttles as accessories to display jewelled rings or taper fingers! The Reformation changed the attitude towards the Virgin Mary, and as people lost the habit of 'Mary-worship' they also forgot the Marian symbols. When the church laces were seized by the nobility for use in dress, the designs of the parrot or the lily-in-a-pot had no more meaning and they were thought of purely as design motifs. By the end of the century, the pot had lost its lily, and was holding fantastic foliage; and still later, the pot had become an urn making an interval between wreaths. The 'potten kant' designs continued in the lace of the Low Countries for the next two centuries. Around this time, too, the 'peasant' laces begin to figure in records.

It is interesting to note that there are traditions of needle lace in all those countries round the Mediterranean where the Venetians had trading connections; as yet, no research has been done on the exchanges between these countries in the stitches and designs. Again, Crete, with its distinctively 'Spanish' C17th influenced designs of needle-lace (coloured or metallic gimp outline and 'sols' motifs) in the ethnic museum, was the home of El Greco, who painted at the Spanish court; here, too, is matter for study.

With improved transport there were conscious efforts made to bring 'culture' to the less populated parts of the known world. It becomes almost possible at this point to trace the missionary paths of the various Catholic orders by looking at the designs of the laces associated with each of the non-European countries. Both the bobbin laces and the needle laces began to separate into distinctive families of styles, while those laces (such as 'sols') with less room for adaptation developed the differences associated with changes of fibre.

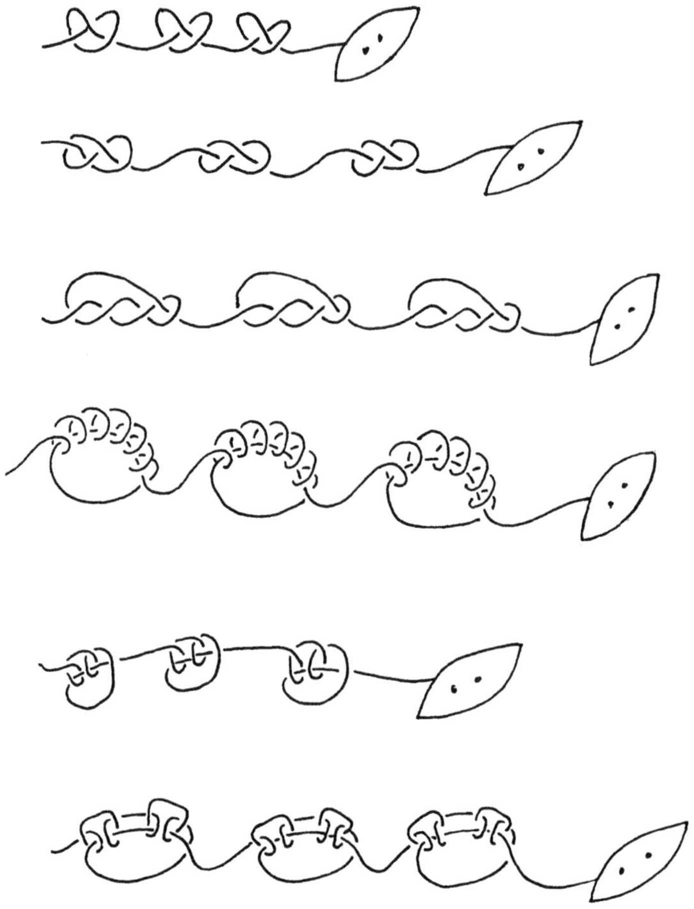

Fig. 3. 'Knotting', and how the 'stitches' evolved into tatting

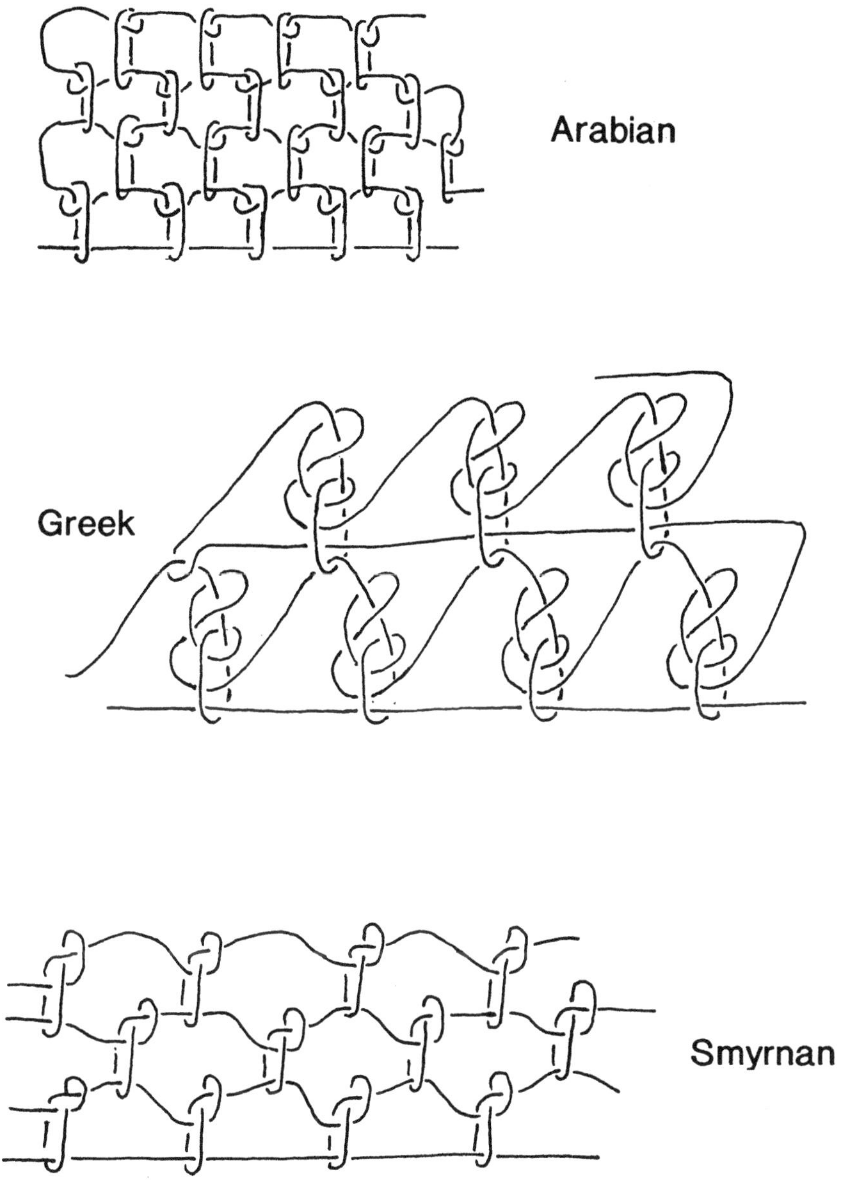

Fig. 4. Knotted lace stitches

The Eighteenth Century

The latter half of the C17th brought the Age of Reason. Minds developed rational habits, losing some of the superstitious clutter. Similarly, houses were more spacious and light. Clothing also changed as people grew aware of the causes of plague, becoming lighter and more frequently cleaned.

King Carlos II of Spain died, willing his Empire to the second grandson of Louis XIV on the Pope's advice; the accession of Philippe Duc d'Anjou to the Spanish throne sparked off the Wars of Spanish Succession (and set the style in wigs and scarves.) In Britain, the dynasty changed with the death of Queen Anne in 1714. George I could not speak English sufficiently well to deal with matters of state, and so the power transferred to the Prime Minister, changing the internal balance of power. The fashions at the Court of St James developed a 'German' flavour, while those further north— supporting the Stuart Pretenders—remained obstinately out-moded in their style of dress. The lace cravats and cuffs, however, concealed more than mere romantic declarations of loyalty; the bunch of blue ribbons tying a jabot (or a cluster of curls in the case of a lady), were a form of livery. (This was the point at which finger bowls were introduced at banquets by the Jacobites, so that when the loyal toast was called they could drink 'To the King over the water' while ostensibly demonstrating their fealty to the House of Hanover.) In mid-century, Frederick the Great began to forge a united country out of the many little German states, France and Britain wrangled over Indian and Canadian territories, and France starved while her monies were thrown at the campaigns. Catherine the Great expanded the Russian Empire, encouraging her subjects to adopt 'German' (in reality, French), fashions at court, while sumptuary laws kept the peasantry in traditional dress. In the last quarter of the century, the American War of Independence broke out.

At this time it is difficult to imagine anyone having the settled conditions necessary for the painstaking work of needlepoint, particularly as so many of the lacemakers had fled from France with the revocation of the Edict of Nantes. The heavy needle-laces fell out of favour, and collecting porcelain took its place. (Tea, or 'bohea' is mentioned in Pope's 'The Rape of the Lock', but despite all the details of dressing-table paraphernalia, lace is not mentioned once!) Lace fans, however, remained in fashion as can be seen from Addison's articles in 'The Tatler'. Silks and cotton threads became more readily available, and the spinning wheel became part of the domestic furniture. In 1733 the flying shuttle was invented so speeding the weaver's work; now it was possible for one webster to use the yarn of six spinsters, each working at a wheel. Consequently fabrics fell in price, homespuns became finer, and there was more cutting involved in dress-making; where previously fabric had been folded, tucked, or gathered so as to make maximum use of it in another garment, shaped panels were being used to give a clearer line and a crisper silhouette.

In Switzerland, those Huguenot lacemakers who had settled in Geneva before the Edict of Nantes was published welcomed those who had fled there upon its revocation; about 2000 lacemakers set up a factory in Neufchatel, smuggling much of their work back to France in 'revenge' for their losses. This helped to enrich Switzerland, and Neufchatel still remains the centre of the Swiss textile industry despite the sumptuary edicts which regulated the use of lace within the state.

At the French court, the vogue for Naples lace embroidered over in silver or gold gradually faded out; blonde and black blonde laces then became popular despite the lack of artistic design, for their use implied a compliment to the court of Spain. The Spanish lace mantillas achieved their best designs at this point, having had French taste to influence the patterns. The fashion developed for lighter laces, particularly as cotton yarns were imported from India. The bobbin laces, being lighter in texture and design, gained importance among the fashionable; spinning-wheels simplified the production of thread, and the comparatively faster technique put the lace within a realistic price-range. The lighter, floral designs of Mechlin, Brussels, and (later in the century) blonde Chantilly became the laces favoured in the summer; while for winter the needle laces of Alencon and Argentan were preferred. Sometimes a compromise was made, the mixed lace having, for instance, flower motifs worked in needlepoint with the ground made in bobbin lace round the motifs.

It is interesting to note that in Russia, despite the court fashions, the only national laces were of peasant origin. Country girls used to work a sampler apron with bands of embroidery and lace, which was worn with 'best' clothes, and designed to show the needlework skills of the wearer. The aprons have alternating bands of coloured designs and drawn or pulled thread work in white, each band showing a different pattern which appears to be copied slavishly from another piece. It is difficult to conceive of such exact reproduction of proportion and colour, particularly as the signficance of the symbolism has been lost; but there is the same precision of copy as in the painting of icons, and this would suggest that the craft pre-dates its records. The dress of the gentry shows little beyond copies (sometimes debased) of the European court dress, with uneasy compromises such as peasant-style sarafans of velvet embroidered with pearls for the royal nursemaids. However one original approach was shown in using split straws as embroidery thread on muslin to imitate the fashion for gold lace on white silk: at arm's length there is no obvious difference.

In the same category of faithful copywork there should be included the Norwegian openwork technique of Hardanger Embroidery, a form in which small rectangular blocks of satin stitch outline open areas of design. Threads are withdrawn from these areas leaving a cartesian grid which may be darned and the spaces filled with lattice-design stitches, as in filet-lacis. The contrast of texture and the strong geometric designs are ideally suited to the even-weave linen that is traditionally used as ground, while the remoteness of Hardanger Fjord may be the reason that the craft has been preserved unchanged. The Fabric of the Age was Indian cotton muslin—fine as silk, (the Indian legends tell of sari-lengths which may be folded into a walnut shell), with the advantage of being easy to wash. These quality fabrics were made more accessible when the spun cotton was imported to Britain for weaving; the 'India muslins' so produced were often 'sprigged' with little sprays of flowers dotted over them in tamboured chains. These light, floral designs were frequently complemented by the embroidered laces: pulled work with raised stitching, drawn-thread work with fine darning, fine embroidery with eyelets and needlepoint fillings, and the like. Once the Spinning Jenny and Spinning Mule machines were invented, it proved faster to import cotton in raw bales, the machine-spun thread was more even, and prices were brought down.

One of the after-effects of the Court Dress (grand habit) introduced by Louis XIV was that clothing fell into three distinct categories for the rich: court-, town- and un-dress. At this time, cotton was generally used for undress (informal attire), town dress often used cotton or wool-cotton unions, and white muslin with rich silk was the fashion for court. The silk industry at Lyons declined with the French Court's craze for the 'shepherdess' look so favoured by Marie Antoinette, and the lighter cotton muslin took over completely. Shawls, scarves and stoles became necessary to the fashionable wardrobe: and for those who could not afford 'real' (bobbin) lace accessories, the netted and knotted laces formed a cheaper substitute. Long scarves and stoles were often made in simple macramé techniques, relying upon the tinsel in the thread to give richness; the fringes for these scarves were often 20–25 cm in length, giving maximum effect for minimum effort. Square and triangular shawls, however, were generally netted (as were some stoles and scarves) again with deep fringes. These netted pieces were generally worked in the plain mesh, with variations in gauge-size round the borders, or clusters of loops to give interest. Filet-lacis as such, however was not used on shawls. Sometimes, too, pieces of embroidered silk with a deeply fringed macramé border were used for shawls. These became more fashionable as Napoleon Bonaparte revived the silk industry of France.

Knotting and fringing (macramé) continued to be favourite domestic crafts; the indefatigable Mrs Delaney wrote in a letter that she had completed the trims for a set of chairs, and was about to begin work on bed curtains to match. Jane Austen mentions netting, and in 'Sense and Sensibility' adds in an aside that the French prisoners-of-war made workboxes for sale to offset their living costs. The taste for fine furniture with matched veneers resulted in a need for cloths to protect the grain of the wood, and netted pieces were very popular for this purpose since they allowed the pattern of the veneer to show through. As a rule, Filet-lacis was used for the bigger pieces, while mat designs were made using the meshes of the net arranged in different ways.

Both in France and in England tambour-work was considered an aristocratic amusement. Portraits and passing references (e.g. in Sheridan's 'School for Scandal') show the tambour- frame as a drawing-room item, prettily decorative although having a proper function, just as the little table-top spinning wheels of the previous half-century were tools for use. However, it is not clear whether the ladies' tamboured work resulted in embroidered muslin or in something more closely resembling lace.

In Britain, the fashion followed the tastes of the royal family, and the embroidered laces of Dresden in Saxony and Hedebo in Denmark were much in vogue. In the latter half of the century, one Luigi Ruffini started a workshop in Edinburgh teaching the techniques of Dresden work to apprentices. His designs were outstandingly good, and he maintained their quality by ensuring that his apprentices over the age of 13 were given the chance to study at the Drawing Academy for a six-months' probationary period; if the student showed promise she was allowed to continue for the full two years. At the turn of the century, the famous Paisley Shawls (Jacquard woven woollen shawls in imitation of the Indian Kashmir webs) were being priced off the market and the Glasgow weavers forced into producing cotton muslin. Ruffini's work had a strong influence upon what followed.

The Nineteenth Century

The century is associated with industry and technology. The ginning machine was invented in 1793, and by the turn of the century cotton manufacture was a fully automated process. The Battle of Waterloo ended fears of further wars and the world settled to a time of consolidation.

The Paisley weavers set up their own tamboured-muslin factories and, using the design principles laid down by Ruffini, sold enough work to offset their losses. In 1814, an Ayrshire lass (Lady Mary Montgomerie) returned from the Napoleonic wars with a French maid, a baby son and a christening gown. The Hon. Mrs Jamieson, an agent for the websters of the Glasgow area, saw the christening gown, learnt the stitch techniques from the French maid, and taught them to the embroiderers of the workshop. This was the beginning of the famous Ayrshire whitework embroidery, a cottage industry where the women sat round the village green exchanging pieces of work and conversation. Each girl specialized in one particular technique associated with the craft: thus all the outlines would be padded by one, stitched by the next, the needlepoint fillings made by several girls—each of whom was expert in one variation,—the cloisons filled by others. The standards were maintained by stringent inspection of the finished pieces; and as the 'Flowerers' were paid by the piece, anyone whose work was below standard was dealt with by her colleagues in the village.In 1840, when the work was at its peak of popularity, a Flowerer could earn as much as 10/- per week. [At this point, a 'very handsome' dress allowance was considered to be 15gns p.a. (=315/-).]

Around the third decade of the century, the lace knitting of Unst in the Shetland Isles was discovered. This was always knitted with wool from the local sheep, which were never sheared but combed by hand ('rooed'). The fine wool was handspun, and the best quality yarn measured 'three-quarters of a mile to half an ounce' (5km length per 50g mass). The method of knitting coupled with the very fine yarn produces a soft, flexible mesh that may be pulled through a wedding ring, and the best quality shawls (often measuring 2m square) were family heirlooms used for wedding veils and christenings (and sometimes called churching shawls.) This proved so popular an item of fashion that by 1840 shawls were being 'exported' to London; but by 1850 imitations were being produced, and the island's brief period of prosperity was over. There are fewer than 35 stitch patterns (the first attempt to catalogue the stitches, in the early part of the C20th listed only ten), some of which are recent inventions. As with the Ayrshire work, the lace knitting was a cottage industry, with an agent to handle the sales. One curiosity of the Shetland knitters is their method of holding the left needle in a belt fastened to the waist, so that the knitting may be produced single-handedly while stirring the porridge or rocking the bairn's cradle.

In England, the finest lace was made in the Devonshire area, and went by the name of Honiton. (Daniel Defoe mentions a lace known as Blandford Point, the finest in the British Isles, in the C18th; but we have no records of the work today.) This industry suffered great setbacks when the bobbinet machine was invented, for the cheaply produced machine nets brought down the price of lace. In an attempt to boost the sales of hand-made lace, Queen Adelaide, wife to William IV, had a dress made of Honiton lace, where the initial letters of the flower motifs spelt her name. Even so, there was a marked decline in sales. There was a fleeting revival in the quality of design when Queen Victoria commissioned her wedding veil: an architect and an artist drew out the design which was then made by lacemakers in Devon. However, all too often the designs were copied off wallpapers or chintzes, and so lacking in balance. The trolley laces of Buckinghamshire and Bedfordshire continued to sell, though lacemakers' prices were forced down; Thomas Lester, an agent, did much to improve the designs in this region.

There was also a new form of lace begun in England at this time. A man called Draygo, fleeing from the Napoleonic wars, settled in the Essex village of Coggeshall, sheltering with a silk manufacturer. In return for hospitality, Monsieur Draygo taught the skills of tambour work to the girls of the village; this lace school had the support of M. Draygo's landlord, and within a quarter of a century a flourishing business was set up, employing around 400 workers who came from a 20-mile radius. Coggeshall lace differs from other tamboured work in that it makes use of the hexagonal mesh of the net in the patterning. Good Queen Adelaide supported this lace craft too, ordering a robe in an attempt to boost the sales. This craft withstood the Industrial Revolution better than most of the other lace forms, being relatively quick to produce. Nevertheless, despite the introduction of little lace mats, lace trims to the dressing-table and lace curtains and bedspreads, machine-made stuffs were so much cheaper that the industry declined.

In Ireland, Limerick lace was developed when some Coggeshall lacemakers were invited over in 1829 to teach tambouring on net. Where Coggleshall uses cotton hexagonal mesh nets exclusively, Limerick was sometimes tamboured on the square meshes of Nottingham machine-made 'filet' nets. The Potato Famine of 1845 resulted in the awareness of the Irish laces as a marketable commodity, and Irish crochet suddenly became a fashion item. The Irish laces are named by region: Carrick-ma-cross, which is muslin applied to net with embroidered fillings; Limerick, a decorated net; Curragh point, a tape lace; Youghal (pronounced 'yawl'), a needlepoint lace in imitation of point plat; and so on.

Tatting was often worked as part of lace-trims in Irish pieces, but had little recognition. Crochet, however, was sufficiently distinctive to be given pride of place nationally. The craft had been taught in convents for as much as a century previously, with its main purpose being for church furnishings; the characteristic method of working, and the patterning in imitation of the heavy Venetian laces made the transition to secular purposes as easy as it was successful. This was such a successful enterprise that other countries began to encourage the making of (peasant) laces as cottage enterprises for the tourist trade, giving us Madiera openwork, or Orvieto lace crochet, to name two of the many available at the time.

In Europe, the production of very fine machine-spun cotton gave Brussels lace a new popularity, for the Bruges flower lace designs could be made more delicate still. The blonde laces remained popular, despite the competition of Leavers and Raschel machine-produced laces. Although these machine pieces made very fine imitations of the design characteristics, the draping qualities were different, so that shawls of real blonde were more comfortable to wear.

The wealthy Austro-Hungarian Empire encouraged art-needlework by setting up a lace school in Vienna. Design was taught side by side with technique, and many of the crochet and knitting designs in existence today originated with the work of students there. The knitted pieces from this region are often characterised by their start from a central point, expanding into circles or squares by carefully positioned increases; this had a strong influence on some of the Victorian knitting patterns, especially those worked in cotton.

With the improved transport and communications, the grand tour of Europe became the objective of everyone who had the means. Souvenirs were an important part of the returning traveller's luggage, and the cottage laces flourished in consequence. C19th novels inform us of the sols laces of Portugal, or the nurse sitting by a bedside tatting with an ebony shuttle. (L.M. Alcott and Susan Coolidge, respectively.) Tourist souvenirs included tatting shuttles of white wood with the local beauty spots transferred under the varnish, or boxed sets of crochet hook and hairpin in a decorated box labelled 'Souvenir de—'; so informing us that the people who worked at these were amongst the leisured classes. Similarly, we learn from novels that knitting, even elaborate lace trims or christening robes, was confined to 'the housekeeper' and netting to the governess. (The Brontes and Mrs Gaskell.) However, lace was becoming a lost craft, and even those women who repaired lace for a living were a dying breed: mechanisation had come to stay.

There were several sorts of machines making lace, with the industry based in the Nottingham area. The style particularly associated with Nottingham is filet design; tablecloths, curtains, mats and even calendars were made in this technique, and proved overwhelmingly successful as novelties. In Switzerland, a chemical form of imitation guipure lace was produced: a sheet of acetate had a design traced upon it, this was embroidered closely in cotton thread and the work then dipped in a bath of acetone which dissolved the base, leaving a stiff lace fabric. This form of imitation lace is still in production, and the trimmings so made are surprisingly durable. It was also in Switzerland that the caterpillar lace experiments were carried out: a design was painted on a varnished wooden board using a 'paint' of succulent leaves pounded into a paste. The board was then propped up against a wall and caterpillars placed along the base; as they climbed up the board eating the paste, they left trails of cocoon-silk which interlaced to form a sturdy mesh. This method of manufacture proved unsuccessful commercially for there were problems of quality control. (The caterpillars would persist in turning into butterflies!)

In the last quarter of the century there was a revival of interest in handiwork. Strongly influenced by the Pre-Raphaelite ideas of John Ruskin and William Morris, the Arts and Crafts movement caused people to re-learn the non-mechanical skills such as pottery, spinning and weaving. Ruskin himself was interested in the production of linen as a cottage industry, and set up a co-operative in the Lake District. Out of this co-op there developed a group of needlewomen who learned to make Reticella/Punto in aria lace using the homespun linen as foundation. With typical Victorian non-correlation of facts and skills, this was first described as 'Greek' lace, but later it became known as Ruskin lace in honour of the man who helped to bring prosperity to the Lakes.

Tatting in this era developed from a series of knots into rings with picots which could be used in conjunction with crochet, needlepoint fillings, or linked to other tatted rings. There was an element of snob value in the craft, for Queen Marie of Romania was known to have written articles on the subject (under the pseudonym of A Lady). It became the fashion to have tatted furnishings on the altars of private chapels. Curiously, tatting permitted the use of colour (generally blue, cream, silver, or gold) in the work while coloured crochet was considered 'not quite the thing' amongst gentlewomen—the housemaid's best outfit in one tale includes a collar of green and yellow crochet lace of shamrocks and buttercups.

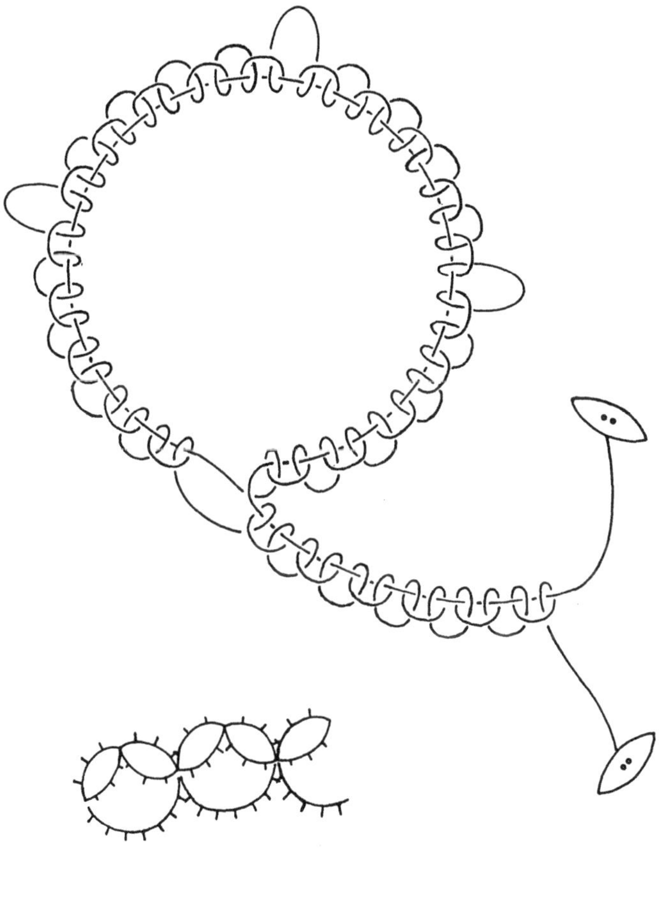

Fig. 5. Construction of Tatting, and typical design

The problem of printing graph-grid patterns was solved in the first decade of the century, but the possibilities were not realised till a German printer (Wittich of Berlin) began to publish charted designs for cross-stitch. This began a craze for counted-thread work on canvas—so passionately was this pursued that for some decades 'embroidery' and Berlin wool work were synonymous. During the same period, this form of printing was being used by a printer in Philadelphia for the publication of filet-lacis patterns. This 'crochet type' did much to make filet-crochet the popular form of lacemaking that it still remains.

Domestic needlework, which included lacemaking, centred round the everyday needs of the household. The elaborately fitted workboxes of the middle part of the century give us an excellent idea of the domestic lace-work; and netting must come high on the list, for many boxes contain netting-needles and mesh gauges, while some boxes are fitted exclusively for netting. Ladies of the era would have used hair nets, plain in the daytime and decorated for evening wear; these would have been made in the home, as were the stocking purses which carried coins. Both these items were subject to hard wear, and would have needed constant replacement, giving the maker a chance to try a new lace pattern each time. Magazines and books on needlework were now available at reasonable prices, and these often carried suggestions as to the suitability of certain designs for some purpose.

After Prince Albert's death, the fashion for mourning resulted in a brief period when hair 'lace' was mounted in jewellery. This was more an open plait than real lace, though the plaiting technique used is a lace stitch.

The Twentieth Century

The Edwardians began the century with a coronation. One of the souvenirs of this was a braid spaced-dyed in lengths of red, white and blue which could be used in all sorts of handiwork. Tape-lace and crochet patterns using Coronation Braid were published, and gave a new interest to the designers. However, the designs were coarse in scale, to accommodate the thick braid, and the emphasis was on the showy rather than on craftsmanship.

A form of tape-lace using thread fillings to hold the design of swirling loops was named in honour of the Battenburgs, and hideous bedspreads were made in this technique. Nor was it a practical design of lace for the tapes lost their crisp form after a few washes; but its ease of construction brought it within the scope of most women.

Needle-lace was still made as a form of embroidery, for there had been a revival of interest in fine needlework thanks to the influence of William Morris. The bobbin-laces, having always been made commercially and now being supplanted by the machine-made pieces, were dying as the lace-makers died.

The designs remained, being preserved in a debased form by the copies made on the Barmen and, later, the improved Raschel machines. In Switzerland, the Schiffli machine with its pantograph made possible the simultaneous production of several lengths of the same design so speeding the process and cutting costs. Modern technology has put a technology spun alginate fabric on the domestic market, making it possible for a needlewoman to attempt chemical lace on the sewing machine in complete safety, for the solvent for this fabric is water.

Tatting developed chains at this time, giving more scope to the design, and strength in use. From a delicate, easily distorted, lace it became a crisp one that maintained its shape even after numerous washes; moreover, it required so little equipment that it was justifiably popular. Knitted and crocheted laces remained general favourites, particularly after the wars, when there was a great need for beauty and little money to spend on the begetting thereof: again the minimal requirements for tools must have helped.

In the 60s, there was a revival of interest; lace bobbins could be purchased once more from specialist needlework shops, and two books were published, one a reprint. The surviving lacemakers were asked to teach their skills, and children in schools began to admit an interest in this 'old-fashioned activity'. In 1976 the Lace Guild was established, one of several international societies, for the study of lace, and in the last 16 years much has been done to further our knowledge of this craft, its traditions and history. Museums specialising in lace have been set up, courses are run all round the world; and, most recently, efforts are being made to help enthusiasts with the design of pieces. Almost all lace diploma courses now include some design exercises, so ensuring that the craft will not stultify. Small developments have occurred in many of the techniques, for modern materials necessitate a change in some areas; but no single new invention has yet come to add to all the accumulated skills of the past.

Written Sources

The Pipe Rolls (mostly C13th) for data on pins, prices and non-existence of lace

Chaucer (ca. C14th)

The Wardrobe Accounts of the Tudor and Stuart Monarchs

The Accounts of Hengrave Hall

The Paston Letters

Sumptuary Laws and other accounts of C11th–C17th England and France

Memoires du duc de Saint-Simon sur le siecle de Louis XIV et la Regence

Memoire du duc de Luynes sur la cour de Louis XV

Pepys' diary
The (life and) letters of Mrs Delany
(Assorted) letters of the Ladies Coke
The Spectator and some of Jonathan Swift's sermons
Jane Austen's letters
Daniel Defoe's *A Tour through the Whole Island of Great Britain*
The History of Lace—Mrs Bury Palliser
Assorted phrases from Shakespeare, Beaumont and Fletcher et al. through to Mrs Gaskell and Anthony Trollope

Visual Sources

Portraits and Fashion Plates
The Ohrid Ikon Gallery (in what was Yugoslavia)
Lace collections, some private

Aural Sources

These come from memories of the ladies of the Gardner family of Paisley; and comments like "my grandmother's aunt made lace, and she told us that ..."

Further Reading

Tatting	Rhoda Auld
Needlelace: Classic and Contemporary Designs and Techniques	Catherine Barley
Lace and Lacemaking	Alice-May Bullock
The Torchon Lace Workbook	Bridget Cook
L'Europe en Dentelles	Rita Carole Dedeyan
Bobbin and Needle Laces, Identification and Care	Pat Earnshaw
Lace in Fashion	Pat Earnshaw
Needle-made Laces	Pat Earnshaw
Youghal and other Irish Laces	Pat Earnshaw
Starting Needlepoint Lace	Valerie Grimwood
Lace, A Visual History	Santina Levy
Flanders Lace: A Step by Step Guide	Mary Niven
The Technique of Bucks Point Lace	Pamela Nottingham
Technique and Design of Cluny Lace	L. Paulis and M. Rutgers
Further Steps in Honiton Lace	Susanne Thompson

Introduction to Honiton Lace	Susanne Thompson
Lace: A Guide to Identification of Old Lace Types and Techniques	Heather Toomer
Introducing Traditional Bedfordshire Lace in 20 Lessons	Barbara M. Underwood
Renaissance Patterns for Lace Embroidery and Needlepoint	Federico Vinciolo
Bobbin Lace in Photographs	Cynthia Voysey
Needle Lace in Photographs	Cynthia Voysey

CHAPTER 17

HERALDIC KNOTS

Charles Warner

Heraldry is a colourful, feudal, picturesque, elitist, decorative, patriarchal, vivid, snobbish, distinctive and artistic way of displaying one's genealogy or allegiances. Amongst its numerous designs can be found some very early clear depictions of the structures of specific common knots. Many books have been written on the subject, often with a small section on heraldic knots; our bibliography references several good introductions to heraldry, particularly its English variant.

Devices or symbols of one sort or another have long been used to represent military or corporate authority and familial or political affiliation. Perhaps the first mention of hereditary family devices is in the Bible, where the Lord told Moses in the Book of Numbers, more than three thousand years ago, that *Every man of the children of Israel shall pitch by his own standard, with the ensign of their father's house.* The use of hereditary symbols became thoroughly systematised more or less simultaneously in feudal Japan (the mon) and in feudal western Europe (heraldry) in the 12th century.

European Heraldry

Heraldic devices are personal to the bearer and are handed on from father to son; wives and daughters occupy a somewhat uneasy subordinate position. The background to the designs is based on medieval armour, and the shield is central to that system. During the Crusades, the heat in Palestine encouraged the use of a light cloth surcoat over the armour, on which designs could be embroidered; this is the origin of the term 'coat of arms'.

Heralds were concerned with the conduct of tournaments, the carrying of messages, and the marshalling of ceremonies; they had to recognise, catalogue and often grant heraldic insignia; because of the hereditary nature of the insignia, they became deeply concerned with genealogy. Tournaments are no longer held, message-carrying has been taken over by diplomats and politicians, and only a few countries still use heralds to conduct ceremonies. But many countries, even those with no trace of feudalism in their republics, maintain an interest, and often some measure of legal control, in the use of the colourful heraldic devices.

The rapid adoption of heraldry throughout western Europe in the early 12th century seems to have had several causes. The feudal leaders returning from the First Crusade (1096–1099) had had their first experience of truly large armies, involving hundreds of thousands of men instead of their accustomed hundreds or thousands. They had found confusion and a need for rapid identification of friend or foe as they faced the highly skilled and organised Saracens [18]. Throughout the 12th century Renaissance, an exuberant delight was taken in personal adornment and colourful decoration; and those military individualists, the armoured knights, with their individual fights and massed tournaments, gave ample scope for heraldic display. Monks, knights, minstrels and scholars wandered over a Europe before the rise of the nation states [14]. All these factors combined in the development and spread of heraldry.

At first, the feudal nobles and knights invented their own heraldic devices, but the need to prevent two people adopting the same device and bring order to the system encouraged the growing power of heralds. Lists or rolls of coats of arms of people associated with particular battles or tournaments were compiled from the early 13th century [2] and a technical jargon was developed to describe or **blazon** the numerous arms. Complex rules were devised to control inheritance, to differentiate the arms of close relatives, or to mark the acquisition of new property by marriage or conquest, or the award of higher rank or honour.

From the 14th century, corporate bodies, such as towns or boroughs, university colleges, craft guilds, merchant companies and cathedrals and churches began to gain coats of arms, and new rules had to be devised [2][6][15]. These arms were all called **impersonal arms**, to distinguish them from the **personal arms** of individual persons.

Different parts of Europe adopted different conventions and rules. Particular designs became much more popular in some regions than in others [2][25]. In the context of this book, the different uses of knots as heraldic devices in England and in Continental Europe is noteworthy. The practice of using heraldic arms has now spread through most of the world, particularly in the ex-colonies of European powers.

Scotland is the only country that still has a functioning court of heraldry with powers of imprisonment and fine. Arms in Denmark are legally protected by registering them as family trademarks. Some countries still have a legal authority to grant particular arms, such as the English College of Arms; other countries have some sort of society that tries to control the assumption of arms; in yet other countries it seems to be open to anyone to adopt what arms he or she wishes, within the bounds of copyright or trademark law.

The earliest heraldic designs were simple geometric patterns [2], but representations of beasts, birds, fishes, flowers and other common or mythological objects soon made their appearance. Indeed, much variety was needed to allow each knight and other bearer of arms to have his own distinctive design. There are now several hundreds of thousands, some say millions, of distinct heraldic coats of arms.

The official design of a coat of arms is given by the blazon, the brief written description. The exact interpretation of blazons has been subject to some standardisation, but the heraldic artist has much freedom in depicting the various objects named in the blazon [26]. This freedom is particularly noticeable with knots; a knot tier will find many different knots or entanglements of cord used by the artists to depict a given blazon.

Knots can be seen in many places in heraldry (Fig. 1), but objectively speaking they are rather rare. Probably less than 2% of arms show any kind of knot or knotted object, whether the knot can be identified or not. Much less than one in a thousand shields depicted in Rietstap's [24] massive set of volumes bear an isolated knot tied in cordage. However, there are few counts of the frequencies of any heraldic devices. Pastoureau [22] has tables of frequencies of major elements (animals, plants etc) and Bolton [4] indexes the individual charges on shields, but neither lists knots.

I now describe the various ways in which knots can be used in heraldry, and the various knots that have been depicted. I begin by giving details of the 'shield', and then discuss designs that are displayed on shields and elsewhere. In the course of this study I have scanned perhaps 150 000 shields and a much smaller number of complete achievements of arms (see below); this seems to be only a small fraction of all arms that have ever been compiled.

The Achievement of Arms

The achievement of arms comprises all the elements of armorial bearings of an **armiger**, a person entitled to arms, usually in display, as in Fig. 1, though sometimes only in writing or blazon. Fig. 1 is a display showing most of the opportunities of using knots; such a collection of knots and knotted objects has never been given to any one person.

The **shield** (14) is the most important element, and is sometimes called the coat of arms, after the surcoat worn over armour and displaying the arms, though that name is at other times given to the whole achievement. Some armigers are entitled to a shield alone, with no other elements; no one is entitled to any other element without also being entitled to a shield. Only the shield can stand alone, except that badges (2) usually, and crests (3) occasionally are displayed, and the regalia of orders and other honours are usually worn, in the absence of a shield (see later). Ladies do not bear their arms on a shield, but use a lozenge, or occasionally an oval, instead, as will be described later.

Fig. 1. An Achievement of Arms

The individual designs displayed on a shield are known as **charges** (for example, in Fig. 1, 9–11, 13, 15). Very many thousands of different charges are known, but few involve knots in any way. Charges involving knots include:

(i) Isolated knots (15) tied in cord, usually imagined as silken, are more common on the Continent of Europe than in England. Several continental designs are shown in Figs. 2 and 3; the only English use of isolated knots that I have seen as charges are some of the badges (Figs. 6 and 7), introduced into impersonal arms a long time ago but into personal arms only this century.

The identity of the knot is usually clear enough, as in Fig. 2, but sometimes as in Fig. 3, (43), there is only an indication of some kind of knot.

(ii) Sometimes animals, birds or mythical creatures, or their tails or necks, are shown tied in a knot, or 'nowed' or 'knowed'. The nature of the knot is usually clear enough, though the term 'nowed' does not usually imply a specific knot, which is often at the choice of the artist. Nowed serpents (13) may be more common as crests than as charges, at least in English heraldry (see Fig. 4); animals or creatures with nowed tails are also perhaps less common as charges than as supporters (e.g., 8, 12) or crests. I have also seen a bird, perhaps a crane, with its neck nowed.

(iii) Sometimes articles are displayed as charges with some kind of cordage attached to them (this is sometimes termed 'stringed'); the interest is in the object, not the knot, which is hardly ever depicted so as to make the identity of the knot clear. The most common knotty articles are: a head wearing a presumably knotted headband (9), often of two colours; a 'stringed bugle-horn' (10), where a cord is tied to a bugle or hunting horn (often a round turn is indicated, rarely a knot) and the cord has some sort of decorative turn or knot (see Fig. 3, 34–38); and a 'garb', or sheaf of wheat or, occasionally, barley or rye, bound with some more stems (11). Quite a number of other objects can be found, with very low frequency, with cords attached and knots indicated; an example is shown in Fig. 3 (32).

Above the shield may be the **helm** (7), except that ladies and the clergy do not use them. Some heraldic traditions have an elaborate system of indicating the rank of the armiger by the kind of helm, the metal (iron, gold) it is made from and the direction in which it is looking. There is no opportunity for knots.

Some form of **mantling** (4) is often shown over the helm, representing the cloth originally used to protect the wearer against the heat of the sun and showed the slashes of many battles. This is often of different colours inside and out, and cut into very decorative shapes. Occasionally there are tassels (6) at the ends of the mantling, and these may be tied in knots, usually a simple Overhand ([1], #515) or Figure Eight ([1], #520) Knot.

The mantling is held in place with a **wreath** (5), which hardly ever shows any sign of knots, or perhaps with a chapeau or cap of maintenance, a medieval hat lined with crimson velvet, which may also be used to line a crown or coronet.

A **crest** (3) rests on the wreath, coronet or bare mantling. Originally, this represented a solid object, often made of wood or leather, which was fastened above the helmet for display, beginning in the 14th century [2]. Very rarely, this is a knot, such as the Wake Knot (Fig. 7, (82)), but knotted objects are less rare. A nowed serpent (Fig. 4) is probably used more frequently than anything else involving knots [11].

The shield is often held up by one or two **supporters**, introduced in the 15th century [2]. A single supporter, such as a displayed eagle with the shield over its chest, is more common on the Continent than England, but over all, paired supporters (8, 12) are the most common. In some heraldic traditions, the use of supporters is restricted to the higher nobility, such as the hereditary peerage in England, though even there some people continue to use supporters that had been adopted before the tradition became established. Supporters are frequently used in impersonal arms, and pressure to grant personal supporters seems to be growing. Occasionally, the supporters may have nowed tails, as shown in Fig. 1.

The supporters are often shown on a sort of mound, called the **compartment** (16), which is occasionally strewn with badges, though I have not seen any examples of knot badges so used. A **scroll** bearing a motto or war cry (17) is usually below the supporters, but occasionally around the helm or crest, depending on the tradition.

Certain **decorations** and parts of the **regalia** (1) of certain knightly orders may be displayed with arms. Medals may be suspended below the shield or, less commonly, around the helm or over the mantling. The 'collars' or other insignia of orders may surround the shield. Knots form part of some decorations and regalia; see Fig. 9.

Badges (2) are separate from the arms and not always included in the official grant of arms; they are much used by some armigers. Some badges, particularly in England, consist of knots (see Figs. 6 and 7). Badges are occasionally displayed on the compartment, beside the crest, on the mantling or worn by the supporters, but this seems to depend mainly on the wishes of the artist.

Lacs d'Amour

This knot (Fig. 2), while relatively common in parts of continental Europe, is unknown in English heraldry, and so has no English name; the French name means, literally, noose of love, but is often translated as true love knot. Pastoureau [22] says that the lacs d'amour is a *figure stylisée, representée par un cordon orné d'un ou plusieurs nœuds, lâches ou serées* (stylised figure, represented by an ornate cord of one or more knots, loose or tight). Rietstap [24] illustrates the knot as in (19), an **alternating** knot, that is, one where if you follow any cord through the knot, the crossings would alternate over and under. But throughout the many volumes of Rietstap [24] and its supplements, several additional types of lacs d'amour are seen (18–31). Not all are alternating knots; indeed some would, if tied in cordage, simply fall apart: they are non-knots.

Rietstap's typical knot (19) is an Overhand Knot with one side extended and twisted; if the twist is in the opposite direction (20), a non-alternating knot is formed; an apparently similar non-knot (21) has other crossings. A plain Overhand Knot (22) yields a similar non-knot (23), or, if the belly of the knot is twisted over the spine, another knot (24) is formed, superficially similar to the Figure Eight, which is itself (18) also used; unlike the previous two, this last is an alternating knot.

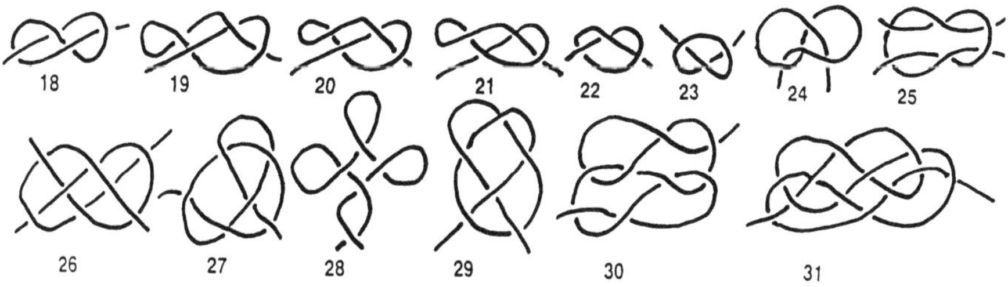

Fig. 2. Lacs d'Amour

Knots (18–24) have all been seen several times each; the following knots, still called lacs d'amour, have been seen only once. The Reef Knot ([1], #1402) is used as a lacs d'amour despite being non-alternating, while the alternating Granny Knot ([1], #1405) does not seem to be. What appears to be (26) an attempt at a Carrick Bend ([1], #1428), is in fact a non-knot. Knot (27) is similar to Anne of Bohemia's knot (Fig. 7, 83), and knot (28) is similar to the Bowen Knot (Fig. 6, 68). Knot (30) is a sort of Double Figure of Eight Knot, listed (#585) but not named by Ashley [1]. Neither the symmetrical knot (29), nor the asymmetrical (31) are mentioned in Ashley. All the knots 27–31 are alternating.

Some other Heraldic Knots on the Shield

Fig. 3 shows a variety of other heraldic knots sometimes used as charges. Knots (39) and (43) are both labelled *entrelacs* (interlaced designs) by Rietstap [24], and are given as charges; the precise knots used in (43) cannot be distinguished. The purse of state (32), with its cords variously knotted has been seen as a charge and borne on supporters [3]. The Pretzel Knot (40) ([1], #2283), also known to some heralds as the Chinese Knot, is listed as a nowed hatband in the 1604 arms of the Company of Feltmakers [15], but Elvin [9] shows (41) as a hatband and (42) as a nowed hatband; (41) would look more attractive as an alternating knot. I have only seen (33) as a crest called a wrestling collar [11]. In the form shown it is similar to the Whatknot ([1], #1208, #1406, #2579); I have no idea how wrestlers used it.

A number of objects in heraldry are shown suspended from a cord; the 'stringed buglehorn' (10) is the commonest. Sometimes the 'string' is shown with a simple turn (34), but quite often some kind of knot is indicated, with one (35), two (36) or three (37, 38) loops showing, and sometimes two ends (36, 38), so that the knot may be considered as either a decorative bend (36, 38) or as a bow (35, 37) tied in the bight. I have never seen the knot shown sufficiently clearly to recognise, though (37) may in fact be the same as (40).

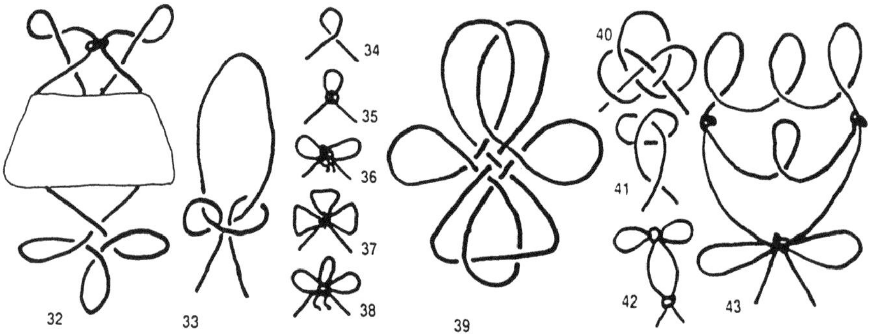

Fig. 3. Some Other Heraldic Knots

Nowed Serpents and Others

Serpents in heraldry seem to live peculiar lives. Either they are shown swallowing ('vorant') a child or their bodies are twined ('nowed') into a curve or knot; sometimes two serpents are twined together. They may appear as charges [9] or as crests [11].

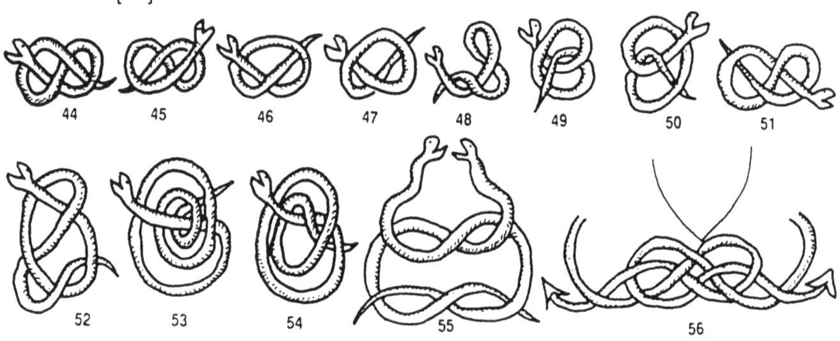

Fig. 4. Nowed Serpents

The commonest formation is the Figure Eight, with the head raised (44); rarely, the knot is tied so that the head points downwards (51), termed 'nowed reversed' [9]. As with other knots, sometimes the crossings are made (45) to give a non-knot [11], or the knot is capsized [7] to give (50). Rarely, the serpent is 'double-nowed' [9][24] as in (52), that is, tied in a Figure Nine Knot

([1], #521). A similar series of knots is based on the Overhand Knot [11], giving (46, 47, 49); the 'nowed embowed' knot [3] is shown in (53, 54), and is a Double Overhand Knot ([1], #516). Sometimes, the serpent's body is simply twisted (48).

Occasionally, two serpents are shown 'tails knit together' [3]; various knots seem to be used; that in (55) looks as if it might have been intended for a Reef Knot or a Granny; the 1571 arms of Gonville and Caius College (Cambridge) have simple turns only [15].

Rarely, eels are shown in similar positions. But the more interesting heraldic objects are animals, most commonly either lions or wyverns (two-legged winged dragons), which may have their tails 'nowed'. These knotted forms appear more frequently as supporters than as charges or crests, but all applications are found. The thin tails of lions are usually tied in a Figure Eight (Fig. 1, 8), the thicker tails of wyverns in Overhands (Fig. 1, 12), or just given a turn or two. A particularly interesting example is seen in the supporters of the arms of the City of Dundee [15], which the 1673 blazon says are dragons (four-legged, winged), though wyverns are often seen, with their tails 'knit together underneath' the shield (56); the knot used is a Carrick Bend ([1], #1428).

Knots for Ecclesiastical Tassels

Throughout the early days of heraldry, there was only one Christian church, now called the Roman Catholic, and everyone belonged to it. Many of the clergy in the higher positions in the hierarchy were of 'noble blood', and wanted or were already entitled to heraldic arms. Being nominally peaceful, they omitted the warlike helm and crest, and often the mantling, but retained the other elements of an achievement of arms. The arms displayed could be those of the individual, those of the parish, see, or other position, or a combination of the two.

Since 1278 [10], an ecclesiastical hat has taken the place of the helm; this has two cords dangling from the brim, ending in tassels. The colour of the hat, the colour of the cord, and the number of tassels are now decided by a complex system that indicates the rank of the armiger [23]. The cords are usually arranged in graceful loops, often crossing between the hat and the shield (Fig. 5) sometimes involving an indeterminate knot at the junction. One or rarely more knots have been shown above the tassels since the 16th century [10]. The nature of the knots seems not to depend on any heraldic rules, but entirely on the artist. Some of the knots seen are shown in Fig. 5; the more complex knots have been seen only in modern depictions. I have not attempted to show any of the knots where I could not follow the crossings in the illustrations seen.

By far the commonest knot used (57) is the Figure Eight; a simple Overhand (60) is rare. Sometimes the turn of the Figure Eight is drawn out and twisted, with the finished knot either horizontal (58) or vertical (59). If the end of (59) is tucked back into the loop a Twisted Braid ([1], #569) is formed (63). Sometimes the knot has the form of a Figure Eight, but the crossings differ (61), giving a non-knot. Two Overhands in line (as in [1], #564) may be opened out to make one enclose the other (62).

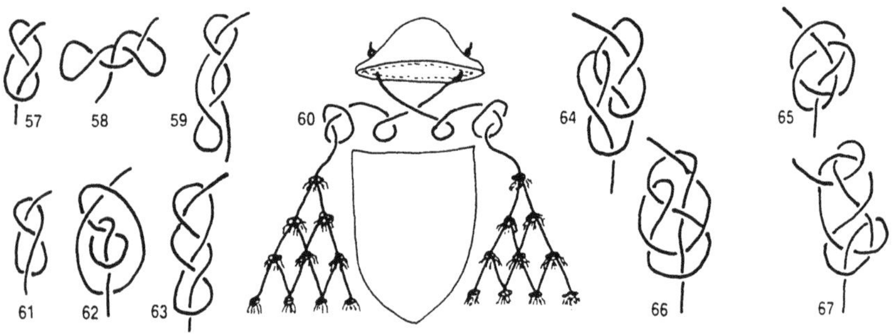

Fig. 5. Knots for Ecclesiastical Tassels

More complex knots used include the Double Figure Eight, seen in (64), as in (30); the Oystermans Stopper ([1], #526) in (65); and two knots not found in Ashley [1], (66, 67); the asymmetrical (66) is used by several modern artists.

Clergy in the Church of England may also adopt arms with ecclesiastical hats and tassels if they so desire [7], but this does not seem to happen often.

Knots as Badges

A badge is a device which, though not part of the coat of arms, could in some cases be used in a similar way; indeed, some badges preceded the adoption of systematic heraldry, and may have been the origin of some heraldic charges. Badges were common in southern Europe [25], but were usually restricted to the personal use of the higher ranks of the nobility, and members of some orders of chivalry. The only knot I have found amongst the Continental badges is the Savoy Knot (Fig. 7, 76), a Figure Eight Knot, often shown vertical (76), but sometimes shown horizontal (75). It has been in use since at least the time of Thomas, Count of Savoy, in the early 13th century [21], and was used by Umberto II (1904–1983), Prince of Savoy and the last King of Italy.

It was in England, however, that the badge, in particular the knot badge, came into common use. They were used personally, on jewellery, fabrics, furniture, carved on buildings and displayed in stained-glass windows. Knights who wished to remain incognito in tournaments often identified themselves with a badge. Retainers and armed followers used the badge of their feudal lord on their livery and banners; local inns displayed the badge; all to indicate allegiance to a faction. The period of greatest popularity of badges coincided with civil unrest, from the 14th to the 16th centuries. Indeed, in the 14th and 15th centuries, the common people were more familiar with household badges than with armorial bearings [13]. The use of badges greatly diminished in the late 16th century, so that by the Victorian age, armigers used their crests instead of badges on servants' buttons etc. However, in 1906 the English College of Arms started granting badges again, and encourages their use.

Anyone at all who so wishes may adopt a device as a badge, provided it is not already protected by copyright or trademark laws, or by a grant of arms to someone else. Recognised heraldic badges are sometimes used (with permission) as badges of local regiments, or they may be incorporated into the arms of local municipal bodies, colleges or corporations [13].

Badges were often poorly recorded; it did not help that some people had many badges. For example, the Earl of Stafford of 1720 had eighteen [16]; all were enclosed in a circle of Stafford Knots (74), though only one badge was the Stafford Knot itself. The total number of badges is unknown. A few are knots (Figs. 6 and 7), or at least arrangements of cordage, and they form the greatest use of specific knots, not just decorative devices of variable form, in heraldry, which I will now discuss. In some depictions, crossings vary, as elsewhere in heraldry, but I have not mentioned these.

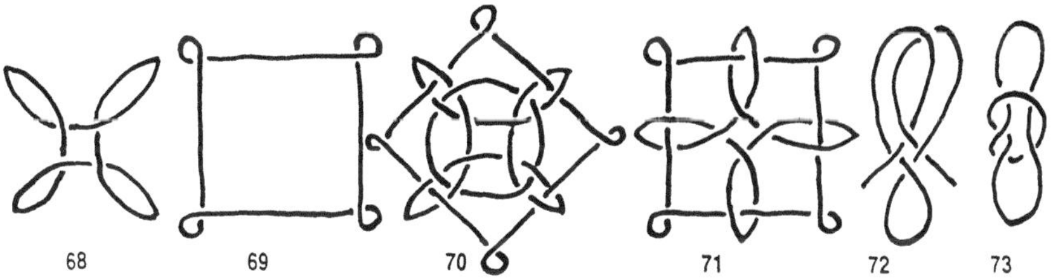

Fig. 6. Some Knots as Badges

The Bowen Knot consists of four loops or bows in a continuous length of cord [7][13]; it may appear in several forms, such as (68, 69).

The Lacy Knot consists of two Bowen Knots interlaced, with usually [7][13] a central loop of cord (70). I have also seen it (71) without the central loop [12][24]. It first appears on the seal of Roger de Lasci (1179–1211) [7], and a slight variant in that of Robert de Laci (died 1193) [21]. It has appeared as a charge of a recent creation, Baronet Lacy, in 1921 [8].

The Heneage Knot [7][13][16] reduces to a Figure Eight Knot (72). It was used by Sir Thomas Heneage, Vice-Chancellor to Queen Elizabeth I in the late 16th century [21].

The Suffolk Knot (73) is one of the rarer knots [20]; it is similar to the

Crows Foot Knot ([1], #1895).

The Stafford Knot (74) is an Overhand Knot. One of the best known of the heraldic knots [7][13][16], it has been the badge of the Earls of Stafford and Dukes of Buckingham since at least 1475 [13], but now occurs also as the badge of the Staffordshire Regiment and of the Scout County. It is used as a charge in the arms of Staffordshire County and several towns [6][15], in the arms of at least two railway companies that grew up in the area, and has recently (1960) been used in the arms of Baron Nelson of Stafford [8].

The Cavendish Knot [13] is a Figure Eight Knot (75); in the vertical form (76) it is used, as already stated, by the house of Savoy, in what is now southeast France, though the Prince of Savoy was for long the King of Italy.

The Harrington Knot (77) is what heralds call a fret, relatively common as a charge or crest. Usually it is thought to represent the lattice-work reinforcing at the back of the shield, and is often so drawn [7]. However, as the badge, it may represent a herring net [7], a pun on the Harrington name and their home seat at Haverington or Herrington. It has been used as a badge from the time of Edward III, mid 14th century [21]. Sometimes it is drawn as if in rope [13].

The Friars Knot (78) is a Girth Hitch ([1], #56) on another rope. It is another rare badge [13].

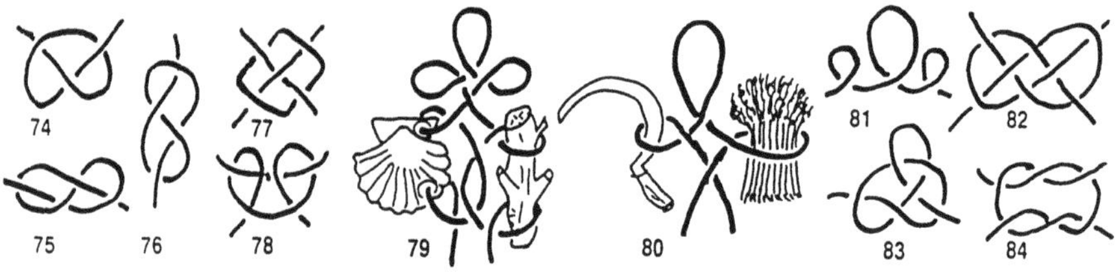

Fig. 7. More Knots as Badges

The badge of the Lords Dacre of Gilsland (79) is a complex one. The rope portion is sometimes called a Dacre Knot [24], but in the badge it unites an 'escallop' from the Dacre arms with a 'ragged staff', which may allude to the barony of Greystock [7]. The badge is said to have been used by Thomas, second Lord Dacre, at the Battle of Flodden Field in 1513 [21].

The badge of the Lords Hastings and Hungerford (80) is similarly complex. The rope portion is called the Hungerford Knot; another version (81) consists of three simple loops [7][13]. In the badge, itself sometimes called a Hastings Knot, it unites the sickle of the Hungerfords with the garb (wheatsheaf) of the Peverells, commemorating the marriage of the two families in the early 15th century [7].

The Wake or Ormonde Knot [13][16] is a Carrick Bend (82). According to tradition, it was the badge of Hereward the Wake, leader of the Saxon resistance around Ely to the Norman invasion of William the Conqueror in the late 11th century [7]. It is now used as a crest by the Wake family since at least 1621 [8], and also by the Isle of Ely County Council. Fairbairn [11] shows it as a crest in the vertical form.

The effigy on the tomb in Westminster Abbey of Anne of Bohemia (1366–1394), queen consort of Richard II of England, shows a curious knot (83), apparently tied in a leather strap [20][21].

The Bourchier Knot [7][13][16] is a Granny Knot (84). According to Palliser [21], who shows it as a Reef Knot, it is shown on a monument to Archbishop Bourchier, who died in 1486; Elvins [10], however, identified the knot as a Granny, both on the monument and on the arms, which showed the cords of the ecclesiastical hat united below it with a Granny Knot.

Knots in the Arms of Ladies

The theory is that women, not being fighters, can't use shields, helms, crests or mantling, but bear their arms on a 'lozenge' (Fig. 8), surmounted by a coronet if so entitled. Now that women take actively combatant roles in the armed forces, they may press for more equality with men in the use of arms, as in other things [2]. There have been a few anomalous instances in the past where women have been granted arms on a shield.

There are complicated systems of uniting the arms of married men and women, depending on rank and armorial history. In some heraldic traditions, sometimes some knots are used outside the lozenge, with particular significance, as in Fig. 8.

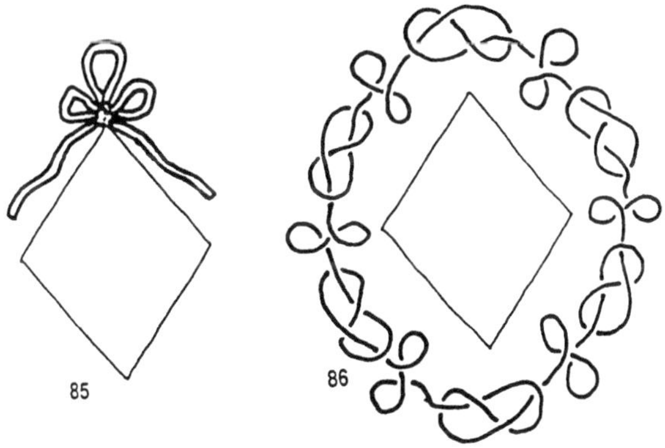

Fig. 8. Knots in the Arms of Ladies

Unmarried women in many traditions sometimes display a blue ribbon above the lozenge bearing their, or their father's, arms. This is tied in a 'true lover's knot'. This knot is usually shown as three loops united at their base, with the ends showing (85), with no possibility of distinguishing the knot;

sometimes it seems likely that the Pretzel Knot (40) is intended. Fox-Davies [16] has pointed out the illogic of not allowing a married woman to display a lover's knot and suggested some legitimate ways of arranging to do so. There are other traditions in which married women surround their arms with a cord tied in true lover's knots; if she becomes widowed, the knots are removed.

In some traditions, widows have encircled their arms with the 'cordelière', a cord tied in alternate Figure Eight Knots and twin turns (86). This practice upsets some heraldic experts [16] on the grounds that this cord is really part of the insignia of the old French Order of the Cordelière, founded by Anne of Bretagne, widow of Charles VIII, in 1498 for widows of noble birth; it should not, they say, be used by those with no connection with that order.

Knots in the Regalia of Knightly Orders

From the 14th century onwards, a number of Orders of Knighthood have been founded as awards of honour and to promote chivalrous ideals. The members wear special robes and display special insignia, some of which may accompany their coats of arms. Some of these insignia have knots incorporated in their designs; some of these knots are shown in Fig. 9 below.

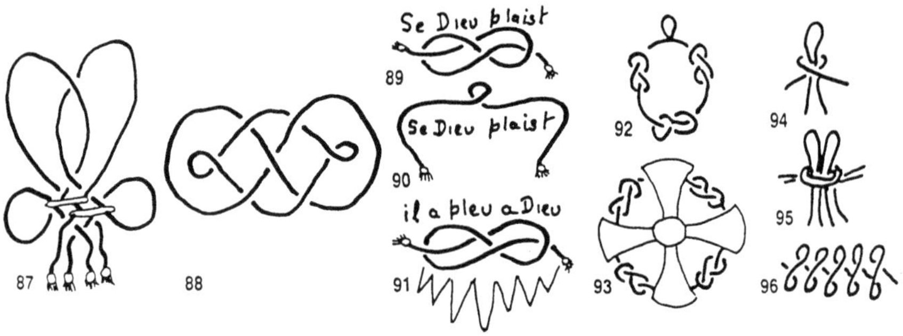

Fig. 9. Knots in the Regalia of the Knightly Orders

Perhaps the oldest such order is the English Order of the Garter, formed in 1348. Members of the Order encircle their shields with the Garter (*Honi soit qui mal y pense*) which is buckled and has the end half-hitched below the buckle. The 'collar' of the Order is a gold chain some of whose links [17] are 'double knots' (87). This is essentially the same knot as the 'entrelacs' (Fig. 3, 39). A similar knot is found in the collar of the British Order of St Patrick; I have not been able to discover its exact structure, except that it has two rather than four ends. The photographs I have seen [17] look different from some of the drawings [19]. I have also not been able to distinguish the structure in the knots formed between 'cables' entwined with sealions and a crown in the collar of the Grand Cross of the Order of the British Empire [17] that was founded in 1917.

The collar of the English Order of the Bath [19], founded in 1399, has some links of 'knots' (88).

One of the most interesting of the knightly orders to the knot tier is the Company of the Holy Spirit of Right Desire, commonly called the Company of the Knot [5]. It existed on mainland Sicily, that is Naples, from 1352 to 1362. Companions wore at all times a Figure Eight Knot (89) tied in about 50 cm of plain white silk cord, sometimes heavily bejewelled, and accompanied with the words *Se dieu plaist* ('if God pleases'). When certain feats of arms and piety were accomplished, the knot could be untied (90). It could then be retied and given ornamentation and a new motto (*Il a pleu a dieu*, 'It has pleased God') (91) if the knight visited the Holy Sepulchre at Jerusalem. A representation of the knot badge, probably in enamel, was also on the pommel of the knight's sword.

The insignia of several other Orders incorporated Figure Eight Knots. The Order of the Collar, formed in Savoy in 1364 and renamed the Order of the Annunciation of Our Lady in 1518 [5] incorporated the knot in several ways at different times in badges hanging from the collar (92). The collar of the Order of the Sword in Cyprus had Figure Eight Knots alternating with the letters R and S.

The Order of the Crown of Italy [17] had the knot between the arms of the cross of the main medallion (93), in much the same way as the knots on the cross of the Netherlands Order of the Lion [17] which were intended to represent the letter W (Fig. 1, 1).

Finally, the French Order of St Michael the Archangel had several designs for its collar between 1469 and 1790 [5]. Some of these incorporated apparent cordage in gold, showing from time to time various knots (94, 95, 96).

References

1. C. W. Ashley, *The Ashley Book of Knots* (Doubleday, New York, 1944).
2. H. Bedingfeld & P. Gwynn-Jones, *Heraldry* (Bison Books, London, 1993).
3. W. Berry, *Encyclopaedia Heraldica* (Sherwood Gilbert, London, 1859).
4. C. K. Bolton, *Bolton's American Armory* (Genealogical Publishing, Baltimore, 1969).
5. D. J. D. Boulton, *The Knights of the Crown* (Boydell Press, Suffolk, 1987).
6. G. Briggs, *Civic and Corporate Heraldry* (Heraldry Today, London, 1971).
7. J. P. Brooke-Little, *Boutell's Heraldry* (Warne, London, 1983).

8. *Debrett's Peerage and Baronetage* (Macmillan, London, 1990).
9. C. N. Elvin, *A Dictionary of Heraldry* (Heraldry Today, London, 1969).
10. M. T. Elvins, *Cardinals and Heraldry* (Buckland, London, 1988).
11. *Fairbairn's Book of Crests* (T. C. & E. C. Jack, London, 1905).
12. J. Franklyn & J. Tanner, *An Encyclopaedic Dictionary of Heraldry* (Pergamon, Oxford, 1970).
13. S. Friar, *A New Dictionary of Heraldry* (A. & C. Black, London, 1987).
14. S. Friar & J. Ferguson, *Basic Heraldry* (Norton, New York, 1993).
15. A. C. Fox-Davies, *The Book of Public Arms* (T. C. & E. C. Jack, London, 1915).
16. A. C. Fox-Davies, *A Complete Guide to Heraldry* (Orbis, London, 1985).
17. A. Jocelyn, *Awards of Honour* (A. & C. Black, London, 1956).
18. C. Kephart, *The Origin of Heraldry in Europe* (Heraldic Books, Baltimore, 1964).
19. J. H. Lawrence-Archer, *The Orders of Chivalry* (W. H. Allen, London, 1887).
20. M. H. Pakula, *Heraldry and Armour of the Middle Ages* (Barnes, New York, 1972).
21. B. Palliser, *Historic Devices, Badges and War-Cries* (Sampson Low, London, 1870).
22. M. Pastoureau, *Traité d'Héraldique* (Picard, Paris, 1979).
23. A. G. Puttock, *A Dictionary of Heraldry* (Gifford, London, 1970).
24. J-B. Rietstap (with H. V. Rolland), *L'Armorial Général* (Vol. 1, Gouda & Van Goor, Zonen, 1884 various publishers to Supplement Vol 8, Heraldry Today, London, 1971).
25. C. A. von Volborth, *Heraldry of the World* (Blandford, London, 1973).
26. C. A. von Volborth, *The Art of Heraldry* (Tiger Books International, London, 1987).

CHAPTER 18

ON THE TRUE LOVE KNOT

Pieter van de Griend

Eljest äro ju knopar övervägande av praktiskt kynne; här er en som representant för poesien. *

Thorild Wullf in a letter to
Hjalmar Öhrvall (ca. 1912)

Introduction

Excepting perhaps the Gordian Knot, there is no other knot name which is so widely known as the elusive True Love Knot. The simple knot-tying act has been one of all peoples and all times. Therefore, in general, it is seldom known exactly when or how a name for a knot emerges. Symbolically a knot implies something 'bound', and hence the action has become at times a spell invoking hindrance or impedance of the actions of other things or people. Conversely the act of loosening a knot has been thought to bring about the removal of the impediment it caused. The belief that magic knots can win or retain a lover is widespread and dates back to extremely ancient times [13].

In China, references from literary works of two post-Han Dynasty States speak of a True Love Knot. Both refer to a 'true lover's knot', believed to have been tied in an endlessly repeating pattern, symbolizing love and affection. The first ruler of the early sixth century state of Liang, Wu Ti, mentions such a knot in a poem [7] about his beloved:

*[The oldest of knots generally have a practical nature; here is one representing poetry.]

I dreamed the silk cords at our waists
Were bound together in a true Lover's knot.

Towards the end of the sixth century a True Love Knot appeared in the official Sui Dynastic history [7], mentioned in a record of a love affair between Yang Ti, son of a Sui ruler, and Hsuan Hua, a concubine of the ruler.

In the 1880's the Mexicans of the Lower Rio Grande had firm faith in the power of a lock of hair, with knots tied in it, to retain a maiden's affection [5]. This comes very close to a practise which Publius Vergilius Maro (70–19 B.C.), known as Virgil in English, describes from his days during the Roman Empire. It is obvious from these examples that knots were already replete with symbolic connotations centuries ago, and certainly not confined to any specific people.

The most-asked question concerning the True Love Knot is about its identity—*What form does it take?* Much has been said on this topic. As the knot's name stands for a symbolic concept, no agreement on any answer can ever exist. There is a demonstrably large number of different knots which somewhere, at one time or other, have been called True Love Knot. Moreover, as Hjalmar Öhrvall noted, the True Love Knot covers several classes of interhuman relationships [36]. In this chapter I shall be concerned with aspects of the emergence of the True Love Knot *phenomenon* . There are many names under which it occurs. In the following, unless quoting, I shall adhere to True Love Knot. Geographically I restrict this study to Europe, where, in a crude chronological ordering, there seem to have been: (i) *Love Knots*; (ii) *Friendship Knots*; and (iii) *Betrothal Knots*.

Classical Love Knots

The earliest traceable development of the knot as a love symbol dates to classical times. Virgil, the great classical writer of Latin poetry, described the efforts of a lovesick Roman maiden to regain the love of her love Daphnis [36]. She does so by means of an image of wax, a coloured woollen cord, the mystic number three, charmed knots and a spell, thus:

Around his waxen image first I wind
Three woolen fillets, of three colours join'd;
Thrice bind about his thrice-devoted head;
which around the sacred altar thrice is led.
Unequal numbers pleasure the gods.—My charms,
Restore my lovely Daphnis to my longing arms.
Knit with three knots the fillets; knit 'em straight;
And say, 'These knots to love I consecrate'

This is from Dryden's translation of Virgil's Eighth Eclogue. It is not more than a literary exercise, but it reflects beliefs held by the common people in Virgil's days [10].

A famous knot in the Greek and Roman culture was the Hercules Knot, also called *nodus Herculis*, *herculaneus* or *herculanus* and *hàmma herakleotikòn*. The origin of this name is not entirely clear and neither is the actual knot. Edmund Saglio and Paul Wolters point out that the Hercules Knot was an everyday (utility) knot [25], [34]. People have chosen to make this Hercules Knot something quite secret and at a certain stage, for some of the Mediterranean peoples, it was identified with the Gordian Knot [36].

Fig. 1. A Vestal statue

Caius Plinius Secundus (A.D. 23–79), in his *Natural History* **5** Book XVIII, he describes (in Chap. 17) the healing powers of a Hercules knot, thus:

> *To thrust an iron nail into the spot where a person's head lay at the moment he was seized with a fit of epilepsy, is said to have the effect of curing him of that disease. For pains in the kidneys, loins, or bladder, it is considered highly soothing to void the urine lying on the face at full length in a reclining bath. It is quite surprising how much more speedily wounds will heal if they are bound up and tied with a Hercules knot: indeed, it is said, that if the girdle which we wear every day is tied with a knot of this description, it will be productive of certain beneficial effects, Hercules having been the first to discover the fact.*

It seems that this belief led to Hercules being recognised as a god who could prevent all illnesses and other evil. Later this caused the emergence of names such as *Nodus Devinctus* for this knot [29]. Traditionally Roman brides tied their belts with the Hercules Knot, symbolising the hope that the couple may be as successful in rearing children as Hercules, who had seventy offspring [12].

On statues from the Vestal Temple we can see how the Hercules Knot looked. Fig.1 clearly shows that the belt harbours a Reef Knot [22]. Another source is provided by Macrobius *Saturnalia*, Book I, chapter xix, sections 16–18. In his description of the caduceus he states that it represents the union of a male and a female serpent as an offering to Mercury, and that they are united by a Herculean Knot (Fig.2) [8]:

> *Another clear proof that it is the sun that we worship under the name of Mercury is the caduceus, which the Egyptians have designed as the sacred staff of Mercury. It shows a pair of serpents, male and female, intertwined; the middle parts of the serpents' coils are joined together as in a knot, called the knot of Hercules; their upper parts are bent into a circle and complete the circle as they meet in a kiss; below the knot their tails rejoin the staff at the point at which it is held, and at that point appear the wings with which they are provided.*

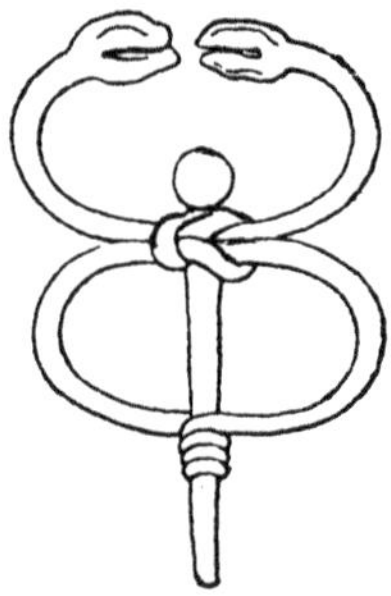

Fig. 2. The Caduceus

Athenogoras says that the wand of Mercury is a symbol of the union of Jupiter and Rhea, whom Jupiter disguised as a dragon, bound to him in the form of a Herculean Knot [13].

As the Roman Empire initially pursued an expansionist policy and came to occupy parts of Europe, including the North, we may assume that they propagated their Hercules-Love Knot culture during that process.

Friendship Knots on Arms and Orders

Early medieval sources for Friendship Knots are given by Gertrude Smola and Ulrike Zischka [27], [35]. They point out that 13th century Southern European religious and secular knighthood orders incorporated the knot as a symbol of friendship in their social identity. Hence from the 13th century the knot reaches us in a new symbolic form. Initially it represented the solemn oaths taken by the members of various religious orders. Franciscus of Assisi (1181–1228) wears a three-knot cingulum (girdle) on a painting from 1222 by an unknown master. The Triple Overhand Knots tied in the end of the cingulum of the Franciscans point to the oaths of poverty, chastity and obedience. To these three Franciscan oath knots the 'Poor Clares' nun-order adds another, which symbolizes their fourth oath: *stabilitas loci*. Cingula with two three-fold knots adorning some doorbeams of Assisian houses in the Central Italian province of Umbria, indicate their inhabitants' relation to the Franciscan Order.

Fig. 3. An Assisian doorbeam decoration

In the 15th century connections between the Franciscan cingulum and the knotted bands on the coats of arms of royal ladies emerge. Around 1750 Hippolytus Helyot made a study of the history of religious and secular orders. In his description of Anne de Bretagne he writes that she founded an order to honour the cords with which our Lord (Jesus) had been tied during his sufferings. Anne de Bretagne had chosen these (Granny) knotted cords, which also return on the *Officium Beatae Mariae Virginis*—a manuscript dedicated to her—as symbols for her worshipping of Franciscus of Assisi. Hippolytus Helyot notes an influence by Duke Francois II de Bretagne, who, also in honour of St. Franciscus, had used these knots on his coats of arms too [20].

The 14th century provides us with many evidences of application of the knot as a friendship-symbol. For example, this interpretation is supported by a statement in the *Index Rerum Notabilium* by Picinellus, where the knot is said to be a symbol of lasting friendship:

Nodus est symbolum amicitiae viciosorum.

This usage also explains the gold knots worn by the members of the secular knighthood Order of Naples, which was founded in 1347 by Queen Johanna of Naples and only existed for a few years. And in 1352 an *Order of the Holy Ghost in Naples* was founded by Ludwig of Tarent, King of Sicily, cousin and spouse to Johanna of Naples: the members of this order wore a knot on their chest, which in the description by Ferdinand von Biedenfeld [2] was called *Zweifelknoten*. To explain this, we shall quote from the Grimm brothers' dictionary [18], which is based on Fischer's *Schwabischen Worterbuch* of 1604:

> ... *People who want to tie and keep their good friend, will send him it* [Zweifelknoten] *in gold or silk love band.*

These knots are signs of mutual friendship, and as such they are mentioned in the Order's constitution. They not only characterised the members, and reminded them symbolically of their Order's constitution, but they were also the motif which represented a group of 300 knights. Ferdinand von Biedenfeld and D. Boulton describe one of the knot tying customs of these Orders [2], [4], thus:

> *If a knight was wounded in battle, or if he had besieged and wounded his foe, then he had to, from then on, bear his untied knot until he had visited the Holy Sepulchre.*

Of such an event, the inscription of an epitaph in Naples witnesses [35]:

> *Member of the order, Colutius Bozzutus, was victorious on the battlefield and untied his knot before travelling to Jerusalem, for to retie it and die there in 1370.*

During the 14th century the use of a knot as friendship-symbol in Central Europe was widely practised. Numerous other secular knighthood orders used it. Knots were used in the chain of the Order of the Garter, which was founded in 1350 by King Edward of England. *The Order of St. Anthony*, which was founded in 1382 at the court of Albrecht von Bayern during a planned crusade against the Turks, has a knot in its heraldic arms. A knot with the same meaning adorns the clothes of the members of the Company of the Knot of St. Esprits, founded by Louis d'Anjou [4]. A Figure of Eight Knot occurs in the chain of the Dell'Annuiziator order, which was founded by Amadeus VI of Savoy in 1360. Around the Savoy insignia, which is on display in the church of Moudon in Switzerland, this Order's chain is wrapped. Amadeus of Savoy also ordered the clothes of his servants and the covers of their horses to be decorated with this knot pattern. This tradition caught on, and characteristics stemming from the Order membership entered family coats of arms. The badges of the English families Stafford, Bouchier, Wake and Ormonde display knots. Although they are not always regarded now as Friendship Knots, their usage in family badges continues. Öhrvall mentions a painting belonging to his family and dating from 1813 [36]. In its background one can see an altar with the inscription *'Sacred be friendship'*. By the side of the altar are figures of a man and a woman. Over them float two doves, which hold a ribbon with a Figure of Eight Knot in their bills. Under it all can be read:

The further we part, the tighter the Friendship Knot is drawn.

How far the expression *The Friendship Knot* in this context equates to the Figure of Eight Knot is not really known. Without doubt, at many times in the past the Figure of Eight Knot has played an important role as a symbol. The knot's property to be secure and yet not be able to lock, so that it can be opened easily, would in many cases have given it its meaning as a symbol.

The connection between Friendship and Love Knots is natural, and before long it became fashionable among noble ladies to have Love Knots in their insignia. According to M. Sebastian Jacob Jungendres, refering to an otherwise anonymous Mr. Trier, the tradition was originated by Anne de Bretagne [23]. The knots, or the lack of them, in the cords on the coats of arms of ladies are indicative of their status, i.e (Love) Knots for married women and none for those who were widowed. This phenomenon had an interesting development. The coat of arms of Louisa de la Tour (ca. 1472), for instance, is divided into her maiden and marital badge surrounded by a knotted cord. Although it displays several statuses, the knot can be considered a Love Knot. When her husband died, Louisa de la Tour got this knotted decoration sewn onto the ceremonial costumes when she founded the Carmelite church in Chalons (France), who kept her knot in the heraldic arms of the church choir. In this way knotted cords became motives in the badges of the clergy. However, this application can hardly be explained as one of a Love Knot, because the number of knots indicates the placing of the bearer in the church's hierarchy.

Betrothal Knots

In our account so far, we have proceeded to the situation of around the 1500s in Southern Europe. We will now turn to Northern Europe and see how parts of its pagan culture reacted to the christianisation process.

At an early stage of its history, the Roman Catholic church supplanted the role of the Hercules Knot as a symbol of love by wedding- and engagement rings. Yet, as we saw in the foregoing section, the symbolic power of knots did not become extinct. In Northern Europe a knot, which (presumably independently) has played an important role, is the Betrothal Knot. Betrothal is the promise to marry, and would be called engagement nowadays. How old this ritual is, can best be seen from the fact that the name was not only used over the entire region we call Scandinavia, but also in Britain where it survived as a remnant from the Danish invasion times until way past the 16th century [30]. In the sixteenth century *trulofa* was still a customary expression for 'to become engaged', which has become *true-love* via some popular etymology. *The English Dictionary* assigns the earliest occurrence of the word *true-love* to the 9th century Cynewulf:

> *Wæs seo treow lufu, hat æt heortum.*

It shows that *true-love*, the concept and the word, is ancient; but not necessarily related to knots some 1000 years ago.

In the Northern parts of Europe attempts by the Roman Catholic church to introduce their engagement customs met with protest. This is described by Troels-Lund [30], thus:

> *In Scandinavia attempts by the medieval Roman Catholic church to introduce the exchange of engagement rings, an originally Mediterranean tradition, met with two kinds of resistance. First of all, we Scandinavians were not willing to let engagement rings take the place of the engagement dowries, even in our churches. Secondly, we had become accustomed to symbolising the relationship between man and woman by different means. Among all gothogermanic peoples, faithfullness in love had been symbolised since times immemorial by either a knot which was tied by both parties (from which came the English* true-love knot*), or by the breaking of a gold coin into two halves after which each party kept one half.*

The first of these traditions is acknowledged by John Brand in *Observations on the popular antiquities of Great Britain* [6]:

> *A knot among the ancient northern nations seems to have been the symbol of love, faith and friendship, pointing out the indissoluble tie of affection and duty. Thus the ancient Runic inscriptions, as we gather from Hicke's Thesaurus, are in the form of a knot. Hence among the northern English and Scots, who still retain, in a great measure, the language and manners of the ancient Danes, that curious kind of a knot, a mutual present between lover and his mistress, which, being considered as the emblem of plighted fidelity, is therefore called a true-love knot: a name which is not derived, as one would naturally suppose it to be, from the words 'true' and*

> *'love', but formed from the Danish verb* Trulofa, fidem do, *I plight my troth, or faith. Thus we read in the Islandic Gospels, the following passage in the first chapter of St. Matthew, which confirms, beyond a doubt, the sense here given*—til einrar Meyar er trulofad var einum Manne &c.; *i.e to a virgin espoused, that is, who was promised or had engaged herself to a man &c. Hence, evidently, the bride favors or the top-knots at marriages, which have been considered as emblems of the ties of duty and affection between the bride and her spouse have been derived. [7, pp. 108–109]*

These notes are of great interest, but give no indication about the Betrothal Knot itself. The (Nordic) literature generally lacks information about its appearance, but we are afforded indications of knots which people believed to be representative. The betrothal rings eventually became acceptable and Love Knots form the crowns in 18 carat gold rings, which emerge in the 14-15th century in Britain and Lubeck, Germany [35].

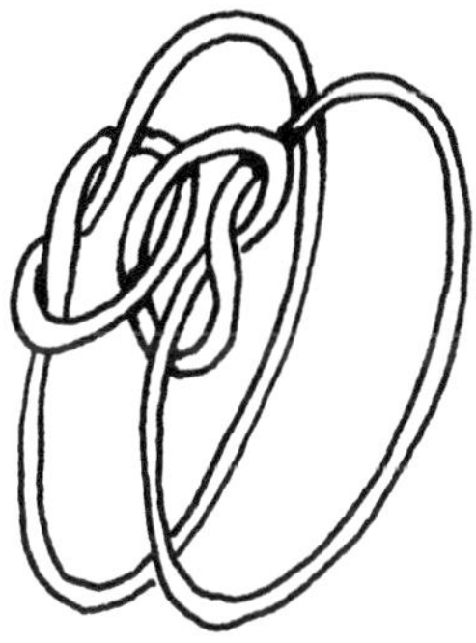

Fig. 4. A knotted betrothal ring

Such expensive traditions existed also in Great Britain [21]: In the inventory of the effects of Henry Howard, K.G., Earl of Northampton, 1614, is recorded:

> *a golde ring sett with fifteene diamondes in a true lover's knotte, with the words "nec astu, nec ense".*

Not only did these rings eventually become more expensive, but also more intricate. The ancient matrimonial jimmal, or gimmal, was a kind of curiously made double ring. There were links within each other, and though generally double, they were, by further refinement, made triple, or even more complicated. Thus Robert Herrick in his *Hesperides* of 1648 can write [21]:

Thou sent'st to me a true love-knot, but I
Return a ring of jimmals, to imply
Thy love had one knot, mine a triple tye.

In this context *Ønskeknuten*, the Wish Knot, may be mentioned. In Scandinavia, about a century ago, a mild superstition among girls was to be found, regarding this. The following example is from Jutland, in Denmark [14]:

The Wish Knot: A magic knot which tells whether a wish will come true. A girl (A) lets another girl (B) pull up five long grass straws, and lets B hold them around the middle. Then A ties the straws, two by two at the bottom and two by two at the top. Finally, the two remaining free ends, one at the top and one at the bottom, are tied together. Fortune will favour A if all straws form a single ring; if two rings are formed, that also is a good omen; three rings are not to be considered hopeful, but that is better than when all rings are separated: then all hope is lost.

According to Öhrvall this custom was also to be found in Sweden, and may still have been found there locally during the first decades of the 20th century [36]. It is remarkable that nothing is mentioned in the event the grass ring should be formed into a Trefoil Knot.

The True Love Knot in the English Literature

The allegorical 14–16th century Love Knots were restricted to an illustrious circle at medieval courts. On the other hand, poetry belongs to an entire people. Moreover, the strong expressive value of the symbolic imagery, relieving poets from saying anything concrete about the True Love Knot's appearance, unleashed an avalanche of True Love Knots in print. Love Knots, as such, began to be mentioned in the English literature in the 14th century. The earliest recorded instance of our phenomenon in the *New English Dictionary* is *Sir Gawaine and the Greene Knight* about 1350. In the 14th century Chaucer's monk in the *Canterbury Tales* had a curious pin of gold, with a *love knotte in the gretter ende* [9].

Thomas Whyatt, the poet at the court of Henry VIII, who reigned from 1509 to 1547, wrote about knots straining around his heart [35]:

The knott whych ffyrst my hart dyd strayn,
Whan that your servant I becam,
Doth bynde me styll for to Remayne
All waies your owne, as nowe I am ....

Such pains of heart were soon to be related to True Love Knots. The concluding stanzas to the fifteenth century ballad of *Fair Margaret and Sweet William* [9]:

Margaret was buried in the lower chancel,
Sweet William in the higher;
Out of her breast there sprung a rose;
and out of his a brier.

They grew up as high as the church top,
till they could grow no higher,
And then they grew in a true lover's knot,
Which made all people admire.

In Shakespeare's (1591) comedy *The Two Gentlemen of Verona* [act II, scene VII] the following dialogue is to be found between Lucetta and Julia who dresses up as youngster:

Lucetta: Why then, your ladyship must cut your hair.
Julia: No, girl; I'll knit it up in silken strings
With twenty odd-conceited true-love knots.

The year 1611 brought a madrigal in six strophes in Davidson's Poetical Rhapsody, entitled *The True Lover's Knot* [6]; and from the 1631 *Merry Devil of Edmonton* stems [6]:

With pardon, sir, that name is quite undon,
This true-love knot cancelles both maide and nun.

John Gay in *Shepherd's Week*, from *Six Pastorals drawn from the English rustic life* (1714), describes the manner of tying the True Lover's Knot by the following lines [15]:

As Lubberkin once slept beneath a tree,
I twitch'd his dangling garter from his knee;
He wist not when the hempen string I drew,
Now mine I quickly doff, of inkle blue;
Together fast I tie the garters twain,
And, while I knit the knot, repeat this strain—
Three times a true-love-knot I tie
Firm be the knot secure,
Firm may his love endure.
With my sharp heel I three times mark the ground,
And turn me thrice around, around, around.

This is a suitable place to quote an early reflection on the origin of the True Love Knot. It comes from Sir Thomas Brown (1602–1682), a man who held an opinion on many a subject. In his *Vulgar Errors* (Works II, p. 82) of 1646, he says the following [26],[33]:

> *The true lover's knot is very much magnified, and still retained in presents of Love among us; which though in all points it doth not make out, had perhaps its original from the* Nodus Herculanus, *or that which was called* Hercules *his knot, resembling the snaky complication in the cadaceus or rod of* Hermes; *and in which form the Zone or woolen girdle of the Bride was fastened, as* Turnebus *observeth in his* Adversaria.

To this Christopher Wren, Dean of Windsor and father to the architect of St. Paul's cathedral, adds the remark [33]

> *The true lover's knot is magnified for the moral signification not esily untyed; and for the naturall,—bycause itt is a knot both wayes, that is two knots in one.*

Whether Wren's note is an allusion to the complexity of love, or whether it is just mention of a mechanical property of a badly jammed utility knot, such as a Reef Knot, we cannot determine.

A comment from Turnebus, written in 1581 [32], supports Brown's opinion, thus:

> *He who wants to know how the Hercules knot looked like should see the Caducea on which the serpents are entwined to form a Hercules knot.*

He refers to the place in the works of Macrobius (from the end of the 4th and start of the 5th century A.D.) which we mentioned earlier.

St. Valentine's Influence

Midway through the 18th century Love Knots appeared in print in more than one way. No longer were they confined to lyrical words, but became embodied in drawings on Valentine cards. In his 1766 novel *The Vicar of Wakefield*, Oliver Goldsmith describes the following customs of his contemporaries [16]:

.... farmers, who tilled their own grounds, and were equal strangers to opulence and poverty ... wrought with cheerfulness on days of labour; but observed festivals as intervals of idleness and pleasure. They kept up Christmas carols, sent true-love-knots on Valentine morning, ate pancakes on Shrove-tide, showed their wit on the first of April, and religiously cracked nuts on Michaelmas eve.

Valentine's Day (February 14th) is the celebration day for the Roman martyr-priest St. Valentine, the patron saint for lovers. The date was chosen to be the one when people believed birds started mating. This custom was recorded in the English literature by Geoffrey Chaucer (ca. 1340–1400) for the first time as the *Parliament for Bryddes*. The original intention of the day was to give (expensive) presents to beloved ones. Halfway through the 17th century the presents became less expensive, and elaborate handpainted knots came to be used as decorations on the gifts. These knots were usually single-stranded, and had endless lines of words written along their lengths—words expressing sentiments of love. In Fig. 5 we show a picture of one of these knots, taken from *Witt's Recreation* (1641). These knots were predecessors of the knots to be printed on later Valentine gifts, such as letters, vases, jugs and cups. The hand-made Valentine objects came to display labyrinths, which were combined with knots to amplify the symbolic power of the Love Knot, and enhanced with pairs of doves and angels. The Valentine objects spread from Britain, and found fertile commercial breeding ground in America. In other countries the custom lapsed on religious grounds [28].

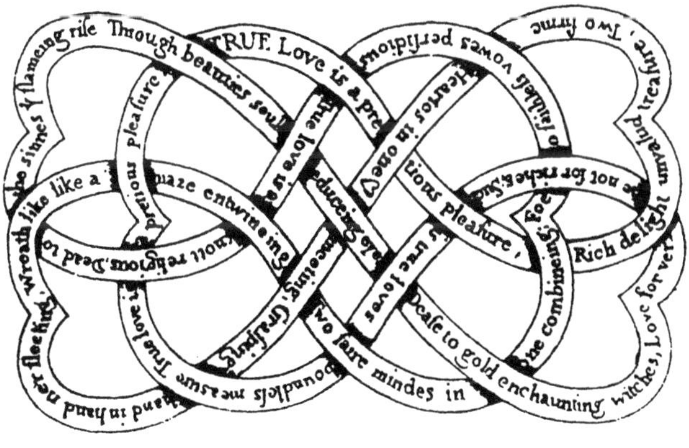

Fig. 5. Knot from *Witt's Recreation* (1641)

The True Love Knot at Sea

As knots were commonly associated with sailors, it is not remarkable that the True Love Knot phenomenon should be found in this pragmatic ropeworkers' world. The question about the True Love Knot's identity seems to have occasionally occupied them too, as a number of knots appear in their lore under that name. As a rule they are based on two interlocking Overhand Knots, but also knots with two loops are to be found [1,#798,#1038,#1414,#2301,#2421], [11, #149].

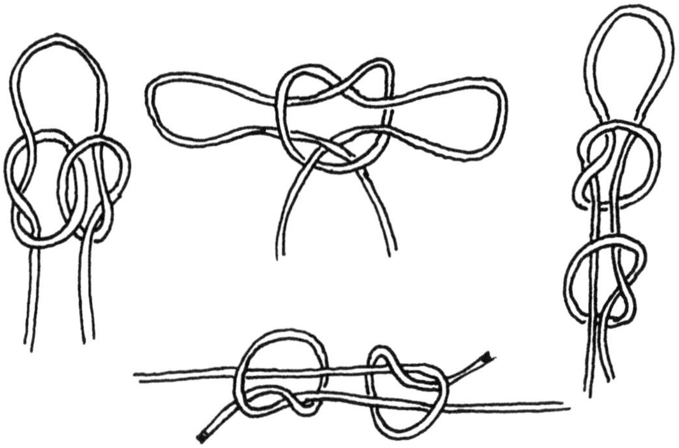

Fig. 6. Some True Love Knots of sailors

There is also a Masthead Knot called *Treble Love Knot* in the English knotting literature [36], but it is uncertain which structure is meant. There exists a quaint (non-Mariner) magazine from 1834 in which there is the following story about a knot, which may have given rise to the foregoing name [38].

> *There is a knot made by seamen, known by the name of* True Lover's Knot, *and it is difficult without inquiring into the uses to which it is applied, to understand in what manner it could have obtained this romantic name. On board ship, it is frequently necessary to raise water by means of a bucket or other vessel, and if this vessel should happen to be without a handle, it would require some ingenuity to find out the means of lowering it into the water: to gain this point, the seamen have invented the knot represented in the engraving. This being the purpose to which the knot is applied, we have now to discover in what manner it has acquired its name. To do this, let us suppose our sailor to be on shore, and paying a*

visit to his relations; his well-known skill is of course, put in requisition, and all the pitchers of his female friends which have lost their handles are brought to him, to be supplied with new ones; and, as a compliment for his exertions, the knot which his ingenuity has formed is called a true-lover's knot.

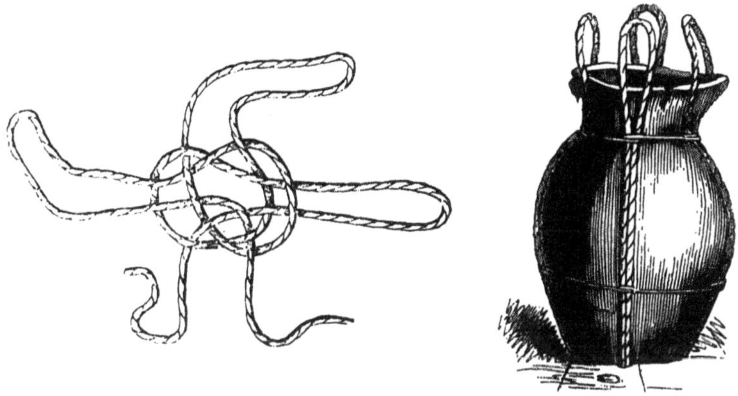

Fig. 7. The True-Lover's Knot; and a use for it

In [19] the story of the Pitcher Knot is continued. Knot lore also has it that such knots were used for hoisting cannonballs on board.

Finally the True Love Knot has had a special significance among seamen. Most of the knots under that name consist of a loop and two Overhand Knots, which can be either separated or pressed close together. In conformity with the symbolism implied by this characteristic of the knot, sailors were said to enclose such knots, tied in silk, with the Overhand Knots separated, in a letter to their sweetheart. If she returned the knot with the knots close together, he would know that she still loved him. If not, another interpretation would be in order. Hjalmar Öhrvall describes this sailor tradition, which was still being practiced around 1860 [36]. This custom also occurred on the coast of Newfoundland [1,#2420]. Remnants of it are reflected by knot names throughout the entire European maritime literature. In Dutch, for instance, there is the Single Fisherman Bend which is called *Twee Zoeteliefjes*, i.e two sweethearts.

Carrick Bends, Wale Knots and Turk's Heads

Since the middle of the 18th century relatively complicated knots turn up in the True Love phenomenon. I wish to mention two examples and take up part of a discussion initiated by Hjalmar Öhrvall.

The first example is a 5 part and 4 bight Turk's Head diagram, which appeared on a Valentine card dated 1741 and made by one Geo Sherwood (Fig.8). It came equipped with the little poem [31]:

There in Knot crosses in Store
But in True Love there Many more

Fig. 8. Geo Sherwood's Valentine motive

The second example occurs on an 1820 German *Neujahrsbrief* from Zittau [36].It shows the patterns which is illustrated in Fig. 9. It was a custom in several European countries when making a call on New Year's Day to leave one's card with a written message conveying compliments of the season. About the 1760's, mainly in Germany and Austria, New Year cards were available with a printed message often in verse and romantic in style. From these emanated the *Freundschaftskarten* , which resembled the Valentine cards of the next century. There are many beautiful examples of such cards by Danish and German artists [37].

It is remarkable that these relatively complicated knots come to have this functionality. Hjalmar Öhrvall conducted the first comprehensive study of the True Love Knot phenomenon and found several interesting results. In the Swedish language *Valknut* is a knot name. At the Swedish Academy's library there is a manuscript for a dictionary by Schultze (ca. 1755). The entry about the valknut reads [36]:

Quasi ut wald knut eller valkad knut, Hårdknut, nodus gordius, nodus adamantinus, inexplicabilis lagneus.
Angl: Wale-knot, adamantine ties.
Germ: Zweifels Knoten.

Fig. 9. Knot on a New Year's letter (1820)

Ihres, in his *Glossarium sviogothicum* 1769 has under Walknut the following, translated from [36]:

> *Walknut, hercules knot, consists of many intertwining bights; called* valknut *from vel which in Icelandic means cunning, superstition. The English call this knot the truelover knot, which has lead Hicke to conjecture that many of the knot decorations, which are chiseled onto our runestones can be considered symbols of the living's love and respect for the deceased.*

From these notes Öhrvall concluded that the Turk's Head Knot could have been used as the ancient Betrothal Knot. He based this on the fact that, to him, *Valknut* was a Turk's Head, yet the meticulous Vilhelm Linders in 1896 also spoke of a *Valknut* , but to him the word quite obviously meant a Carrick Bend [24]. Linder's Turk's Head is called *Valknop* . Öhrvall was aware of this discrepancy. Furthermore he also knew about John O'Gilvie's *Imperial Dictionary* (1833) in which is found an illustration of the Carrick Bend among the entries on the True Love Knot. In the English knotting literature the Carrick Bend structure has also been named True Love Knot [1, #2426]. To Öhrvall it seemed plausible that the Hercules Knot, which most likely was the Reef Knot, could have been made more decorative by interweaving the parts, so as to give a Carrick Bend. Further interweaving would lead to more complicated structures such as Turk's Heads.

Öhrvall set out trying to find evidence supporting his conjecture that the Turk's Head had been used as a Betrothal Knot in Sweden. After a literary search he found that the Swedish name had been used to denote a symbol of love and friendship. In a work of 1825 by Törneros one can read [36]:

On the loose sands of word friendship rests,
which stretches out of days and late times
although silent it speaks of faithfulness
our 'valknut-*bond' goes over mountain and valley.*

It is obvious that here is mention of the knot being used in a symbolic sense. Öhrvall also found that Sparre, in his 1847 novel *Standaret*, used the word *valknut* in the sense of Love Knot [36].

Two things are worth noting. First of all it is remarkable that the Swedes chose to use a customary knot name, i.e. *Valknut*, to actually identify a concept like the True Love Knot. I have been unable to trace anything similar in Denmark or elsewhere.

The second thing on which I wish to say a few words, are the relationships between the Carrick Bend, Wale Knot and Turk's Head. The first of the noteworthy links is that in Scandinavia the Carrick Bend has an ancient Icelandic connection. The Viking descendant fishermen of the Vestmanna archipelago, which is situated off the southwest coast of Iceland, traditionally have made extensive use of the Carrick Bend to connect the segments of their gillnets [17]. In the Scandinavian knotting literature the Carrick Bend goes by the name *Grønland Stik*. This is due to its frequent use in the Greenlandic gillnetting fishery since 1500 onwards.

The second link is that the Carrick Bend relates to the Turk's Head and vice versa. The Carrick Bend forms a basis for knots like the Doubled Diamond Knot and the Footrope Knot [1, #695, #697]. These knots strongly resemble the four-stranded Wale Knot, which in turn conceals a Turk's Head pattern. The Wale Knot name has been applied indiscriminately to many of the Wall and Crown derivatives [1, #845].

Furthermore, there are distinct peoples, widely separated geographically and in time, such as the Bushoong and Tchokwe people in Central Africa, who draw fascinatingly complex planar graphs in the desert sand which more than superficially resemble braid diagrams (Fig. 10a). Closer to Europe the Celts have produced similar curves (Fig. 10b); admittedly not in desert sand.

Fig. 10. Tchokwe and Celtic Motives

The Polar Inuits have known the Diamond Knot, and thus implicitly the Carrick Bend. This may imply that the Viking settlers, who colonised Greenland around year A.D. 1000, knew it too. In 1910 Hjalmar Öhrvall studied some Viking knots excavated at Mossle Jössagård in Sweden. He said one of them was a 3-part Turk's Head [36]. All these facts together seem to indicate that such complex knots as Turk's Heads were around at a very early date, and similar patterns were known to many distinct peoples. It is therefore quite well possible that one (or more) of these complex patterns may have been used for a Betrothal Knot in Europe, perhaps even before any Roman entered the scene. Who can tell?

Conclusions

It appears that there have been numerous distinct knots which were called and used as a True Love Knot. It is pointless to inquire which one of them was the real True Love Knot. As we have seen, the Reef Knot probably qualifies as the original one for Southern Europe. At a later date other (more complicated) knotted structures were used to symbolise the True Love concept. During the medieval revival of the knot as mystic love symbol, the True Love Knot found fertile grounds in the hands of poets. Finally the commercialisation of the True Love Knot phenomenon in the Valentine tradition seems to have been stronger in the Anglo-Saxon world than elsewhere. As a result of this, more knot remnants are to be found in its literature and art. Hence one gets the (incorrect) impression that during the past one thousand years True Love Knots were more popular in the English-speaking world and in Northern Europe generally than elsewhere.

This paper has greatly benefitted from several days' quite intense discussion with Dr. Charles Warner, in the Australian summer heat of Christmas 1994. He is, however, not to be held responsible for its final form.

References

1. C. W. Ashley, *The Ashley Book of Knots* (Doubleday, New York, 1944).
2. F. von Biedenfeld, *Geschichte über Verfassung aller geistlichen und weltlichen, erloschenen und blühenden Ritterorden,* (Weimar, 1841).
3. J. Bostock and H.T Riley, *The Natural History of Pliny* (London 1856) 299.
4. D. J. D. Boulton, *The Knights of the Crown* (Boydale Press, Suffolk 1987) 222–226, 258–261, 427–431.
5. J. G. Bourke, The Medicine Men of the Apache, *Ninth Annual Report of the Bureau of Ethnology* (Washington, 1892) 550–580.
6. J. Brand, *Observations on the popular antiquities of Great Britain* **II** (London, 1854) 108–109.
7. L. Chen, *Chinese Knotting* (Echo, Taipe, 1982).
8. P. V. Davies, *Macrobius: The Saturnalia* (Columbia University Press, New York, 1969).
9. C. L. Day, Knots and Knotlore—A Study in Primitive Beliefs and Superstitions, *Western Folklore* **9** (1950) 229–256.
10. C. L. Day, *Quipus and Witches' Knots* (Kansas University Press, Lawrence, 1967).
11. C. L. Day, *The Art of Knotting and Splicing* (Naval Institute, Annapolis, 1955).
12. W. J. Dilling, Girdles: their origin and development, particularly with regard to their use as charms in medicine, marriage and midwifery, *Caledonian Medical Journal* **IX** (Glasgow, 1913–1914) 337–357.
13. W. J. Dilling, Knots, *Hasting's Encyclopedia of Religion and Ethics* **8** (1915) 747–751.
14. H. F. Feilberg, *Ordbog over Jysk almuesmål* (Copenhagen, 1904–1910).
15. J. Gay, *The Poetical Works of John Gay* **1** (Lawrence and Bullen, London 1893) lines 109–118.
16. O. Goldsmith, *The Vicar of Wakefield* (Folio Society, London 1971) 28.
17. P. A. v.d. Griend, 'The Faeroese Way', *Knotting Matters* **23** (1988) 18–19.

18. J. and W. Grimm, *Deutches Wörterbuch* (Leipzig, 1914).
19. P. N. Hasluck, *Knotting and Splicing Ropes and Cordage* (London, 1904) 81–83.
20. H. Helyot, *Ausführlige Geschichte aller geistlichen und weltlichen Kloster und Ritterorden für beiderlei Geschlecht* (Leipzig, 1753–56).
21. W. Jones, *Finger-ring Lore* (Chatto & Windus, London 1877) 312, 414.
22. H. Jordan, *Der Tempel der Vesta und das Haus der Vestalinnen* (Berlin, 1886).
23. M. S. J. Jugendres, *Einleitung zur Heraldik* (Nürnberg, 1729) 139–145.
24. V. Linder, *Lärobok i Sjømanskap* (Stockholm, 1896).
25. E. Saglio, Nodus, *Dictionaire des Antiquités Grecques et Romaines* **4** (Paris, 1907).
26. C. Sayle (ed), *The works of Sir Thomas Browne* (London, 1904) 266.
27. G. Smola, Das Grabgewand Herzog Ernst des Eisernen, *Festschrift 150 Jahre Joanneum* (Graz, 1969) 335–355.
28. F. Staff, *The Valentine and its Origins* (Lutterworth, London, 1969) 7–39 .
29. J. C. Svabo, *Dictionarium Færoensis* (Copenhagen, 1966).
30. Troels-Lund, *Dagligt Liv i Norden i det Sekstende Aarhundrede* **IX** (Copenhagen, 1910).
31. J. C. Turner, 'Will the real True Lover's Knot stand up, please?' *Knotting Matters* (**34** (1991) 4–5.
32. Turnebus, *Adversariorum* **III** (Basiliae, 1581) 588.
33. S. Wilkin (edr.), *The Works of Sir Thomas Brown* (London, 1835) 165.
34. P. Wolters, *Zu Griechischen Agonen* (Museums der Universität Würzburg, 1901).
35. U. Zischka, *Zür sakaralen und profanen Anwendung des Knotenmotivs als magischer Mittel, Symbol oder Dekor* (Tuduv, München, 1977) 55–74.
36. H. A. Öhrvall, *Um Knutar* (Bonniers, Stockholm, 1916).
37. *The Oxford Companion to the Decorative Arts* (edr. Harold Osborne, Oxford University Press, 1975).
38. *The Saturday Magazine* **4** No. 126 (Guernsey, June 1834) 236–237.

ABOUT THE AUTHORS

Twelve authors have contributed to the making of this book. Their nationalities, ages, interests and careers range widely, and this is clearly reflected in the variety of the topics and treatments of the essays which they have submitted.

The authors were asked to supply a few paragraphs about their lives, to inform on their backgrounds and qualifications, and to indicate how they became involved with the subject of knots. Although a very small sample of people is involved, the following collection of their replies constitutes a written contribution on the sociology of the current knotting world. At the very least, it supplies evidence that such a world exists!

Robert G. Bednarik

Robert Bednarik is an Australian, born in Vienna in 1944. He has produced some 400 refereed scientific publications in nine languages, mostly in cognitive archaeology. He edits three archaeological journals and a series of monographs. As Permanent Convener and Editor of the International Federation of Rock Art Organisations he plays a leading role in the world-wide study of palaeoart, and he has conducted extensive field work in most continents. In 1983, Robert Bednarik founded the Australian Rock Art Research Association, of which he remains the Editor and Secretary, and which has become the flagship of the discipline of global palaeoart studies. Since 1990, he has also been the Editor of the Archaeological and Anthropological Society of Victoria. He is an independent publisher and scientific writer.

Geoffrey Budworth

Geoffrey Budworth was born on 22 February 1937 in Oakham, a rural market town of Rutland, England's smallest county; but from the age of seven, he grew up on the South coast, in Bournemouth, and there acquired a love of swimming and boats (with both of which he remains besotted).

He completed 25 years with the Metropolitan Police, doing undercover work with the London West End's vice squad, nine years afloat in Thames Division (London's river police), and several stints lecturing criminal law and social studies at the renowned Hendon police training school.

Ex-Inspector Budworth then began another career, managing public sector swimming pools and leisure centres; and for the past six years he has been a full-time lecturer of sport and recreation management in a large South London college of further education.

His fascination for knot tying was triggered when, as an 11-year-old Sea Scout, he asked his troop leader about a bellrope he'd seen—and was told: 'Don't bother with that, son. It's too complicated for you.' In 1982 he co-founded the International Guild of Knot Tyers and for seven years edited their publications, becoming the knotting writer that other writers consult. He promoted the systematic identification of knotted clues discovered at scenes of crime—and the subsequent presentation of evidence about them in courts—now widely practised by criminal investigators and forensic scientists.

Geoffrey has written several books (including two on knot tying). The latest, nearing completion, is an authorised 200-year history of Thames Division.

He is married, with two grown-up daughters, and lives in Tonbridge, Kent. His current pastimes include cruising tideways in his 12 foot inflatable sailing dinghy, riding unicycles and juggling. He is also reading biology as a third-year undergraduate student with the Open University.

John Roger Carter

Roger was born in 1932, in the industrial Midlands of England—about as far from the sea as one can get in Britain. He joined the Royal Navy as a Boy Seaman, at the age of fifteen, and served for twelve-and-a-half years in all manner of ships, from a battleship to a small wooden minesweeper.

Leaving the Navy, he spent some time ashore (making nylon), before becoming Skipper of an esturial tanker working on the River Severn. After a few years of that, he emigrated to New Zealand, to join the Pilot Staff of the Wellington Harbour Board. His next twenty years were spent in pilot boats and tugs, as well as on other floating plant; his jobs included nautical surveying and Berthing Master. His final ten years before retirement were spent as supervisor of the Boat Harbours and Marinas in Wellington Harbour.

Roger is a Member of the International Guild of Knot Tyers, and President of its New Zealand Chapter. His knotting interests stem from his nautical background, and extend from sailor's work; he is expert in both practical and decorative knotting. In recognition of the debt our cultures owe to cordage and knots, he has developed strong interests in the history and etymology of the subject.

Married, with two sons, he now lives on the sea-front overlooking Cook Strait, about twenty miles north of the New Zealand capital. There he engages in numerous maritime crafts, including knotting; for example, he makes and restores ship models, and puts ships in bottles. He is a Member of the Ships in Bottles Society of America, and is on the National Committee of the New Zealand Ship and Marine Society. He has had several articles published in various Nautical Publications.

Europa Chang-Dawson

Europa Chang was born in 1944, in Shanghai, China. Her childhood education was garnered from nine or ten schools, scattered through five countries. After her VIth form at the Perse School for Girls, in Cambridge, England, she read Mathematics at Somerville College, Oxford. There she gained a B.A. in 1967, and M.A. in 1971. Recently she has achieved City & Guilds Diplomas in Embroidery and Lacemaking; her formal studies of the latter craft are still continuing.

Until her marriage to John Dawson in 1981, she taught mathematics full-time. Nowadays she is a widow, devoting much time to caring for her aged parents and relatives. She teaches mathematics part-time, takes a couple of Tatting workshops each year, does research for a small museum, and enjoys her nephews and nieces. She also enjoys doodling with string, or cord, or yarns, or thread, or, if desperate, with pencil and paper! Her favorite TV programmes are the Tom and Jerry cartoons. ('I seem to bumble through life, not achieving very much but having a happy time.')

She has a detailed memory, with a lengthy story, about how she first became interested in knots. Around the age of five years, in Hong Kong, an Uncle gave her two linked elastic bands, and two separated bands, to play with. She played with them for several days, and by various cutting and rejoining maneouvres discovered for herself rules relating twists and mirror-symmetries. She tried to explain these ideas to her Uncle, and asked him how to tie the linked rings. When he laughed, and said it wasn't possible, Europa threw a tantrum; as punishment for this (from her Grandparents), she was made to learn how to splice two ropes together—and the Uncle had to learn how to do that, too, in order to teach her. Small wonder that Europa Chang grew up to be a great enthusiast about Chinese Knotting and Lacemaking!

Lydia H. S. Chen

Born in 1940, Lydia Chen holds a B.S. degree in Agricultural Chemistry from the National Chung-hsing University in Taiwan. When first exposed to the Chinese art of Knotting in the early 60s, she was immediately captured by its elegance and subtle delicacy, and has since then devoted her time to the study of Chinese decorative arts in general and knotwork in particular. Since 1981 her crafts and knotting work have appeared in eighteen exhibitions throughout the world—for example in Taiwan, Japan, Korea, Germany and the United States of America.

Ms Chen is a research fellow at the National Palace Museum, Taipei, and is the author of many monographic publications on the arts and crafts of China. Two books of hers which teach the knotting craft are *Chinese Knotting* and *Chinese Knotting 2*, published by Echo Craft Books; they contain many beautiful colour photographs and pictures of the decorative knots being demonstrated.

Antje Christensen

Antje Christensen (née Kaselowsky) was born in 1969, in Stolzenau, Germany. She began studying mathematics at the Ruhr-Universität Bochum in Germany in 1989. As a student of Professor Heiner Zieschang, a renowned topologist and group theorist, she became involved with knot theory. She finished her 'Diplom' in Mathematics in February 1994. After that she started working for a Ph.D. degree, again in knot theory and with Professor Zieschang. In the Spring term of 1993, she went as an exchange student to Århus University, Denmark; there, at the Institute of History of the Exact Natural Sciences, she followed a course in history and philosophy of Mathematics, given by Kirsti Andersen. During this course she became interested in quipus, and wrote an essay on them, which is the basis for her contribution to this book.

She spent the academic year 1993/94 at the Université Paul Sabatier in France, studying with Professor Zieschang. Currently (mid-1995) she is a research visitor at the University of Melbourne in Australia, staying for three months; early in 1996 she will continue her knot studies at Osaka University in Japan, again for three months.

Recently Antje married Thomas K. Christensen, from Denmark and also a mathematician.

Sally Jo Cunningham

Sally Cunningham was born in Louisiana in 1961. She has several very diverse interests, as is demonstrated by the degrees she has earned: a B.A. in Humanities (Asian Studies), then B.S. in Computer Science, and minor in Mathematics from the University of Tennessee at Chattanooga; and a Ph.D. in Computer Science, with minors in Chinese History and Library and Information Science, from Louisiana State University. During all her schooling, she maintained a childhood hobby—crocheting. This was sometimes difficult, as she had to hide her doilies from her major professor (who felt that she should be more 'productive')!

After graduating from LSU, she became a lecturer in Computer Science at the University of Waikato (Te Whare Wananga o Waikato) in Hamilton, New Zealand. One of the most appealing aspects of the position is the very large amount of wool available throughout New Zealand, and the number of talented members of the local Spinners and Weavers Guild (who have taught her to spin lumpy yarn). She particularly appreciates the support of her colleagues in her studies of crochet and computing. Despite the distance from her home in Louisiana, she remains a Southern belle.

John Christopher Turner

John Turner was born in 1928, in Goole, Yorkshire, England. He is married, with five children and eight grandchildren to date.

His career, after obtaining an Honours degree in Mathematics at Leeds University in 1954, included positions as Scientific Officer in Armaments Research in Kent, and College and University lectureships in Kenya, Sierra Leone, Huddersfield and Leeds. In 1970 he became Reader in Mathematics and Statistics in the University of Waikato, New Zealand. His 23 years of service there included a four-year spell as Foundation Dean of the School of Computing and Mathematical Sciences; he retired in 1994.

He served a term as President of the New Zealand Mathematical Society; and three terms as President of the New Zealand Federation of Classical Guitar Societies.

His academic achievements include the invention and development, with W. Rogers, of a statistical computing language (STATUS) in the 1970s; publication of several mathematics textbooks; numerous papers on number-tree graphs and related number theories; and, with Georg Schaake since 1987, an extensive series of books and pamphlets on Schaake's new theories of braiding processes.

His interests in Knots began in 1978, when he started a study of operations on knot-graphs; this work led to a D.Phil. degree in 1984. The main discovery was a powerful twist-spectrum invariant—a polynomial which was able to detect amphicheirality and distinguish handedness; but unfortunately it only worked with alternating knots.

John Turner's other interests lie mainly in music. He plays 'cello for the Waikato Symphony Orchestra; and piano in a jazz duo. A fun retirement hobby is juggling with balls and clubs, which he enjoys with his son-in-law William Rogers.

Pieter van de Griend

Pieter responded as follows:

'I was born in 1959, in Airdrie, Scotland, and am of Scottish-Dutch descent. I have enjoyed playing with knots ever since I can remember. Indeed, I believe I acquired a love for Celtic knotting from the traditional Arran sweaters worn in our family, and knitted by my mother. My first memories of these were implanted when I was three or four years old.

Knots have certainly played a major part in my short life so far, bringing me salty sea-adventures on racing yachts off-shore, and deep-sea fishing experience on trawlers on the North Atlantic Ocean. Knots have also earned me a cand. scient. degree in mathematics from Århus University. I am convinced that they will demand the rest of my time, and I certainly do not mind happily exploring the richness of the subject in the years to come. In fact, I feel very privileged to live at a crossroads in time when the combination of Physics and Knot Theory is enjoying a renaissance, with a consequent surge forward in several mathematical fields.

Currently I am teaching mathematics at a grammar school in the Netherlands. For recreation I enjoy long distance running, racing large yachts offshore, and listening to Pink Floyd whilst playing with knots ...'

Gerre van der Kleij

Gerre responded as follows:

'I was born on 15 September 1962, in the town of Assen, province of Drenthe, the Netherlands.

After my B.A. (four years) in Classics at the State University Groningen on 1984, I graduated as an M.A. in Mediterranean Archaeology at the same University in December 1987. By that time, my interests had already veered towards prehistory and more anthropologically oriented archaeology; and in 1988 I started graduate studies in archaeology and anthropology at the State University of New York at Binghampton (USA). I graduated there as an M.A. in anthropology in 1990.

In the Fall of 1991 I moved to Denmark, and started Ph.D. work in prehistorical archaeology at the Department of Prehistory of the University of Århus specialising in the Northern Funnelbeaker Culture. Financial problems forced me to cut these studies short in the summer of 1994, and I moved back to the Netherlands. There, I worked, and still work, as a volunteer assistant of the municipal archaeologist of Zutphen, while at the same time trying to obtain paid employment. Besides excavation and find-processing, my archaeological activities centre round bringing archaeology closer to the public through courses, columns, lectures, articles etc.

I became involved in the *History and Science of Knots* project through Pieter van de Griend, whom I met at the University of Århus. Being compatriots, we talked a lot together, to keep our Dutch in shape; and in the course of these conversations Pieter suggested I might add an archaeological perspective to HSK. This I thought was a good idea, and a nice challenge, being totally ignorant of the world of knotting.

I don't drive, don't smoke and don't drink, but one can get me to do almost anything by tempting me with Bach, Mahler or Palestrina.'

Charles Warner

Charles Warner is an Australian, born in Sydney in 1920. He first learned about knotting in the Scouts, and has since applied those skills in a variety of outdoor sports, including climbing, canoeing, caving, camping and sailing. He has used and developed them, too, in activities such as pioneering (field engineering) and rescue work, and in the home manufacture of much of the equipment needed for these operations. He is the author of a small but comprehensive book, *A Fresh Approach to Knotting and Ropework*, self-published in 1992.

His profession was in research science, with CSIRO Australia; and he has been interested in prehistoric rock art and in heraldry since youth.

Willemina Z. Wendrich

Willemina Wendrich was born in September 1961, in the city of Haarlem, in The Netherlands. Her University education led in 1988 to an M.A. degree in Theology, specialising in ancient religions. After that she enrolled as a Ph.D. student at the Research School Centre of Non-Western Studies, her subject being the social context of ancient Egyptian basketry. She expects to complete her Ph.D. degree in 1996.

Since 1994 she has been co-director of the excavations at the ancient harbour town of Berenike, on the Egyptian Red Sea coast. In 1995 she joined the staff of the Netherlands Institute for Archaeology and Arabic Studies in Cairo, and teaches archaeology to advanced students.

Her interest in knotting was aroused by a friend, a textile archaeologist, who advised her not to specialize in textiles, but in basketry. She found that little attention had been paid to the knotting and cordage that is found on ancient baskets: and from 1987 onwards she has joined a large number of excavations in order to study that aspect of basketry. She soon discovered that: 'To identify the construction, function and use of knots is a major task which requires a thorough inventarization of the knots occurring in excavations, and at the same time a broad general knowledge of knots in order to interpret the ancient knotting techniques.'

When not pursuing her professional studies, which she finds absorbing and exciting, she seeks relaxation in reading, dancing, scuba-diving and horseback-riding.

Willemina is married to a medical doctor, Hans Barnard, who is doing research into Diabetes mellitus type II.

INDEX

INTERNATIONAL GUILD OF KNOT TYERS

The International Guild of Knot Tyers is a worldwide association of folk with a common interest in knots. For more information, please contact the Secretary:

Nigel Harding
3 Walnut Tree Meadow
Stonham Aspal
Stowmarket
Suffolk, IP14 6DF
United Kingdom